JUN 2024

JAPANESE

DESIGN

JAPANESE DESIGN

A SURVEY SINCE 1950

Kathryn B. Hiesinger and Felice Fischer

PHILADELPHIA MUSEUM OF ART

in association with

HARRY N. ABRAMS, INC., PUBLISHERS

Published on the occasion of the exhibition
Japanese Design: A Survey Since 1950
Philadelphia Museum of Art
September 25 to November 20, 1994

The exhibition and catalogue are made possible by generous
grants from the E. Rhodes and Leona B. Carpenter Foundation;
The Pew Memorial Trusts; the National Endowment for the Arts,
a Federal agency; The Japan Foundation; and the Japan–United
States Friendship Commission; and a grant from The
Commemorative Association for the Japan World Exposition
(1970) to promote world peace.

Produced by the Publications Department of the Philadelphia Museum of Art
Editor and coordinator: George H. Marcus
Copy editors: Jane Fluegel, Sherry Babbitt
Translators: Minoru Endo, Felice Fischer, Rosemary Morrison
Designer: Mitsuo Katsui, Katsui Design Office Inc.
Printer: Toppan Printing Company

Clothbound edition published in 1995 by Harry N. Abrams,
Incorporated, New York
A Times Mirror Company

ISBN 0-8109-3509-0 (Abrams); 87633-092-8 (Museum)

Library of Congress Catalog Card Number: 94-67907

Printed and bound in Japan

Contents

A comprehensive survey of Japanese design since 1950 accords so naturally with the history and collections of the Philadelphia Museum of Art, as well as with the enthusiasm and knowledge in their respective fields of two colleagues at the Museum, Kathryn B. Hiesinger, Curator of European Decorative Arts Since 1700, and Felice Fischer, Curator of Japanese Art, that it is astonishing to discover that this is the first such exhibition to be organized anywhere in the world. The Museum's origins in the Centennial Exposition of 1876 in Philadelphia, when works by contemporary Japanese craftsmen were shown and admired by a broad American public for the first time, suggest a reason for its early and sustained interest in Japanese art, and the large airy gallery containing a seventeenth-century temple and a twentieth-century traditional tea house, surrounded by bamboo and permeated by the faint sounds of water dripping into a stone basin, has been among the most popular in the Museum since it opened in 1957.

The Museum has also maintained a long tradition of showing and documenting contemporary design, exemplified by two major exhibitions, *Design for the Machine* of 1932, and *Design Since 1945*, organized by Dr. Hiesinger in 1983. It was in the course of her research for the latter project, which included a few works by Japanese designers but focused its attention upon four decades of work in Europe and the United States, that Dr. Hiesinger hit upon the idea of an exhibition devoted exclusively to the achievements of Japan during the same period. Since 1989 Dr. Fischer and Dr. Hiesinger have been zealous collaborators in a new venture, together exploring this field in which Japan's ancient traditions, its grasp of modern technology, and the innovative imagination of its artists have combined to infuse so many contemporary Japanese products with distinctive visual values.

This has been a truly collaborative international project, in which the generosity of the many lenders to the exhibition and the enthusiasm and cooperation of our Japanese colleagues who have written essays for the catalogue or given advice on so many aspects of the exhibition and publication have been invaluable. It is wonderfully appropriate that the design of the spectacular installation of the exhibition in Philadelphia has been conceived by Kisho Kurokawa, the distinguished architect and theorist, and that this striking volume, which presents the objects in the exhibition in printed form, has been the design of Mitsuo Katsui, an artist whose work has been at the forefront of graphic design in Japan for several decades.

The funding for this complex undertaking has also been international in character. An initial planning grant from the E. Rhodes and Leona B. Carpenter Foundation enabled the curators to conduct extensive research and travel in Japan, making contact with the impressive circle of colleagues and designers who have assisted us so willingly. Funding from government foundations and agencies in both Japan and the United States has been gratifying in its extent. Handsome support from both the Museum Program and the Design Arts Program of the National Endowment for the Arts has been balanced by most welcome grants from The Japan Foundation, the Japan–United States Friendship Commission, and The Commemorative Association for the Japan World Exposition in 1970. The Pew Charitable Trusts, which has assisted the Philadelphia Museum of Art in the research and realization of so many major international exhibitions, was again crucial to our ability to undertake this one, and the Carpenter Foundation generously followed its early planning grant with invaluable support for implementation. Thanks are also owed to the efficiency and skill of Nippon Express, which has coordinated the complex packing and transportation of loans between Japan and Philadelphia.

The staff of many departments in the Museum have played vital roles in this project. None has worked harder to bring it to fruition than Suzanne F. Wells, Coordinator of Special Exhibitions, and Robert Morrone, Vice President for Operations. In the Departments of East Asian Art and European Decorative Arts, all have devoted themselves to the cause over the last several years, while the Registrar's Office, the Division of Education, and the staff in Publications, Public Relations, and Installations sprang into action to assure the best possible presentation to our public.

It is the fervent hope of the curators of the exhibition, and of their colleague George H. Marcus, Head of Publications at the Museum, who has overseen the creation of this splendid book, that both visitors and readers will be struck by the extraordinary range and variety of Japanese design presented here. Themselves necessarily only a fraction of the impressive number of imaginative and efficient designs produced in Japan over the last five decades, each of the 255 objects were selected for both their individual quality and their ability to add to a composite picture. In the process of their work on this project, the curators reflected upon the mix of qualities that help define the special character of recent Japanese design: compactness, craftsmanship, simplicity, asymmetry, and humor are among the most striking. Whether it be a witty paper package in the shape of a banana (no. 137), a Sony Walkman (no. 132), a graceful bent-plywood stool (no. 27), or a spectacular variation on the traditional form of the kimono, each object partakes of at least one, and often several, of such qualities, and in so doing establishes its place in a visual world that is both centuries old and undeniably modern. Japanese traditions infuse contemporary Japanese design in the most unexpected ways: Hanae Mori may translate a technique for making lacquerware into the built-up beaded surface of a dress (no. 202), while the serene concentric patterns of raked sand gardens of Zen temples may reappear in a printed furnishing fabric (no. 90). The controlled dramatic violence of Noh theater may have echoes in a brilliant, aggressive poster, or the pictogram of the word "celebration" may find itself transformed into a piece of furniture (no. 228). The lyrical asymmetry that characterizes the venerable art of flower arranging may reappear in the design of a piece of ergonomic equipment that proves satisfyingly shaped to the user's needs. In few twentieth-century cultures do past and present seem so intricately interwoven, or does the work of the most gifted designers (in the most diverse mediums and styles) bring these subtle connections so vividly to the surface.

This exhibition and book are envisioned not as the definitive word on a vast subject but rather as the beginning of an exploration. Neither would have been possible without the exemplary commitment over the past four decades that Japanese government agencies, private companies, and individual designers have made to the effort to infuse everyday objects, clothing, and machines with that individual attention to their shape, color, and appearance—one may almost say their spirit—that makes Japanese design so distinctive despite its diversity. It is to the designers above all to whom we owe a profound debt of gratitude, for their enthusiasm and their cooperation with this complex project, and for their extraordinary creative energy, which inspired this undertaking from its inception.

Anne d'Harnoncourt
The George D. Widener Director
Philadelphia Museum of Art

Japanese Design: From Meiji to Modern

The Japanese, by geographic disposition and aesthetic temperament, have historically exhibited an extraordinary talent for learning and adapting from outside sources without sacrificing their age-old traditions and beliefs. When, with the introduction of Buddhism in the sixth century, the Japanese encountered Chinese civilization, they deliberately set about experimenting with aspects of China's sophisticated culture. They adopted Chinese dress at court, and elaborate Chinese models dominated the architecture of Buddhist temples and even entire city plans. The successive eighth-century capitals of Nara and Kyoto were laid out on a grid plan copied from China's early-seventh-century T'ang dynasty capital of Ch'ang-an. But the most profound foreign adoption was that of the Chinese system of writing, based on ideographs, or characters, which is still used in Japan today. By emulating the Chinese example, Japan sought to become China's equal, and Chinese civilization remained the model for Japanese culture for over a thousand years.

This talent for cultural accretion became even more pronounced during the period of Japan's encounter with Western civilization. In 1853 the American commodore Matthew Perry first brought his ships into the bay at Uraga and began the process of "opening" Japan to American and European trade after more than two hundred years of virtual isolation under the military leaders, the shoguns. The treaties signed with the United States and various European nations in the 1850s came at a time when internal opposition to the shogunate was growing, and they ultimately contributed to the overthrow in 1867 of the last shogun, who had ruled from his castle in Edo in eastern Japan in the name of the reigning emperor, whose residence was in Kyoto. The fifteen-year-old prince Mutsuhito ascended the throne, ushering in the reign period known as Meiji (Enlightened Rule), which started in 1868 and lasted until his death in 1912. With the new emperor's accession, the government returned to the lessons of Japan's own history: emulate what was perceived as superior models to become their equals without discarding existing strengths.

The telling difference in the nineteenth-century encounter with the West was the matter of time. While Chinese religion, literature, and political philosophy had been learned over the course of centuries, the lessons of modernization, industrialization, and imperialism were absorbed in a matter of decades. This led to an often-confusing telescoping of Western ideas and methodologies introduced simultaneously, and sometimes only the adoption of outward forms, devoid of their social, philosophical, and aesthetic underpinnings. What is amazing, however, is how quickly and how well Japan succeeded.

In 1869 the seat of the imperial government was moved from Kyoto to Edo, renamed Tokyo (Eastern Capital), and the emperor, depicted in photographs with a beard and in Western-style military uniform, became the center and symbol of the drive for modernization. Japan's entry into the contemporary world came with dizzying speed. Establishing a pattern of support for industry that persists even today, the government actively decreed and financed new ventures, starting with strategic areas such as mining, shipbuilding, and munitions. The first telegraph line was laid in 1869 and the first rail service began in 1872, running between Tokyo and Yokohama. That same year fire destroyed most of the Ginza district in Tokyo, and its over 950 wood houses were rebuilt in brick, following the English model. As with Chinese influence in the past, architecture was one of the most evident aspects of foreign culture emulated, one that could be used to impress visitors with Japan's progress toward becoming a thoroughly modern nation. Some of the early attempts by Japanese carpenters working in the traditional wood idiom to replicate Victorian buildings led to eccentric hybrids, but examples such as the Kyoto National Museum, designed by Tokuma Katayama in the Second Empire style and built in 1895, stand as ample proof of how quickly and completely new architectural styles were mastered (fig. 1).

In keeping with the aim of the Charter Oath of 1868 that "knowledge shall be sought throughout the world so as to strengthen the foundation of imperial rule,"[1] the Meiji government dispatched its first official delegation to the United States and Europe in November 1871. The Iwakura Mission, named after its leader, Tomomi Iwakura (1825–1883), minister of foreign affairs, consisted of some one hundred officials, translators, technical

Fig. 1 Kyoto National Museum, 1895
Fig. 2 Coffee service, c. 1875, signed *Kichizan*. Porcelain. Philadelphia Museum of Art. Gift of Visi Aromiskis
Fig. 3 Design for a cloisonné vase, by Yasuyuki Namikawa (1845–1927). Ink and colors on paper. Philadelphia Museum of Art. Gift of Clayton French Banks, Jr., in memory of his grandparents, George W. and Mary French Banks

experts, and students, the last sent primarily to learn about European manufacturing techniques and to collect examples of European goods. Before their return to Japan in September 1873, some of the members attended the Weltausstellung in Vienna, the first world's fair in which the Meiji government itself participated as an exhibitor.

The phenomenon of international expositions as a stage for displaying the best and latest wares of the industrialized nations began in London in 1851. Japan did not exhibit at the 1851 London exposition, but several of its feudal domains had participated in subsequent expositions, in London (1862), Paris (1867), and San Francisco (1871). At Vienna in 1873 Japan was represented by 6,668 catalogue items in 25 sections, which included displays on mining, agriculture, the chemical industry, textiles, leather, basketry, lacquerware, metalwork, and ceramics.[2] Between 1873 and 1910 the Japanese participated in twenty-five such events, which provided valuable bellwethers for assessing trends abroad and directing development at home.

Preparations for and participation in the Vienna world's fair provided many useful lessons for Japan's infant industries. One telling result was soon manifested in the Japanese language itself. Until the Meiji period there were no equivalents for the categories of *Schöne Kunst* (Fine Art) and *Angewandte Kunst* (Applied Art) that appeared in the announcements for the Vienna exhibition, for the Japanese did not see distinctions between art made to be looked at and art made to be used. In order to work within the Western model, new words were coined. For fine art, the term *bijutsu*, which initially referred only to Western-style oil painting and sculpture, was introduced, while *kogei* was the word used for applied art, encompassing both handmade crafts produced with traditional techniques and objects of daily use made with new manufacturing processes.

In selecting the objects for exhibition in Vienna, the Japanese government relied heavily on the counsel of a German chemical engineer, Dr. Gottfried Wagener (1831–1891), who in 1868 had been one of the foreign advisors in different fields hired by the Japanese. Wagener, whose specialty was ceramics, recommended that the displays for Vienna concentrate on high-quality traditional crafts, such as ceramics and textiles, rather than on machinery or industrial products, and indeed, the craft items that were exhibited there proved very popular and sold extremely well.

Following this success, the Japanese government, which badly needed funds for the national coffers, decided to encourage the production of crafts for export. Under the supervision of the potter Kaijiro Notomi (1844–1918), who had been to the Vienna exposition, such ventures as the Edogawa ceramic plant in Tokyo were started (1874) to experiment with European-style plaster molds for ceramic production. These molds were useful in producing new shapes, for example, cups with handles and coffee services, suitable for the Western market (fig. 2). Notomi was also involved in the Kiritsu Kosho company, founded the same year in Tokyo with government assistance, to produce ceramics, lacquerware, metalwork, and cloisonné (fig. 3). By commissioning painters and craftsmen to create original designs for manufacture, it became the primary source for the decorative arts sent by the Japanese to subsequent world's fairs. By 1882 the company had retail shops in Paris and New York, and its workshops, which employed some sixty craftsmen, became an important training ground for young artists.

Textile production was another field in which foreign technology was successfully introduced. In Kyoto, the traditional center for textiles, the first government-sponsored plant was established in 1874, and it experimented with aniline dyes and new weaving techniques. Yanosuke Date (died 1876), one of the country's leading weavers who had been sent with the textile exhibits to Vienna, subsequently purchased the first Jacquard loom, which was brought back to Tokyo and set up in 1875. Regional centers such as Kiryu were not far behind the larger cities, and they too soon became important producers of machine-made textiles.

The government also organized a series of domestic industrial expositions—the first in 1877—to further competition and to display products from all parts of the country. The official policy of supporting mass production and exports soon proved almost too successful as dozens of private companies sprang up to share the profits from the overseas trade. By the mid-1880s Notomi and others were criticizing the poor quality and lack of creativity displayed by these goods, and new organizations were founded to promote and control standards of quality in production, among them, the Japan Textile Association (1885) and the Japan Lacquer Association (1887). Notomi also advocated the establishment of technical schools near regional production centers at which students could learn design and craftsmanship along with new industrial techniques. He opened the first of these schools in Kanazawa with support from the Ishikawa prefectural government in 1887. Notomi's school admitted women, and offered revolutionary courses such as Western dressmaking, as well as textile weaving and dyeing, metalwork, and papermaking. While such efforts were being made to promote and improve goods made by mass production, there were those who feared that the traditional handicrafts were in danger of being overwhelmed by commercialization and industrialization. The one-of-a-kind examples of ceramics, lacquer, textiles, and metalwork being made were conservative in conception and tended to emphasize technical virtuosity while the cost of fabricating them increased as the market declined.

Within twenty-five years of initiating policies for rapid modernization, the government found that it had to take measures to ensure the survival of traditional crafts. This in part reflected the general mood of the Japanese in the latter half of the Meiji period. The excitement and euphoria of change and Westernization began to wear off and a conservative reaction set in, fueled in part by the opinions of Westerners, such as the American Ernest Fenollosa (1853–1908), a much respected teacher and government advisor in Japan, who urged that traditional Japanese art be preserved.[3] The Technical Art School founded by the government in 1876 to teach Western painting and sculpture was closed in 1883. When its successor, the Tokyo School of Fine Arts, opened in 1889, the curriculum established with Fenollosa's advice emphasized traditional ink painting and decorative arts, which had been neglected at the earlier institution. The later domestic industrial expositions had separate categories for industrial crafts and "art crafts" (*bijutsu kogei*), and both were shown in the Japanese exhibits at the Chicago World's Columbian Exposition in 1893.

In 1888 eighteen artists were named members of the Imperial Household Arts and Crafts (later the Imperial Academy of Fine Arts), and to provide financial support for them, their works were purchased for the imperial collections. This system raised the prestige of traditional crafts and assured their survival into the twentieth century; similar measures were taken after the disastrous earthquake of 1923 and during World War II, when materials were scarce and the studios of many craftsmen were destroyed. The postwar designation Holder of Intangible Cultural Property (Living National Treasure) for artists in traditional crafts is a descendant of this system.

Works by artists named members of the Imperial Household Arts and Crafts were shown at the Exposition Universelle in Paris in 1900. This international exhibition proved to be another landmark in the evolution of the Japanese design community, when the elaborate and intricate crafts were admired for their technical virtuosity but were criticized for their old-fashioned designs (fig. 4). If the result of the Vienna exhibition had been to make the Japanese conscious of Western methods and techniques of production, the Paris exposition made them aware of new ideas about design.

A number of artists and craftsmen who had studied in Paris and returned to Japan around the time of the exposition tried to introduce changes that would bring Japan into the international art community. Among them were men like Mataichi Fukuchi, founder of the Japan Design Association (1901), and first senior professor of design at the Tokyo School of Fine Arts. He organized the first Japanese exhibition of Art Nouveau (1902), introducing the style that had taken Europe by storm. A group of teachers and students from the Tokyo College of Industrial Arts formed the Greater Japan Design Association around the same time. The fruits of their labors were displayed at the Domestic Industrial Exhibition of 1903, where the Art Nouveau style predominated in ceramics, textiles, and graphics.

Graphic design significantly reflected the artistic trends in early twentieth-century Japan. Changes in graphics, following a pattern similar to those in other mediums, were two-fold: technical and aesthetic. The technical developments had already started in the 1870s with the introduction of Western machinery and printing techniques. Lithographic-engraving technology advanced quickly, and four-color lithography made its debut in 1881, when it was used for advertisements by a tobacco dealer. The tremendous expansion of the book-publishing industry in the early Meiji period encouraged the rapid changeover from manual woodblock printing to the use of moveable type and modern machinery, and during the 1870s and 1880s, innumerable translations of Western books, from technical manuals on shipbuilding to the complete works of Shakespeare, were printed on the new equipment. Gas- and electric-powered presses were in use by the late 1880s, as were rotary presses for newspaper printing. Daily newspapers (the first, the *Yokohama Mainichi*, appeared in 1871) were another new phenomenon, as were their advertisements and the agencies that sold advertising space. Some manufacturers set up their own printing facilities to produce advertising and packaging, notably those established by rival tobacco companies in Tokyo and Kyoto. It also became fashionable to use foreign words and lettering for brand names, and among the new tobaccos were those dubbed Peacock, Telephone, and Pin Head.

New aesthetic possibilities of graphics began to be explored after the turn of the century, quickly reaching great sophistication in advertisements for kimonos. In 1904 Echigoya, one of the dry goods stores that were the traditional purveyors of kimono fabrics, converted its shop in Tokyo into the first department store modeled on the Western prototype (better known by its later name, Mitsukoshi). The firm had started a design department in 1909 and hired artists like Hisui Sugiura (1876–1965) to create advertisements for their kimonos. Perhaps because the

Fig. 4 Box carved with quails and millet, c. 1900. Ivory. Philadelphia Museum of Art. Gift of Mrs. Henry W. Breyer, Sr.
Fig. 5 Mitsukoshi department store, Tokyo, 1914

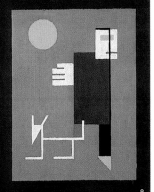

kimono was a traditional dress form, the aesthetic inspiration for advertisements came from the heritage of eighteenth-century *ukiyo-e* woodblock prints, particularly *bijinga* (pictures of beauties). In a poster issued to commemorate the opening of the new, imposing Mitsukoshi store, built in 1914 in Tokyo (fig. 5), Sugiura translated the *bijinga* into a contemporary idiom (fig. 7), depicting his beautiful woman amidst modern Art Nouveau motifs, butterflies on her kimono, the carving on the table, the picture on the wall behind her, and even the style of calligraphy for the store's name. Mitsukoshi, which featured Western-style furniture and interior design, introduced the department store art gallery, which became, and still is, an important venue for exhibitions of both traditional and contemporary art in Japan. Mitsukoshi and other firms also regularly sponsored competitions for new textile and graphic designs, exhibiting the prize winners and publishing catalogues to spur interest in them (fig. 6).

Sugiura himself sought to raise the prestige of commercial art and designers as the profession was growing in Japan; he founded the graphic art research and study group Shichininsha (Group of Seven) in 1924, and published the magazine *Affiches* (from 1927) to introduce the best of graphic art from Japan and abroad. A spate of new art and architectural magazines, such as *Mizue* (started 1925) and *Kenchiku Shincho* (New Wave Architecture, founded 1929), also began publication. Young graphic artists like HIROMU HARA and TAKASHI KONO absorbed the lessons of French Cubists, Italian Futurists, and Russian Constructivists, producing posters and magazine illustrations that reflected these influences (fig. 8), and in every field the decades between the two world wars proved a rich and fruitful period for Japanese artists. An increasing number of young artists were studying abroad, while Westerners were coming to Japan, among them, the American architect Frank Lloyd Wright in 1905 and the English potter Bernard Leach in 1909.[4]

Two influential German institutions, the Deutscher Werkbund and the Bauhaus, had a great effect on young Japanese of that generation. Their impact was probably so strong because their philosophies and aesthetic ideals coincided with much that was an integral part of the Japanese concepts of beauty and functionality, and their most important contribution was perhaps to remind the Japanese of their own rich heritage, and how it could be applied in the contemporary world. At the Bauhaus school and workshops established in 1919 at Weimar, its founder, Walter Gropius, called for the reintegration of fine art and applied art, an attitude the Japanese had held before the distinction was introduced from the West. Japanese architects had visited the Bauhaus as early as 1922, but the first to enroll as a student was Takehiko Mizutani (1898–1969), who studied there from 1927 to 1929. Mizutani taught at the Tokyo School of Fine Arts after his return, and one of his students, the architect Iwao Yamawaki (1898–1987), and his wife Michiko (born 1910) left to study at the Bauhaus in 1930. Both were won over by the school's curriculum, which included theoretical courses as well as studio classes in drawing and painting, and workshops in metalwork and textiles. All three became principals in establishing and teaching at the New Academy of Architecture and Industrial Arts in Tokyo, which was modeled on the Bauhaus. The driving force behind the school was the architect Renchichiro Kawakita (1902–1975); although he had not attended the Bauhaus, nevertheless he devoted his career to its educational ideals. The New Academy gave courses in architecture, painting, stage design (taught by Kawakita), dressmaking, and weaving, the last two courses being taught by Michiko Yamawaki, who had studied at the Bauhaus under Josef Albers (fig. 9). Among her students was the dress designer Yoko Kuwasawa (1910–1977), who founded the Kuwasawa Design School in Tokyo on the Bauhaus model in 1954. Another New Academy graduate was graphic designer YUSAKU KAMEKURA, who applied the lessons learned there in his work for *Nippon*, Japan's first multilingual graphics magazine, published for international distribution by the Nippon Kobo (Japan Workshop) from 1934.

The model to which Japanese industrial designers looked in the 1920s was that of the Deutscher Werkbund, an association founded in 1907 by a group of designers and manufacturers to improve the quality and design of industrial goods in Germany. An early advocate of the Werkbund ideals of integrating craftsmanship and industrial technology was Chikutada Kurata (1895–1966), a lecturer at the Tokyo College of Industrial Arts. A handful of Kurata's students joined him in 1928 in establishing the Keiji Kobo (Form Workshop), which according to its 1928 brochure, was "a conscious attempt to give contemporary form to the architecture and industrial arts that are part of our daily lives. [Its] focus is crafts for interiors, and it aims to produce pieces that are simple and economical for as large a market as possible."⁵ Members planned to produce furniture, lighting, ceramics, and glassware, but in fact furniture became their main product. The lack of manufacturers willing to experiment with new products without a proven market, especially in the depressed economy of a Tokyo still recovering from the great Kanto earthquake of 1923, was an obstacle to the mass production of their work, although it was the shortage of housing and goods after the earthquake that had first stimulated this generation of architects and designers to plan low-cost housing and furnishings.

Keiji Kobo was the longest lived and most successful of the idealistic groups that sprang up in the 1920s and devised plans to use industrial design to better the lot of the urban masses. Unit furniture, emphasizing standard sizes for ease and economy of production, was introduced, beginning with designs for Western-style chairs. KATSUHEI TOYOGUCHI designed a series of low-cost chairs (1927–34) with wood veneer for the frame and woven hemp for the seat (fig. 10). To bring members' work to public notice, Keiji Kobo launched a series of exhibitions, accompanied by lectures, held in Tokyo in 1928 and 1930 at the Kinokuniya bookstore gallery and in 1934 at the Takashimaya department store, where the complete line of furniture was shown in two model rooms. The final exhibition, featuring its series of childrens' furniture, was held at the Itoya department store in 1937. Keiji Kobo members published their designs in architectural magazines, and in 1930 began propagating their ideas and marketing their furniture through articles and advertisements in the women's magazines, such as *Fujin no Tomo* (The Housewife's Companion). The group was dissolved in 1937 under pressure from the militarist government, which disapproved of its international outlook.

Ironically, the ideals that Kurita, Toyoguchi, and their generation espoused were flourishing at the same time under the aegis of the government in Sendai at the Industrial Arts Institute. In 1928 the government had recruited Kitaro Kunii (1883–1967) to direct the Institute for the Promotion of Industrial Arts of Northern Japan. Established to promote small- and mid-size manufacturers in the six northern prefectures of Japan by introducing mass-production techniques and encouraging the manufacture of new products for export, the institute published a monthly journal, *Kogei Shido* (Industrial Art Promotion), and held an annual open house exhibition to show new products, for example, experimental bamboo furniture in 1931. Kunii was sent to Europe, America, and Southeast Asia for six months to study manufacturers and markets. In 1933 the government opened a Tokyo office of the institute (later renamed the Industrial Arts Institute) to coordinate regional craft production nationwide. The magazine was renamed *Kogei Nyusu* (Industrial Art News) and was distributed nationally from an editorial office in Tokyo. The venue for the annual exhibition was also moved to Tokyo, where it opened in September 1933 at the Mitsukoshi department store.

Of all Europeans and Americans who visited Japan in the first half of the twentieth century, the German architect Bruno Taut (1880–1938) left the most indelible impression on young architects and designers. His is the one name that is invariably mentioned when Japanese industrial designers of that generation are asked about influences on their work. Taut had practiced architecture in Berlin from 1904, and after World War I became involved with utopian city planning for community housing projects. His interest in Japanese architecture prompted him to join the International Architectural Association of Japan as an overseas member (as Walter Gropius had also done), but he was unable to attend the association's first meeting in Osaka in 1930. In the spring of 1933 Taut fled the Nazi regime and made his way to Japan, where the architect Isaburo Ueno met him when he arrived on May 3. The following day, Taut's birthday, Ueno took him to see the Katsura Detached Palace in Kyoto, the paradigmatic example of Japanese traditional architecture (fig. 11). Taut was deeply impressed by its "modernity": its spare, plain architecture, clean lines, perfect proportions, and the logic of the modular *tatami* mats that were the standard unit for interiors. He expressed his enthusiasm in his book *Nippon: Yoroppajin no Me de Mita* (Japan: Seen with European Eyes),⁶ which became a best-seller in Japan and was instrumental in helping to turn the attention of the Japanese back to their own architectural heritage.

Taut worked on several architectural projects in Japan, but his ideas were most concretely realized in the area of product design. In 1933, when Taut visited the exhibition at the Mitsukoshi department store that showed experimental works of the Industrial Arts Institute, he criticized the tendency of the institute's designers to produce imitations of European products rather than to use the inspiration and resources of Japan's own rich craft traditions. Kitaro Kunii subsequently invited Taut to Sendai as an advisor (fig. 12). During his stay, from November 1933 to March 1934, Taut introduced his ideas for the design and production of consumer goods, based

Fig. 10 Chair, designed by Katsuhei Toyoguchi for Keiji Kobo, 1934. Wood and hemp. Museum & Library, Musashino Art University, Tokyo
Fig. 11 Katsura Detached Palace, Kyoto, 1636
Fig. 12 Bruno Taut and Kitaro Kunii in front of the Industrial Arts Institute, Sendai, 1934 (Courtesy of Kenmochi Design Associates)
Fig. 13 Datsun Deluxe Sedan DB, made by Nissan Motor Company, 1948
Fig. 14 Standing lamp, designed by Bruno Taut, 1936. Lacquered wood and paper. The position of the shade can be adjusted along the entire height of the frame (Photograph by Osamu Murai)

1 For the complete Charter Oath of 1868, see Ryusaku Tsunoda, Wm. Theodore de Bary, and Donald Keene, comps., *Sources of Japanese Tradition* (New York, 1958), p. 644.
2 See Herbert Fux, *Japan auf der Weltausstellung in Wien 1873* (Vienna, 1973).
3 See Felice Fischer, "Meiji Paintings from the Fenollosa Collection," *Philadelphia Museum of Art Bulletin*, vol. 88 (Fall 1992).
4 Frank Lloyd Wright (1867–1959) was first exposed to Japanese architecture at the Chicago World's Columbian Exposition of 1893. He first visited Japan in 1905, and began collecting woodblock prints there. His most important building in Japan, the Imperial Hotel, was completed in Tokyo in 1922. Bernard Leach (1887–1979) first came to Japan from England in 1909 and stayed eleven years. During that time he worked with the potter Shoji Hamada and with the founder of the folk-crafts (*mingei*) movement, Soetsu Yanagi. Leach introduced the Japanese climbing kiln and pottery techniques to European studio potters.
5 Gruppe 5 and Katsuhei Toyoguchi, eds., *Keiji Kobo kara: Toyoguchi Katsuhei to Dezain no Hanseiki* (From Keiji Kobo: Katsuhei Toyoguchi and a Half-Century of Design) (Tokyo, 1987), p. 64.
6 Bruno Taut, *Nippon: Yoroppajin no Me de Mita* (Japan: Seen with European Eyes) (Tokyo, 1934).
7 Kogyo Gijutsuin Sangyo Kogei Jikkensho, ed., *Sangyo Kogei Jikkensho Sanjunen-shi* (The Thirty-Year History of the Industrial Arts Institute) (Tokyo, 1960), p. 283.
8 Ibid., p. 53.

on Deutscher Werkbund practices. Young designers at the institute, such as ISAMU KENMOCHI and RIKI WATANABE, were impressed with Taut's program of building and testing full-scale models of furniture and other products before sending drawings off to the manufacturer. Above all, they recall his insistence on *Qualitätsarbeit* (quality workmanship) and his call for a return to the craftsmanship still evident in the works of some of the artisans, such as ROKANSAI IIZUKA, whose studio he visited. In 1934 Taut met Fusaichiro Inoue, who asked him to become an advisor to his workshop at Takasaki, in Gumma Prefecture, where Inoue wanted to revive local craft production. During a two-year stay at Takasaki, Taut designed over six hundred items, including furniture, lighting, trays, and salad bowls, which Inoue sold through the Miratesu craft shop in Tokyo beginning in 1935. In his standing lamp of that period, with its cylindrical handmade paper shade and lacquered wood frame, Taut borrowed directly from Japanese tradition (fig. 14). Taut stayed in Japan until 1936, his sojourn coinciding with a time when Japanese designers were receptive to the message of modernism and to quality craftsmanship, and they saw him as the manifestation of those ideals. Nearly thirty years later, Isamu Kenmochi recalled Taut's days in Sendai: "For that short period, the Modern Movement was alive directly in Japan, and planted the seeds of a tradition that we could not deny in our later careers."[7]

The Industrial Arts Institute subsequently invited other foreign designers to consult and lecture. The most notable was Charlotte Perriand (born 1903), a French architect who had worked with Le Corbusier in Paris before the war. She came in 1940, and brought much the same message as Taut, noting the lack of character in modern Japanese crafts in contrast to the beauty of traditional works. In 1941 Perriand organized an influential exhibition in Tokyo, "Tradition, Selection, Creation," which included her own pieces inspired by Japanese crafts. The activities of the institute continued through the war, although the National Mobilization Act of 1938 had decreed restrictions on uses of such materials as copper, iron, rubber, and leather. The institute worked on finding substitute materials, and its exhibition in 1940 featured items such as shoes and handbags made from fish skins. Experiments also included work with molded plywood, principally for making decoy airplanes, technology that would find peacetime applications in the field of furniture manufacture. During the years from 1939 to 1945, the government recruited the best and brightest of the young designers for the institute, and the employee roster reads like a who's who of leaders of postwar industrial design in Japan: Isamu Kenmochi, Katsuhei Toyoguchi, MOSUKE YOSHITAKE, IWATARO KOIKE, JIRO KOSUGI, Iwao Yamawaki, and Masaru Katsumie (1909–1983).

The arrival of the Americans brought Japan its first experience of occupation by a foreign power. The Japanese showed a remarkable resilience in the face of another tidal wave of new ideas and demands, not dissimilar to that shown nearly one hundred years before. Once again, they resolutely sought to learn as much as possible as quickly as possible. The Industrial Arts Institute became one of the focal points of learning and recovery when, in 1946, American General Headquarters (GHQ) ordered furniture and equipment for 200,000 new housing units for its personnel. In all, thirty types of furniture were designed with an eventual output of 950,000 pieces.[8]

The Industrial Arts Institute coordinated the production of electronic goods for occupation housing, working with firms such as Toshiba, which made washing machines and coffee percolators, and Mitsubishi, which supplied refrigerators. Japanese industry was in effect given a jump start by GHQ orders: this included such industries as glass manufacturers, which suddenly had to produce hundreds of thousands of tableware items for the Americans. Other industries experienced a quick move toward recovery. By 1947 Nissan and Toyota had put their first passenger cars on sale (fig. 13). By 1949 electric fans, autobikes, and cameras appeared on the market, and exports, aimed primarily at the United States, were making headway as well. The Americans were surprised and impressed by the energy and abilities of the Japanese as they once again demonstrated their openness to new and even opposing ideas and ideals.

The history of the first century of Japanese design might be defined by the aesthetic principal of *wa* (harmony). The great originality of Japanese civilization lay in its ability to harmonize disparate elements, seeking not so much to conquer the new or exotic as to accommodate them, to allow them to coexist and find their own place in the rich and varied blend of tradition and modernity that characterizes the culture of Japan.

Within a decade and a half after the Second World War, Japan had emerged from utter devastation to become one of the world's biggest industrial producers, and its finished products—motorcycles, cameras, radios, and television sets—were the visible measure of that success. Whereas immediately after the war and until the occupation formally ended in 1952, the words "Made in Japan" had meant goods that were cheap and poorly manufactured, they increasingly came to indicate a quality product. The American magazine *Popular Photography* reported in 1957, for example, that Japanese cameras were playing a significant role in building consumer confidence abroad for Japanese goods "as word of their quality and precision spreads," exports having already increased by a "fantastic 650-fold" between 1947 and 1954.[1]

The history of postwar design in Japan is tied to Japan's rise to prominence as a great manufacturing nation, and the government, in cooperative relationships with business and industry, guided its course with the aim of promoting exports and building a domestic market. The institutions it established or revised for this purpose in the 1950s remain largely in place today. The Japan External Trade Organization (established in 1951 as the Japan Export Trade Research Organization and known as JETRO), an operating arm of the Ministry of International Trade and Industry, provided the government with information on foreign markets by sending students abroad to study design and inviting foreign experts in design to visit Japan. The Industrial Arts Institute (established in 1928 as the Institute for the Promotion of Industrial Arts and known as IAI until its reorganization in 1969), a testing and research institute attached to the Ministry of International Trade and Industry, had as its objective to develop and promote exports by serving as a research and development resource for Japan's medium- and small-size industries—experimenting with and testing materials and processing techniques, and manufacturing sample products. During the 1950s, the IAI expanded the range of its research to include packaging (to prevent claims against exports and to promote economy in materials usage) and promotional activities. It participated in trade fairs and in exhibitions at trade centers abroad, as well as in the international exhibitions held at Hälsingborg, Sweden, in 1955 (figs. 1–2); the Triennale in Milan in 1957 and 1960; and at Brussels in 1958, to which it contributed widely admired model rooms. It also continued to issue its periodical, *Kogei Nyusu* (Industrial Art News, established 1932), which remained the most important source of information about design activities in Japan until it ceased publication in 1974.

Through the Ministry of International Trade and Industry (commonly known as MITI)[2] and the Ministry of Finance, the government encouraged growth industries by guiding and controlling the import and use of foreign technologies and the availability of foreign currency to firms, most famously in the case of SONY CORPORATION for the transistor device it licensed from the American Western Electric Company in 1953 and developed for use in Japan's first transistor radios and succeeding generations of semiconductor products. In an attempt to counter design piracy problems that plagued Japanese exports in the 1950s, MITI created in 1957 the Good Design Selection System with its G-Mark award to encourage innovative design and the Design Promotion Council to operate it;[3] in 1958 it established a design section in its international trade bureau. That year MITI sponsored amendments to Japan's design law of 1921 requiring, for the first time, a statement of "originality" in copyright applications,[4] and in 1959 it sponsored the Export Commodities Design Law requiring designs to be officially registered. The industrial designer MOSUKE YOSHITAKE reported in 1958 "with embarrassment" that he had been refused permission to visit factories in Sweden and was shown the reason why—some thirty Swedish products and their nearly identical Japanese copies.[5] In view of the serious consequences of the piracy problems, among them American restrictions on some imports, in 1958 JETRO opened its first overseas offices and trade centers to provide public relations and advertising abroad for Japan's industries.

The government's efforts to regulate design imitation recognized the side effects of its own trade policies, which encouraged importing technology and mimicking the competition to achieve the widest possible markets, this by a nation historically conditioned to assimilate and value foreign influences. The importance of the

financial and technical assistance provided by the government to industry and the overall protectionism of MITI's policies might suggest that the history of modern Japanese design has been economically determined along Marxist lines. Yet Japan is not a socialist state and most economic decisions affecting product design have been made within the company unit. Moreover, the government has proved itself vulnerable to organized resistance (for example, the notable failure of MITI's efforts to concentrate the automobile industry around the firms of Nissan and Toyota in the late 1960s) and criticism (in the 1950s, from designers and manufacturers who, fearing the anticompetitive nature of MITI's G-Mark system, successfully lobbied against the public exhibition of G-Mark products, which had been part of the original program). According to a 1958 editorial in *Kogei Nyusu*, members of the professional Japan Industrial Designers Association (established in 1952 and known as JIDA) and certain manufacturers criticized the G-Mark selections for favoring consumer products at the expense of industrial equipment (which was finally made eligible for G-Mark selection when the categories were expanded in 1984).[6] JIRO KOSUGI, a JIDA member and the first Japanese industrial designer to exhibit his work independently,[7] thought that the G-Mark should be eliminated entirely, arguing that a largely nonprofessional selection committee could not really understand the design process.[8]

During the 1950s as Japan forged the institutions of its high-growth economic system, design became a modern profession. For the first time designers were hired as consultants by manufacturers to improve the function and appearance of their products and to create new ones, as well as to advertise them; among them were Jiro Kosugi by Toyo Kogyo (later Mazda) in 1949; SORI YANAGI by a number of firms including the furniture producer TENDO MOKKO and the Tajimi Porcelain Institute in 1956; and YUSAKU KAMEKURA by NIPPON KOGAKU (later NIKON CORPORATION) in 1954. The need for professionalism in design and for recognition of the designer's status was reflected in the organization of Japan's professional design societies, most notably, the Japan Advertising Artists Club (JAAC) in 1951, Japan Industrial Designers Association (JIDA) in 1952, Tokyo Art Directors Club (Tokyo ADC) in 1952, Japan Craft Design Association (JCDA) in 1956, and Japan Interior Designers Association (JID) in 1958. JIDA's founding members consisted primarily of designers associated with the IAI (including KATSUHEI TOYOGUCHI and ISAMU KENMOCHI), which served as a training ground for industrial designers, many of whom later opened their own independent offices.

It was of great significance that the United States occupied Japan after the war, not only for the country's larger political and economic history but also for the history of design. The consultant-design profession in Japan was modeled on the practice of American industrial designers, who had been charged by manufacturers with stimulating consumer interest and multiplying sales during the Depression and early postwar years, and whose well-publicized successes firmly linked their practices with the advance of economic recovery. As the design critic Masaru Katsumie related, interest in design was a by-product of the American occupation: "To the eyes of the general public of Japan, which had suffered from want of materials throughout the war . . . all items that came across the ocean from the United States looked startlingly new and were regarded as symbols of civilization."[9] Yusaku Kamekura described his own experiences in the early postwar years: "Many rectangular boxes were discarded on station platforms by American soldiers. They were the empty containers for combat rations, and they were decorated with some abstract designs in blue . . . . I picked up some of them and took them home. Displaying them on shelves, I felt as though a fresh air of civilization and culture was suddenly filling my room . . . . I thought: This is civilization, this is design."[10] In a newly democratized, parliamentary Japan, American imports reached all segments of the consumer market, and it was this broadened social base that the government wanted to capture in its plans for the nation's economic recovery and growth. Government and industry alike looked to American designers for expertise, and it was no accident that among the first to be invited to Japan was Raymond Loewy, America's best known and most flamboyant designer, whose prolific creations—including the Coldspot refrigerator (1934–37) and Lucky Strike cigarette package (1940–42)—pervaded almost every aspect of American life. Loewy came to Japan in 1951, where he designed the Peace cigarette package for the Japan Tobacco Monopoly and received a then-astronomical fee of four thousand dollars, which instantly increased the prestige of the design profession (fig. 3). His autobiography, *Never Leave Well Enough Alone*, considered a kind of "how-to" manual, was translated into Japanese in 1953. Other American designers followed, among them, in 1957, George Nelson and Freda Diamond, who like the Europeans Bruno Taut and Charlotte Perriand before the war, were invited to review Japanese furniture and other manufactured goods for their suitability as exports. In an informal report to MITI on exporting to the American market, Diamond emphasized the growing importance of housewives as consumers and the need for multipurpose products adapted to small living spaces that were "not only functional but attractive and decorative as well."[11]

If American industrial designers served as role models for their Japanese counterparts, Britain's Design Centre, established by the Council of Industrial Design in London in 1956, set the example for Japan Design House, established in Tokyo by JETRO to promote higher standards of design in industry. One of the British Design Centre's missions was to exhibit what it considered to be well-designed products, initiating in 1957 an awards program to recognize them (as MITI did the same year). From 1960 until 1969, when the activities of Japan Design House were taken over by the Japan Industrial Design Promotion Organization (JIDPO), Japan Design House similarly exhibited articles of "good design . . . in various products for export,"[12] and published a bimonthly news bulletin and the *Japan Design Annual* in English. Likewise in 1960, the Osaka Design House (now the Osaka Design Center) was established under the auspices of the city and prefecture of Osaka, to promote good design in western Japan with its own permanent exhibition hall and "good design" certificates of recognition.

What "good" design was, according to MITI and JETRO, was a subject of much discussion among designers in the 1950s. The phrase itself was an American import, named after the "Good Design" exhibitions and sales sponsored jointly by the Museum of Modern Art in New York and the Merchandise Mart in Chicago between 1950 and 1955, and which Isamu Kenmochi, one of Japan's first designers to travel abroad after the war, saw in New York in 1952 and described as a correspondent to the *Kogei Nyusu*.[13] The belief in absolute aesthetic values and spirit of consumer service—to make beautiful products that serve their purpose and are appropriate to need—gave good design its moral imperative both in the West and in Japan, linking the emergent Japanese design profession to the higher interests of the international design community. In 1953 Kenmochi, along with Sori Yanagi, RIKI WATANABE, Yusaku Kamekura, the architect Kenzo Tange, Masaru Katsumie, and others, formed the International Design Committee (renamed the Good Design Committee in 1959, and finally, the Japan Design Committee in 1963). The aim of the committee, according to Katsumie, was to unite Japanese architects, designers, artists, and critics for the "realization of international good design," by communicating with design organizations abroad and participating in international conferences and exhibitions, and by undertaking "all other matters requisite to the promotion of good design."[14] One of the committee's first, continuing activities, beginning in the fall of 1955, was the selection of well-designed articles for the Matsuya department store's Good Design Corner (fig. 4), established some six months earlier with the help of Yanagi and Watanabe. The committee took its role seriously: "There were not so many articles from which we could select good ones," Katsumie later related, "so we even separated ourselves into several parties and visited a number of wholesalers to seek good articles."[15]

In the book *What Is Modern Design?* by Edgar Kaufmann, Jr. (who as director of the Museum of Modern Art's design department helped make the selection for the "Good Design" exhibitions), good design was defined as "a thorough merging of form and function revealing a practical, uncomplicated sensible beauty."[16] The book, translated into Japanese in 1953, increased "the public interest" in good design, according to Katsumie.[17] But the criteria approved by MITI in 1958 for its Good Design (G-Mark) products selection expressed concern for aesthetic values differently, and for economic ones more strongly, as befitted the government of an important

Fig. 4 Good Design Corner, Matsuya department store, Tokyo, 1961
Fig. 5 Cover of Masaru Katsumie's *Good Design*, Tokyo, 1958

manufacturing nation: "The form and function of the product should work together in a way that is unique; the product should be able to be mass-produced economically using advanced technology; the product design should enhance the natural qualities of the material that it is made of and should be in touch with human nature."[18] Even as government policy these last criteria are distant from the functionalist aesthetics that have informed Western notions of good design, expressing not an idealized relationship of form to function based on industrial methods but a craftsmanlike sensitivity to material and personal content of design. Through his book *Good Design*, published in 1958 (fig. 5), Katsumie, then a member of MITI's G-Mark selection committee, became Japanese good design's chief spokesman; in it he argued against the "abstraction" and impersonality of "functionalist" values and advocated a national version of good design that recognized Japan's cultural traditions.[19]

The movement to create a "good" modern design style that was uniquely Japanese had been initiated by Isamu Kenmochi on his return from America in 1952, when he directed the design division of the IAI. In America, Kenmochi rediscovered Japan's aesthetic traditions through the admiration expressed for Japanese forms by his American colleagues, among them Edward Wormley, T. H. Robsjohn-Gibbings, and Paul McCobb. McCobb was described later by Kenmochi as a designer whose works "would fit into any Japanese house because they are beautifully and naturally shaped and have a Japanese intelligence, smell, and taste, even if he is unaware of it."[20] "We have cared too much about what is foreign," Kenmochi told *Kogei Nyusu*, "and forgotten our own origins. We still need to learn some techniques from the West, but in terms of concepts, we have no reason to follow."[21] This, after all, was the lesson taught by Taut and Perriand to the IAI before the war: that industrialization should not blot out traditional Japanese culture and that Japanese designers should find their modern concepts of design and a new vitality first in Japanese aesthetic principles and then in those of the West.

While all styles and movements of the contemporary West were automatically considered modern, younger Japanese designers could see no reason why a new viable tradition could not be created outside the strict canon of Western modernism. What they discovered in the process of creating a "Japanese Modern" style in the 1950s was that many formal design elements considered prerogatives of Western modernism matched their own ancient visual traditions, among them the use of simple, basic forms and modules, exposed materials, undecorated surfaces, and open spaces. As Katsumie noted in an article about the Swiss designer Max Bill in 1954: "We Japanese had great love for honest form and function. Clarity, purity, and simplicity—these were the aesthetic ideals of Old Japan . . . . Now we must look back upon our traditional treasures and bring [them] to life again."[22] At the same time there were differences: the respect for craft processes, the love of incompleteness, irregularity, and asymmetry, and the use of pattern, description, representation, and quotation. Breaching Western modernism's ingrained taboos on historical references, Japanese Modern became the first postmodern style, well before postmodernism as a movement existed in Europe or the United States.[23] Moreover, in the wide range of its aesthetic expression, complexity, and decorative sensibility, Japanese Modernism displayed the artistic as well as intellectual character of postmodernism. Experiencing history in an active way, Japanese designers could look to the past even when designing objects without historical precedent like electrical appliances, as, for example, ZENICHI MANO, who borrowed the grid pattern and detailing from the architecture of the seventeenth-century Katsura Detached Palace in Kyoto for the housing of his National radio (no. 10), winning in the process a prestigious award initiated by the Mainichi newspaper to promote industrial design (1953). Some manufactured goods such as packaging, for example, remained tied to ancient craft techniques like paper folding. Others retained the tradition of handcraft in engineering as attention to detail and close definition of differences, factors that contributed to the quality and commercial success of Japanese production in the 1950s, as did the adoption of the "quality-circle" management system and standards of professional practice that Japanese manufacturers learned from the Americans J. M. Juran and W. Edwards Deming on their visits to Japan around 1950 and from Juran's book *Quality-Control Handbook* (1951) and Deming's *Sample Design in Business Research* (1960).

Combining traditional design elements and craft techniques with modern industrial ones, Japanese Modern established enduring values and parameters for Japanese design that have provided its best examples with a clear national identity. The issue of identity was of particular importance to designers as they became internationalized in the 1960s and 1970s, discovering their own aesthetic achievements in recognition from abroad, and their competitive strengths on world markets in certain high-technology products requiring miniaturization and precision.

Some designers went abroad to study and work: among them, there were MOTOMI KAWAKAMI, TOSHIYUKI KITA, and MASANORI UMEDA, who worked in Italy; FUJIWO ISHIMOTO and KATSUJI WAKISAKA, in Finland; and Toshihiro Katayama, in the United States. Other designers owed their exposure to the international design community to three international events held for the first time in Japan between 1960 and 1970: the World Design Conference (WoDeCo), which took place in Tokyo in 1960; the Eighteenth Olympic Games, held in Tokyo in 1964; and Expo '70, the world exposition held in Osaka in 1970. Subsequently, there were meetings in Japan of the International Council of Societies of Industrial Design (ICSID) in Kyoto in 1973 and in Nagoya in 1989 (fig. 6), where a World Design Exposition took place at the same time. To each event the government applied a characteristic task-force approach, appointing committees for their planning and production. Graphic designers dominated the WoDeCo and Olympic committees to address the urgent task of creating a graphic program for conferencees and visitors, who were unlikely to understand Japanese.

The WoDeCo committees included HIROMU HARA, YOSHIO HAYAKAWA, Yusaku Kamekura, TAKASHI KONO, and RYUICHI YAMASHIRO, as well as Katsumie, Kenmochi, and other members of the Good Design Committee. The sweeping idea of "visual communication," which lay behind such a program (and also included environmental design), was new to most Japanese designers, and was cited by Katsumie as the most important lesson of the conference.[24] Katsumie himself was named design coordinator and art director for the Olympics and used his conference experience to develop with teams of designers (including KOHEI SUGIURA, IKKO TANAKA, KIYOSHI AWAZU, MITSUO KATSUI, and YOSHIRO YAMASHITA), the first symbol system in Olympics history (no. 69). This system in turn influenced other international graphic programs, including Expo '70 and the Sapporo Winter Olympic games of 1972. "Japan has a long tradition in the function of emblems," Katsumie wrote at the time. "The symbolism and design of Japanese family crests, for example, is one of the most sophisticated uses of visual language in the world. I saw the Olympic Games as a splendid opportunity to capitalize on our experience in this design field . . . . I was secretly counting upon the remarkable possibility of design in Japanese tradition."[25]

Under the theme Progress and Harmony for Mankind, Expo '70 at Osaka presented a view of Japan as one technocratic nation among others. As at other world expositions, there were separate buildings for each country, but there were more buildings representing commerce and industry than at any other exposition. This was "a reflection," one reviewer wrote, "of the importance of industry to the Japanese economy and Japanese life,"[26] as well as an opportunity to display the latest in manufactured goods and concepts. The most distinguished architectural achievement at the exposition was Kenzo Tange's Festival Plaza (fig. 7), a vast techno-expressive space frame at once entirely modern and regional in character, which sheltered futuristic mobile robots, strobe lights, and sound machinery.

Outside the exposition site visitors could see evidence of the newly prosperous Japan that the rapidly growing Japanese economy had produced over the past decade, including new highways and new railways built in connection with the Olympics, most notably the high-speed Bullet Train (Shinkansen) linking Tokyo and Osaka (fig. 8). In addition, new high-rise buildings ripped the skylines, and television antennas sprouted everywhere from tile roofs (television was first broadcast in 1953, and by 1962, 45 percent of all households had owned a set). Televisions were among a number of manufactured goods that the new urban Japanese sought to possess, as Edwin Reischauer and Albert Craig have described: "The 'three sacred treasures' of Japan, anciently the mirror, the jewel, and the sword, became in the late 1950s the television, the refrigerator, and the washing machine. In the early 1960s a new set of consumer 'treasures' were designated the 'three C's,' the car, the color television, and the 'room cooler,' and by the late 1960s there were the 'three V's,' the villa, the vacation, and the visit to a foreign country. The Japanese media described the new privatistic orientation toward home and possessions as 'my-home-ism' and 'my-car-ism'."[27] The government's widely publicized plan of 1960 to double the Japanese income

1 "The Japanese Photo Industry," *Popular Photography*, vol. 41 (Apr. 1957), pp. 136–37.
2 Formerly the Ministry of Commerce and Industry, established 1925, and reorganized as the Ministry of International Trade and Industry in 1949.
3 The Good Design Selection System was operated in its first year by the Japanese patent office and from 1958 to 1974 by MITI, and since 1974 has been under the Japan Industrial Design Promotion Organization (JIDPO).
4 This dealt with patent and copyright issues like the previous Meiji legislation of 1888, 1899, and 1909; see Takashi Takada, "Ishuho no Kaisei ni Tsuite" (Amendments of Design Laws of Japan), *Kogei Nyusu* (Industrial Art News), vol. 26 (Jan. 1958), pp. 13–18.
5 See Mosuke Yoshitake, "Sutokkuhorumu no Soshun: Dezain Toyo Koso Senso ni Tsunagaru" (Stockholm: How Miserable I Was When Scolded for Japanese Imitation), *Kogei Nyusu* (Industrial Art News), vol. 26 (Aug. 1958), pp. 19–21.
6 "Guddo Dezain no Sentei ni Tsuite" (Controversial Problem over "The Good Design Selection of MITI"), *Kogei Nyusu* (Industrial Art News), vol. 26 (Dec. 1958), p. 2.
7 Tokyo, Bridgestone building, Oct. 18–23, 1954.
8 Kogei Zaidan, ed., *Waga Indasutoriaru Dezain: Kosugi Jiro no Hito to Sakuhin* (My Industrial Design: Jiro Kosugi, the Man and His Work) (Tokyo, 1983), pp. 104–5. Although they were in the minority professionally, architects and designers were included in the first G-Mark selection committee of 1957: Junzo Sakakura, RIKI WATANABE, MOSUKE YOSHITAKE, IWATARO KOIKE, MASAKICHI AWASHIMA, Masaru Katsumie, SORI YANAGI, KATSUHEI TOYOGUCHI, YUSAKU KAMEKURA, ISAMU KENMOCHI, and Kenzo Tange.
9 Masaru Katsumie, "Pro et Contra," in *The Graphic Design of Yusaku Kamekura* (New York, 1973), p. 15.
10 Yusaku Kamekura, in ibid.
11 Freda Diamond, "Fureda Diaimondo Joshi no Hokoku" (Report by Mrs. Freda Diamond), *Kogei Nyusu* (Industrial Art News), vol. 26 (Aug. 1958), p. 45.
12 Japan External Trade Organization, *Japan Design Annual* (Tokyo, 1963), preface, n.p.
13 Isamu Kenmochi, "Tobei—America Tsushin" (Trip to America), in Kenmochi Isamu no Sekai Henshu Iinkai, ed., *Kenmochi Isamu no Sekai 4: Sono Shiteki Haikei Nenpu / Kiroku* (The World of Isamu Kenmochi 4: Historical Background—Chronological History / Record) (Tokyo, 1961), vol. 4, pp. 118–19.
14 Masaru Katsumie, in Japan Design Committee, *Dezain no Kido: Nippon Dezain Komittei to Guddo Dezain Undo* (The Way of Design: The Japan Design Committee and the Good Design Movement) (Tokyo, 1977), pp. 17–18.
15 Ibid., p. 19.
16 Edgar Kaufmann, Jr., *What Is Modern Design?* (New York, 1950), p. 9.
17 Katsumie, in *Dezain no Kido*, cited above, p. 18.
18 "G Maaku o Eta Seihin" ("Good Design" Approved by Japanese Government), *Kogei Nyusu* (Industrial Art News), vol. 26 (May 1958), p. 54.
19 Masaru Katsumie, *Guddo Dezain* (Good Design) (Tokyo, 1958), n.p.
20 Isamu Kenmochi, "Pauru Makkabu no Hito to Sakuhin" (Paul McCobb: His Profile and Works), *Kogei Nyusu* (Industrial Art News), vol. 23 (June 1955), p. 30.
21 Isamu Kenmochi, "Tobei—America Tsushin," cited above, p. 119.
22 Masaru Katsumie, "Makkusu Biru no Zokei" (Plastic World of Max Bill), *Kogei Nyusu* (Industrial Art News), vol. 22 (Apr. 1954), p. 15.
23 On postmodernism in the context of design, see Kathryn B. Hiesinger and George H. Marcus, *Landmarks of Twentieth-Century Design* (New York, 1993), pp. 277–83.
24 Masaru Katsumie, "World Design Conference (WoDeCo)" [1960], in "Katsumie Masaru: 24 Messages," *Graphic Design*, no. 100 (Mar. 1986), p. 82.
25 Masaru Katsumie, "On Problems of International Symbology" [1965], in ibid., p. 83.
26 J. M. Richards, "Expo '70," *Architectural Review*, vol. 148 (July 1970), p. 67.
27 Edwin O. Reischauer and Albert M. Craig, *Japan: Tradition and Transformation* (New York, 1978), pp. 28–29.
28 "Expo ABZ," *Architectural Design*, vol. 40 (June 1970), p. 273.
29 "Sony TR Terebi 8-301 Kei" (Sony TR Television 8-301), *Kogei Nyusu* (Industrial Art News), vol. 28 (Feb. 1960), p. 79.
30 J. Gordon Lippincott, *Design for Business* (Chicago, 1947), p. 10.
31 "Design Policy for the 1990s," *Design Quarterly Japan*, no. 1 (1989), p. 18.
32 Ibid.
33 Shinya Iwakura, in John Heskett, "Made in Japan," *Industrial Design*, vol. 31 (Jan.–Feb. 1984), p. 18.

within a decade (it was surpassed in seven years) produced in Japan Asia's first modern consumer society, although living conditions still remained poor by Western standards (a reporter for the British magazine *Architectural Design* noted in 1970 that an extended family of eight with three television sets and three cars was living together in Kyoto in a small, six-room traditional wooden house without indoor plumbing).[28]

The manufactured goods that drove and reflected Japan's transformation into a world economic power were developed by a generation of entrepreneurs, Sony Corporation and HONDA MOTOR COMPANY among them, who were encouraged in their expansionary instincts by the government with lower interest rates and elastic credit lines. The entrepreneurs came late to design: in the 1950s their first products were designed largely by engineers, as at Sony, or in imitation of European designs, as at CANON, where widely admired German Leica cameras were copied, and at Honda, where its first automobiles of the 1960s were inspired by the British MG and Austin-Healy. Pushed by the government, the entrepreneurs responded to the problems imitation raised by hiring designers, Zenichi Mano becoming the first full-time product designer at a Japanese corporation when he joined MATSUSHITA ELECTRIC INDUSTRIAL COMPANY in 1951. By the 1960s most large firms had in-house design divisions (Canon's was established in 1965), whose team achievements in the innovative function and appearance of their products drew public attention. *Kogei Nyusu*, for example, reported enthusiastically, in a feature article of 1960, the technological and stylistic advantages of the world's first all-transistor television made by Sony (no. 44).[29] With its irregularly spaced knobs and grill slots and projecting hood that overhung the screen like a Japanese roof to reduce glare, the television exemplified the aesthetics of Japanese Modernism. Improving on the principle basic to American commercial competition of restyling products annually to make them appear new ("The American consumer *expects* new and better products every year," wrote J. Gordon Lippincott, in *Design for Business* in 1947[30]), Japanese electronics firms pushed rapid technological innovation with semiconductor devices to make their products new, better, and different while becoming ever smaller and increasingly multifunctional. Since manufacturers could copy and bring to market quite complex products within a few months, innovation and rapid introduction of models remained a key to staying ahead.

Oil crisis, recession, and competition from rapidly developing countries in the 1970s made vividly clear to the government Japan's dependence on the outside world, and its need to respond to changing global economic conditions with new paradigms for design. In 1981, with the support of MITI and the city and prefecture of Osaka, the Japan Design Foundation was established to promote international exchange in design activities, including an international design competition held biannually in Osaka, and a pan-Pacific design exchange program.

In the years around 1990, when the recession in the West spread to Japan and all leading industrial countries of the world were in recession simultaneously, MITI published a "Design Policy for the 1990s" to "encourage the creation of innovative design of which we can be truly proud."[31] The policy statement set broad guidelines for the study, practice, promotion, and international exchange of design in Japan and stressed the enduring values of design as a cultural aspect of industry, "deeply . . . interwoven with society . . . to be handed down to our descendants as the heritage of our age."[32]

Japanese design is still a field largely unknown today in the West and one whose visual forms are less understood than the technology they incorporate. One of its most enduring characteristics since 1950 has been its ability to make use simultaneously and without bias of both the traditional and the modern. For Shinya Iwakura, director of research and development at Honda in 1984, Japan's long history of craftsmanship and aesthetic values entered the modern age in engineering and technology: "Historical tradition is therefore the base, proving a capacity for sensitivity to detail and a clear definition of small differences. Aesthetic standards are implicit in Japanese culture, in much the same way that classical standards were once a norm in Western countries."[33] Since the West has lately begun to invest cultural meaning in design and to broaden its aesthetic standards by referring to historical examples and adopting quotation as an approach, one wonders whether this development can be under the influence of Japan.

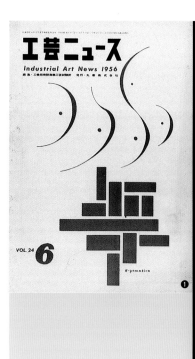

In 1958 the Dezain-ka (Design Section) was established in the Ministry of International Trade and Industry (MITI), and I was appointed its first section chief. This was not the beginning of the government administration of design in Japan; matters concerning design had previously been handled by several sections in MITI, but the significance of a section established to handle design matters exclusively, and more important, one with a foreign name, Design—used without translation—was a landmark and can be considered one of the first steps in the evolution of design in Japan.

The Design Section was under the Bureau of International Trade Administration. If design had been considered a creative activity concerned with life-style and culture, its administration would have been under the jurisdiction of a department concerned with the production and distribution of household goods. But just as the tendency today among developing nations is to view design as something that will help to promote exports, so it was with Japan in the late 1950s.

Although the economic white paper for 1954 declared the end of the postwar period, economic policy at that time focused on inflation control, and the economy was finally overcoming shortages and showing signs of recovery. The seller's market was becoming a buyer's market, and design was beginning to be recognized as a powerful means to stimulate sales. The Japanese economy was restricted by the amount of foreign currency available, and the acquisition of foreign currency through exports was an issue of highest priority. However, there were also frequent claims of design piracy from abroad. This began to affect exports in general, and measures to cope with the problem became urgent, which was one of the issues that led to the establishment of the Design Section. Its first task was to deal with this problem. In 1959 the Export Commodities Design Law was enacted to prevent design piracy. It provided that the minister of international trade and industry would designate certain products (Specially Designated Commodities) that required the approval of the agency designated by the minister (Specially Designated Agency) before they could be exported.

Among the Specially Designated Commodities were fountain pens, toys, household furniture, cameras, fishing reels, cookware, Christmas tree decorations, and tape recorders. Among the Specially Designated Agencies were the Japan Ceramics Design Center, Japan Merchandise Design Center, and Japan Machinery Design Center. For textiles, the Japan Textile Design Center, a nongovernment organization in Osaka, was already following self-imposed controls. With these measures, design piracy problems gradually decreased and Japanese design advanced. Now the situation is reversed, and Japan has itself become the victim of design piracy by other Asian nations.

The phenomenon of design piracy in the 1950s fell into two areas. One concerned the small- and medium-size enterprises that produced goods exactly as specified in the orders of foreign buyers; while these firms bore the liability for design piracy, most of the profits were being made by buyers abroad. The second, which should not be overlooked, was the failure of Japan to recognize that an intangible product called design had its own economic value. That climate supported the rise of piracy, and the rectification of this had to be the starting point of the advancement of design.

Such measures as preventing design piracy and maintaining the integrity of design were not enough to promote the advancement of design, but clearly they were the minimal requirement. Positive examples of the role that the government played in design development are exemplified by the Industrial Arts Institute (IAI), and the

Design Promotion Council. The Industrial Arts Institute was established in Sendai in 1928 as a regional undertaking; in 1932 it became an agency of the Ministry of Commerce and Industry (later MITI). It was moved to Tokyo, and as the core agency for design development in Japan, it played a leading role in the advancement of design in Japan for almost half a century. *Kogei Nyusu* (Industrial Art News), which was a publication of the institute (fig. 1), contributed greatly to pointing design education in the right direction. Of great value were the superior talents that the IAI generated, including those of ISAMU KENMOCHI, KATSUHEI TOYOGUCHI, MOSUKE YOSHITAKE, IWATARO KOIKE, and many others. These civil servants, together with nongovernment designers, were the pioneers who made Japanese modern design what it is today.

The Design Promotion Council was established in 1957 in the patent office, an extraministerial office of MITI, but it was transferred to the Design Section after that had been established. As advisor to MITI for design administration, it has submitted reports on design promotion measures such as the Good Design Selection System and the establishment of design promotion agencies. Under the Good Design Selection System, products are selected by specialists from among those recommended, and confirmed and publicized by the government. These products are then entitled to use the special G-Mark logo for sales display. Much discussion about whether the government should be involved in deciding what is good design took place, but in making improvements in the system to conform with changing times, the G-Mark still performs a useful function and remains one of the important activities of the Japan Industrial Design Promotion Organization.

The Japan Design House was established in 1960 in the Japan travel bureau building opposite Tokyo's central station through the cooperation of the Japan External Trade Organization for the purpose of promoting design (fig. 2). The first director was Shinji Koike, who was then dean of the engineering faculty at Chiba University in Tokyo. One of its projects was the plan for promotion of Japanese craft exports to the United States. Under the plan, called the Wright Plan because the project was proposed by the American designer Russel Wright, American designers and businessmen were invited to Japan, where they gave advice, studied Japanese design, and scouted for goods that were suitable for export. Japanese designers responded by redesigning their products in line with the recommendations. Although the Wright Plan was not as successful as had been anticipated, it contributed greatly to the awareness of design in many localities.

Other projects were also implemented, such as foreign design-study programs, gathering of samples of design from foreign countries, exchange of merchandise between foreign and Japanese department stores, design-study tours to the United States, and invitations to foreign designers to visit Japan. The leading role that the United States played in the advancement of Japanese design during that period cannot be overemphasized.

Although government intervention was successful in regulating design piracy and maintaining design integrity, the high value placed on Japanese design today is the result of both government administration and civilian efforts. While the public was finally becoming aware of a profession called design, and design organizations were formed by various groups, the position of the designer in the commercial world of product manufacturing was only secondary. But the mere fact that a Design Section was established by the central government made Japanese businessmen begin to value design, promoted design awareness, and provided greater opportunities to further the activities and involvement of designers.

Fig. 1 Cover of *Kogei Nyusu* (Industrial Art News), designed by RYUICHI YAMASHIRO, 1956
Fig. 2 Cover of *Japan Design House*

Design and Marketing

For a century after the beginning of Japan's modernization in the mid-nineteenth century, Japanese companies adopted (and the government supported) a relatively paternalistic view of product development. Companies made the kinds of products they thought would be "good for society" and consumers accepted what was offered. Freedom of choice was not an issue.

But values changed as Japanese society became more affluent. Even if a mass-produced item is of good quality and reasonably priced, it has little cachet now. The Japanese used to scorn handmade clothing, feeling that it was the last resort for people who could not afford anything better; now fashion has come full circle: those who wear handmade clothes are seen as being able to afford the ultimate luxury—an individualistic life-style. Manufacturers must come to grips with these shifting values. The ideal marketing strategy today is to produce a large variety of small product lots, which are seen as more individualized, instead of mass-producing a single item in large quantities. This demands a high degree of production-engineering skill and flexibility, as well as impeccable quality control. Achieving these new skills has meant a complete revamping of Japanese corporations.

Early marketing techniques relied heavily on feature-benefit comparisons. At first, marketers emphasized models with more features or those with comparable features at lower prices. "Compare and save," was the watchword of the day. As time passed, however, Japanese industry enjoyed improvements in the general level of manufacturing technology, and distinctions between products became more and more blurred. Consumers could rely on virtually all manufacturers to supply products that had similar features and were of equally high quality.

At the same time manufacturers were engaging in a fierce cost-cutting war that resulted in virtually identical cost factors across the board. A good example is the home appliance industry: while large manufacturers had originally been able to offer products at lower prices by taking advantage of large production runs, by the late 1970s even small producers were matching them, and price, too, was eliminated as a basis of comparison for consumers. The average consumer was left with no definitive motivation for purchasing, and marketing strategists had to develop a new way to project product appeal.

That appeal is now largely based on image—be it product design, brand image, corporate image, or some other undefined aspect. In Japan consumers choose products on the basis of a good reason to buy rather than a good price. Japanese mass-merchandisers differ from their American counterparts in several ways, one of the most significant being the use of store brands. Until recently, both Sears and J. C. Penney in the United States relied heavily on private-label merchandise. In contrast, Japanese merchandisers did not have great success with private labels, although the American concept had been widely imitated. According to one estimate, store brands account for only 20 percent of Japanese general merchandise sales. Why? Japanese are more image conscious. Japanese consumers will choose a national brand over a store brand even when the store brand is of comparable quality and represents substantial cost savings.

In order to counteract this, merchandisers have begun to turn the private label into a life-style brand of its own. Seiyu's No-Brand brand was the pioneer—a house brand packed and merchandised as if it were a distinctive national label. This resulted in explosive sales. Renown, a major Japanese clothing manufacturer, used the same technique for its [ixi:z] brand. Renown started by creating a master brand image for [ixi:z], then filled out the product line over a ten-year period with a wide variety of supplier-made items, from sport shirts to luggage. All were narrowly targeted to male college students between twenty and twenty-two. Renown's carefully targeted marketing strategy was a principal asset in product development, and through the narrow targeting of this label a distinct brand image was quickly established. Not surprisingly, it sold very well, but what was surprising was that despite the narrow targeting, its customers actually ranged from teenagers to fifty-year-olds.

Such umbrella brands can unify a great variety of products under one familiar label. The key is to build the information value of the brand first, and only then to insert the products. Thus, image marketing has reversed the conventional wisdom of building a brand on the strength of one successful product. Image marketing works

1

Fig. 1 Bridgestone corporate logo
Fig. 2 Original Mazda logo
Fig. 3 Redesigned Mazda logo

the same way on the corporate level. Consumers are motivated to purchase products because the corporate umbrella brand is appealing. "If I buy a Honda," thinks the Japanese consumer, "I will be cool."

Internationalization is another way that the marketing of design has been spurred in Japan. Japanese companies have traditionally relied upon tacit understandings that define their positions in the domestic marketplace. This was possible because the Japanese society is so homogeneous. However, the continuing expansion into world markets has brought a need for clearer definitions.

Words borrowed from other languages (and frequently used in ways far removed from their original meanings) make up a large share of Japanese product names. One source estimates that between four and five thousand foreign words have already been incorporated into everyday Japanese, an exotic assortment of homegrown English, French, and German. These foreignisms have also made their way into corporate names. Bridgestone is one example (fig. 1). The founder's name was Ishibashi, which is written in Japanese with the characters meaning "stone bridge." He turned the words around and used them in English to coin Bridgestone, a name he felt connoted forward thinking.

But the use of international borrowings does not always work outside Japan. The stylized M used in Mazda's old logo was read as hn by many Americans, although the Japanese saw the M as intended (fig. 2). The problem was that Westerners have a conventional understanding of how the letter M should look—change it too much and it no longer looks like an M. To the Japanese, on the other hand, the Roman alphabet is less familiar, and for them an M can be defined any way the designer likes. Originally, Mazda's trademark was written with the three Japanese phonetic symbols for *ma-tsu-da*. The most common association with the name was the old putt-putt three-wheel motorbikes that had been Mazda's mainstay in the early postwar years. Although the company had ceased production of three-wheelers and had repositioned itself as an automobile manufacturer, the public still clung to its old putt-putt image of the Mazda, which was largely associated with the old trademark (no. 42). This was hardly an asset in entering the cutthroat international automotive marketing arena. Thus the company's mark was changed to the Roman letters MAZDA, and a visual identity system (VIS) was developed for it (fig. 3). A set of standard signage designs and showroom layouts was created for different types of dealers. These were customized by an automated system for each site, according to VIS standards, a system that presaged what we now call CAD / CAM (computer-aided design / computer-aided manufacturing).

Toyota provides a similar story. Toyota's early corporate mark used the Japanese phonetic symbols for to-yo-ta in a circle. In the company's early stages of international marketing, dealers frequently displayed the symbol backward at sales counters, even at dealership opening ceremonies, while service technicians attached the symbol plate to cars backward, because to Americans who do not read Japanese it looks right no matter which way it faces. The Toyota mark was not visually international and had to be changed.

Many industries must now face rapid and profound changes at a time when the demand for traditional product categories has stagnated. These businesses must diversify into new areas and then differentiate themselves in their new positions. The best chances for success are among the companies that carve out and name entirely new business categories rather than those that move into established areas.

Inax expanded into a new area when it opened its luxurious Xsite showroom in 1986 atop Tokyo's newest and tallest "intelligent" building in the Ark Hills multi-use complex. Although it is the world's largest bathroom-fixture showroom, with products from thirty different manufacturers from ten countries, Xsite (pronounced "excite") sells more than products. It sells the concept of bathrooms as a life-style choice. By opening Xsite, Inax broke out of manufacturing into the service sector and undertook a major new marketing step.

The case of the Matsuya department store in Tokyo's central Ginza shopping district illustrates the process of marketing as practiced in Japan. Matsuya's troubles can be traced back to the immediate postwar years. Because it was the largest store in the Ginza, Matsuya was requisitioned for use by the United States occupational forces as

2

3

the Tokyo PX. It was not returned to private enterprise until 1952, giving Matsuya a seven-year handicap as compared to other downtown stores. In the 1970s Matsuya retained the Paos design agency to develop a strategy for the main Ginza store. By that time it had already divested its Yokohama and Funabashi stores, leaving only the Ginza store and a smaller facility in Tokyo's quaint Asakusa "old town." Matsuya's troubles went deeper than floor plans; the whole department store sector, in fact, had conceptual problems that limited its profitability. Yet no one was addressing them. How should a department store be run in this modern age? Why do people shop at department stores? What does it mean to be a Tokyo department store? And what is the role of a Ginza department store? These basic questions were not even being considered then.

At that time most Japanese department stores were expanding at a fast rate, opening new branch stores to keep pace with the mass-merchandising chains. As Matsuya had already shrunk to a comparatively tiny, two-store chain, its managers feared that they had lost the battle for survival. But Paos suggested a different approach and showed that there would still be hope if they could elevate shopping from a chore to an experience, from a needs-based to a wants-based activity.

Paos demonstrated that a department store with only one or two sites could use its size as a mark of distinction. Works of art derive part of their value from their rarity; custom-produced products are more valuable than mass-produced items, their very scarcity making the difference. And Matsuya, too, would offer more originality—and more value—by having only one main store, in the Ginza; this was renamed Matsuya Ginza.

The first step suggested was to discard Matsuya's old-fashioned pine and crane symbol because it was synonymous with the old Matsuya (fig. 4). Most of the old-guard Japanese department stores used a circle-based mark in their logos, denoting the rolls of kimono fabric that had been their mainstay. Matsuya was no exception, but this tradition-rooted symbol had to go in the quest for a more exciting shopping experience. Positioning was the point of departure in developing Matsuya's new mark. Youth was a major theme in retailing of the day, typified in Tokyo by the stores at railway terminals in trendy areas of the city. Matsuya's Ginza location meant that a youth identity was out of the question, but it could appeal to a slightly older, more sophisticated, and better heeled group: Japanese yuppies, with their keen sense of quality and style. Paos concentrated on making the store design appeal to this group, choosing sophisticated colors like slate blue and chocolate instead of the hot shades of red and yellow typically found in Japanese store environments.

One aspect of Matsuya's identity campaign that made an especially deep impression was the decision of the designer Masayoshi Nakajo to use the name Matsuya Ginza in Roman letters (fig. 5). Almost all department stores were using Japanese characters in a circle to identify themselves, and Romanization seemed like a breath of fresh air. In the choice of fresh colors and Romanized letters, the idea was to demonstrate that Matsuya was a store unlike any other, in its merchandise, its display, and its service. The strategy succeeded, leaving Matsuya in good shape after the department-store war. Matsuya continues to keep up with its targeted audience through its Tokyo Life-style Research Institute, founded in the fall of 1985. The institute is at the core of Matsuya's single-focus development, a fountain of creative ideas and innovations, and the combination of its research and market data is fed back into the store's merchandising policies.

Japan has attained its goal of world-class standing in the economic and technological spheres; our corporations must do the same in culture. Many corporations are more comfortable dealing with quantitative data than with aesthetic considerations. Beauty is unfamiliar, uncomfortable, subjective, transitory, even personal, and it cannot be measured. Management has to understand more about the mechanism of beauty in order to make a reasonable investment in aesthetics. Japan's headlong rush to modernize over the last century and a quarter has alienated us from some of our cultural roots. Now that Japan is searching to define a new international role for itself, we must return to our aesthetic and cultural origins.

Today's company must find a balance between quantitative management, humanistic management, and aesthetic management. In other words, the corporation must be an economic machine that provides aesthetic as well as social value. Those companies able to synthesize these diverse spheres will have the edge in the century to come.

4

5

Fig. 4 Matsuya department store's crest of pine and crane
Fig. 5 Redesigned Matsuya Ginza logo

Six professional designers associations have been recognized by the Ministry of International Trade and Industry (MITI) as having the status of incorporated bodies: Japan Industrial Designers Association (JIDA), established 1952; Japan Craft Design Association (JCDA), established 1956; Japan Interior Designers Association (JID), established 1958; Japan Package Design Association (JPDA), established 1960; Japan Jewelry Designers Association (JJDA), established 1964; and Japan Graphic Designers Association (JAGDA), established 1978 (see fig. 1). Their goal is to perpetuate the cultural traditions by supporting the livelihood of their craftsmen and designers as they draw on their life experiences and philosophies to accommodate the design of objects, cities, and the environment to the future needs of the world. At the same time they want to bring international recognition and understanding to the underlying philosophy of the Japanese life-style and its culture.

Since the 1970s these organizations have worked closely to further the cause of design, receiving government support and sponsorship for their joint activities. For the 1973 meeting in Kyoto of the International Council of the Societies of Industrial Design (ICSID), for example, these organizations formed the nucleus of the executive committee. The government designated 1973 as Design Year in Japan, and various design-related groups participated in organizing it. The ICSID theme, "Soul and Material Things," was the proposal of the design organizations and showed that they understood that the industrial society in which objects were measured in quantity and size was changing into a society that questions their communicative values.

Thus the function of design has expanded from material objects to the environment, and international contacts are becoming more active. Teaching and research of design in Southeast Asia are conducted as a joint project by the six organizations, and foreign research students assigned through the Japan International Cooperation Association (JICA) also participate in this undertaking.

The 1989 ICSID conference in Nagoya was held in conjunction with the Design Year designated by the government, and the six organizations again became the nucleus of the executive committee. The theme, "Emerging Landscape: Order and Aesthetic in the Information Age"—agreed upon after more than a year of deliberations—was, as the words indicate, a confirmation of a new purpose and direction for the designers of Japan. From what viewpoint should designers approach the relations between man and his environment, man and the city, and man and the planet, and how can they work for the betterment of society?

The six organizations, with the Japan Industrial Design Promotion Organization (JIDPO) as its hub, form a Design Organizations Council. This council functions effectively for international events as well as national ones that encompass a wide scope. Information is disseminated to all members accurately and quickly through regularly scheduled symposiums and exhibitions of new products from the various specialized fields.

In 1993, in an interim report of the MITI Design Promotion Council entitled "A New Design Policy to Cope with Changes in a New Age," a joint proposal of council members invited from the six organizations enunciated three concrete plans for action: establishment of a system to support the training of promising designers, promotion of design in small and mid-size enterprises and in various regions, and expansion and strengthening of international cooperative design projects. As the new global age approaches, the key to the future lies in giving true value to design as a common language.

Fig. 1 Logos (from top) of Japan Industrial Designers Association (JIDA), Japan Craft Design Association (JCDA), Japan Interior Designers Association (JID), Japan Package Design Association (JPDA), Japan Jewelry Designers Association (JJDA), and Japan Graphic Designers Association (JAGDA)

The development of design education in Japan followed two separate paths. Before World War II, it was shaped by developments in Germany, particularly the Deutscher Werkbund and the Bauhaus, while after the war, the American industrial design approach, which had made extensive inroads in close association with industry, had a significant impact.

While Japan's art education was reformed under the influence of the Bauhaus before the war, design education was not immediately established as a specialized discipline. Indeed, it was not until the early postwar period, during the 1950s, that specialized design education at the high-school and university levels had their real beginnings in Japan.

The 1950s marked the dawn of Japanese design itself. This decade witnessed a new beginning, heralded by the adoption, in the early 1950s, of the word "design" itself, a direct borrowing from Anglo-Saxon usage. Before the war the Japanese concepts denoting design had all been terms written in Chinese characters, such as *zuan* (design / sketch / pattern) and *isho* (design / idea). In Japanese, words of foreign origin (loan words) are transliterated in a phonetic syllabic alphabet known as *katakana*. The word "design" (*dezain*) presented in this phonetic transliteration was a new concept also in the sense that it stood for change, a revolutionary start with the promise of new hope after the war. The use of the *katakana* writing itself suggested a break with the old, offering a new image and inspiring new expectations of peace, democracy, freedom, and economic betterment. The model for this was of course American culture, the American life-style with its material prosperity and affluence and its familiar adjuncts—products and goods offering efficiency and convenience, such as household electrical appliances and private automobiles.

In the general quest for this type of affluent life-style and economic growth, Japan's industrial and educational sectors—and with them the design field—went through a short period of dramatic change in the first half of the 1950s. The overseas study program for basic research played a major role in the postwar development of design education and research and the progress of design activities in Japan's industry and corporate world. This program, from 1955 through 1966, was instituted under the Ministry of International Trade and Industry (MITI) and administered by the Japan External Trade Organization (JETRO). These efforts were launched as an attempt to improve industrial design and promote exports. During the eleven years that the system was in effect eighty-two candidates from the design area were selected for overseas study in art and design schools and colleges, and internships in design offices for at least one year. While most of the candidates went to the United States, some were also sent to other countries, including Denmark, Sweden, Finland, West Germany (particularly to the Hochschule für Gestaltung at Ulm), Italy, the United Kingdom, France, and Switzerland. Most who went to the United States studied at the Art Center School (now the Art Center College of Design) in Pasadena. Other institutions that played host to Japanese students were the Pratt Institute in New York, Illinois Institute of Technology in Chicago, and the Cranbrook Academy of Art in Bloomfield Hills, Michigan.

Kogei Nyusu (Industrial Art News), a monthly publication issued by MITI's Industrial Arts Institute that was published from 1932 to 1974, also played an important role. With its informative articles and its papers on design methodology and research above and beyond the scope of practical design activities, it made a significant contribution to the development of design education and research in Japan, especially after the war.

During the fifteen years between 1956 and 1971 the Industrial Arts Institute also furthered the teaching and propagation of design through its lectures and workshops featuring design experts invited from abroad. Between 1956 and 1960, the institute invited well-known designers and design educators, mainly from the United States, who brought information about the current design trends and practices abroad. From 1961 to 1963, the institute organized events primarily with the participation of leading authorities, again mainly from the United States, in the areas of consumer education and human-factors engineering. From 1964 to 1967, the institute invited designers primarily from European countries, and from 1968 to 1971, leading design critics and experts, mainly from Europe, came to present their views and theories as the dawn of the information age approached.

Until the 1950s, when design activities in Japan had not yet penetrated sufficiently into the private sector, the Industrial Arts Institute was a haven for design, unique for the considerable freedom it offered for design experimentation and research activities. Many of those who played an important and pioneering role in the development of postwar Japanese design as designers, critics, and educators were attached to the institute, which they used as the ideal workshop for their design activities.

The universities began to search for an industrial design education system early in the 1950s. These efforts began in 1951 with the establishment of the department of industrial design within the faculty of engineering at Chiba University in Chiba City and the founding of the design department at the Tokyo National University of Fine Arts and Music. The establishment of an industrial design department within a faculty of engineering was a watershed in Japanese design education. The mainstream of design education until then had been shaped by the Tokyo College of Industrial Arts (forerunner of the department of industrial design at Chiba University, founded in 1921) and the Tokyo School of Fine Arts (forerunner of the Tokyo National University of Fine Arts and Music, founded in 1887). Both are national institutions with long and proud traditions. They represent the historic dichotomy of design education in Japan, one as part of the faculty of engineering and the other as part of the fine arts curriculum. Each of these approaches has its own features and its own standing and status tacitly endorsed by the students as they pride themselves on having followed one pathway or the other.

The theoretical basis of the department of industrial design at Chiba University was established by Professor Shinji Koike (1901–1981), later the founder of the Kyushu Institute of Design (1968). The spiritual father of design education at the Tokyo National University of Fine Arts and Music was Professor IWATARO KOIKE, an industrial designer who was also the progenitor of the GK design group and an activist in Japanese design, making significant contributions to the development of the Japan Industrial Designers Association and the Japan Society for Science of Design (JSSD, founded in 1953).

There are historic precedents for government-sponsored design schools with unique approaches. The curriculum of the Kyoto University of Crafts and Textiles, founded in 1899, fuses traditional techniques, including such textile processes as machine weaving and fabric dyeing, and modern technology. Equally outstanding and unique are the Kanazawa University of Arts and Crafts and the design department of Tsukuba University (formerly Tokyo University of Education).

As an alternative to these national design institutions, there was yet another endeavor to create a design school, a unique attempt worth noting here. In 1954 Yoko Kuwasawa (1910–1977), a clothing designer in Tokyo's Aoyama section, founded a professional school known as the Kuwasawa Design School. Her school was the basis of a new design movement that sought to find answers to such challenging issues as the nature of modern design and the way in which design should be taught. Later, in 1966, Kuwasawa established the Tokyo University of Art and Design.

In the 1930s the architect Renshichiro Kawakita had opened a short-lived school in the Ginza section of Tokyo in which he sought to establish a curriculum of design education based on the methodological approach and principles of the German Bauhaus. He invited the architects Takehiko Mizutani and Iwao Yamawaki and the textile designer Michiko Yamawaki to join the school; these three designers shared a common background as former students of the Bauhaus in Dessau. The graphic designer YUSAKU KAMEKURA and Sofu Teshigahara, initiator of the avant-garde Sogetsu school of flower arrangement, studied there, as did Yoko Kuwasawa, who first came under the influence of the Bauhaus at this school.

In June 1954 Walter Gropius, founder of the Bauhaus, came to Japan and paid a visit to the Kuwasawa Design School (fig. 1). Preeminent among the many leading designers who taught at the school was Masaru Katsumie (1909–1983), who as a design critic and long-time editor of *Graphic Design*, an international journal of the highest standards, played a significant role in the promulgation of modern design in Japan. As president of JSSD he also made a major contribution to the promotion of academic design studies and to the advancement of design as a serious academic discipline.

In the arts field, private universities such as the Musashino Art University (fig. 2) and the Tama Art University occupy an independent position in relation to the government-supported universities and engage in design education on this basis with greater freedom to experiment. For example, during the 1960s, a decade of rapid economic growth for Japan, a high-level mass-consumer society ushered in an ever-finer division and subdivision within the domain of design, and design education followed this trend with increasing specialization and compartmentalization. In the face of these developments Musashino Art University decided to engage in a new attempt, an experiment in design education (fig. 3).

The academic establishment offers an already defined range of departments in the design field. Within this established spectrum of visual communication—industrial design, interior and craft (ceramics, metal, textile, plastics, wood) design, scenography, display, and architecture—the searching question was what new course of study might be added to this and what the nature of its educational content should be.

We have not yet established a design theory that can be shared by, or be universally applied to, the various fields of design. I am convinced of the need to build the philosophical foundation of design and reassert the notion of what the Greeks called *poiesis* in its complex meaning, denoting the creativity of design.

Instead of breaking down the design field further, it is important to embrace design in its totality. The department of science of design came into being at Musashino in 1967 to seek the overall synthesis of design and to search for an approach that would emphasize interdisciplinary theories and educational methods to educate a new type of specialist. In Japanese this scientific department is called Kiso Design-gaku (Fundamental Study of Design), a discipline that aims to discover and develop new design topics and provide potential for creative activity in such varied fields as design planning, research, education, and criticism. Amid the turning tide toward the information-based society, we are now witnessing an even greater need for this type of educational approach to train a new generation of design specialists with unique capabilities encompassing numerous areas of design.

1 For an attempt to search for the relationship between the system of notation used in written Japanese and Japanese culture, see Shutaro Mukai, "Characters That Represent, Reflect, and Translate Culture in the Context of the Revolution in Modern Art," in Yoshihiko Ikegami, ed., *The Empire of Signs: Semiotic Essays on Japanese Culture* (Amsterdam, 1991).

Fig. 1 Walter Gropius at the Kuwasawa Design School with Yoko Kuwasawa (second from left) and Masaru Katsumie (right), 1954
Fig. 2 Museum & Library, Musashino Art University, Tokyo
Fig. 3 Student presentation, Musashino Art University

Recently, we have seen the continued founding of design universities and faculties in all parts of Japan. There are now some forty universities in the area of art and design, and some sixty junior colleges offering two-year courses, as well as over a hundred professional and related schools. In addition, there are more than forty universities of education (Teachers Training Colleges) with art and design courses for high school teachers.

This may reflect, in many ways, the importance attached to regional society and local industries. The Kobe Design University in western Japan and the Tohoku University of Art and Design in northern Japan are examples of regional schools. While the English names of these two universities differ, the Japanese titles, except for place names, are identical. Written in Chinese characters, they both read Geijutsu Koka Daigaku; literally translated, this means Art Engineering University, a name that may be rendered as University for Engineering of Art or University for Integration of Art and Engineering. As the concept of art engineering (*geijutsu kogaku*) indicates, the way in which design is being understood or interpreted implies a synthesis of art and technology.

This approach was first proposed by Shinji Koike. While the English name of the school he founded is Kyushu Institute of Design, the Japanese title, in Chinese characters, is Kyushu Geijutsu Koka Daigaku (literally, Kyushu Art Engineering University), a first attempt at proposing the synthesis of art and engineering. Here a specialization in a hierarchical order is eschewed in favor of a juxtaposition of the two concepts of art and engineering in a horizontal side-by-side ideal of an education based on an overall synthesis. The objective of this education is not to train the specialist but to create the generalist, at home in a wide range of specialties.

In this sense, the educational orientation of these universities has much in common with the concept of the Musashino Art University's department of science of design, opened in 1967. This is a testimony to a new quest for the notion of an integral design education. We are witnessing today a new call for a holistic system of design education and an educational system that opens a broad spectrum of possibilities.

For the Japanese, the postwar adoption of the word "design" in syllabic transliteration (*katakana*) ushered in a change in our life-style. While this term has firmly established itself in general usage, the art and design universities and faculties have retained the old concepts molded in the form of Chinese characters for their names and for the description of the courses they offer. In part, this may be explained by the Ministry of Education's known dislike for foreign loan words that have found their way into Japanese. We have here a problem deeply rooted in the phenomenon of design itself.

As the meaning of design as we know it is constantly being reshaped and regenerated toward a perceived ideal, we are looking for a new concept in the guise of new Chinese characters, which may take the form, as we have seen, of Kiso Design-gaku (Fundamental Study of Design) or Geijutsu Kogaku (Art Engineering).[1]

With the spread of present-day multimedia communication, we are now considering what the next stage of design education should be within a perspective that makes us aware of how we should view these problems. The direction in which new reform is pursued points to a new, more open, educational system offering a broader range of possibilities.

One of the central issues of Japanese design education in the future may be the process of selection that will assure the survival of the fittest, and the structural reforms and changes needed to achieve this end.

Industrial Design

Shiichiro Ishikawa

In Japan, as in Europe and the United States, an independent design movement was initiated before the Second World War, drawing its influence from modern design in the West. Architects and designers such as Iwao (Gan) Yamawaki, his wife Michiko, and Takehiko Mizutani went abroad to study at the Bauhaus design school in Germany, while in 1933 Renshichiro Kawakita founded the New Academy of Architecture and Industrial Arts in Tokyo, where he instituted a new method of design education based on the Bauhaus model. In addition, the government invited the German architect Bruno Taut and the French designer, and associate of Le Corbusier, Charlotte Perriand to Japan in an effort to introduce modern means of production and industrial materials to Japanese design and consequently, to Japanese styles of living, while retaining, however, traditional aesthetic elements.

Parallel to this were the goals of the Japanese *mingei* (folk crafts) group centered around Soetsu Yanagi (fig. 1), who found in typical, unsophisticated Japanese folk crafts similarities in attitude toward function (*yo*) and beauty (*bi*). These people and those who worked or studied with them believed in the concept of functional design, following the precepts of modernism. Just as they were about to succeed in giving form to these ideals, however, their work was interrupted by the war.

When at the end of the war, the Japanese regarded the devastations of their country, they were apprehensive about the ability of the new government to bring them back to normalcy. To the designers working amid this instability, America extended a helping hand, offering goods, materials, and instruction in methods of design and manufacture, indirectly by example and directly with visits of designers such as Walter Gropius. It would not be an exaggeration to say that the direct result was the revival and ultimately the enormous expansion that has come to mark Japanese design today. Japanese designers envied American designers, who had been able to see their dreams become reality, from the late-nineteenth-century architect Louis Sullivan, who coined the phrase "form follows function," to the founders of the industrial design profession in the 1930s, Henry Dreyfuss, Norman Bel Geddes, Raymond Loewy, and Walter Dorwin Teague.

Among these, Raymond Loewy was the one who probably exerted the greatest influence on Japanese designers, starting with his conception of the Peace brand cigarette package (fig. 2). Its simple and sleek image conformed to the standards of modern design that the young generation of Japanese designers sought; moreover, the high fee Loewy received for his work emboldened Japanese designers in the pursuit of their profession.

Loewy's book *Never Leave Well Enough Alone* was translated into Japanese by Aiichiro Fujiyama, the foreign minister, and published in 1953, becoming a sort of designer's bible in Japan. His phrase "from lipsticks to locomotives," which was written on the jacket of the Japanese edition (fig. 3), served as the best introduction to the broad possibilities of the designer's job, and is still used to explain what it is that a designer does.

In the early 1950s many Japanese designers also went to study in the United States, at such schools as the Art Center College of Design in Pasadena, Pratt Institute in New York, Cranbrook Academy of Art in Bloomfield Hills, Michigan, and the Illinois Institute of Technology's Institute of Design in Chicago. Most of them were supported by the Japan External Trade Organization (JETRO).

European modern design was reintroduced to Japan in the postwar period first through books, such as Herbert Read's *Art and Industry* (fig. 4) and Nikolaus Pevsner's *Pioneers of Modern Design* (fig. 5), both published in Japanese, and Max Bill's *Form* (fig. 6). From these texts, designers were able to assess the history, theories, forms, and philosophies of modernist design.

At the same time Japanese manufacturers were striving to make their own products match the high quality of European and American goods, which they achieved by mimicking not just their forms but also foreign technology, facilities, materials, and so forth. In products such as automobiles, motorcycles, cameras, and household electronics, especially, they had the backing and the support of JETRO; these products were held in higher regard than others, and incentives were given to advance their development.

It has always been recognized that Japanese cameras were based on German models—Nikon on the Contax,

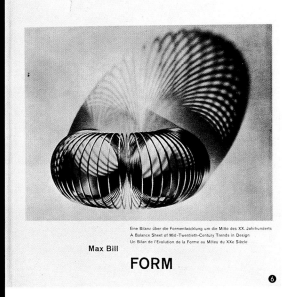

Canon on Leica, and Zenza-Bronika on Hasselblad—in which the imitation of external forms reflected the attempt to imitate their technology, while in the automotive field, Nissan worked with the English Austin and Hino with the French Renault. In broader terms the household electronics maker MATSUSHITA ELECTRIC INDUSTRIAL COMPANY cooperated technically with the Dutch firm Philips. These concerns were all truly devoted to their work, putting in long hours of overtime, and they pursued their goals using the industriousness and technology that Japan's best minds could produce.

The Japanese did not question this process of imitation as copying. Historically, they do not feel the same pride in the creation of an original work that the modern Western spirit does. Nor do they derive the same sense of joy that such creation seems to bestow on Westerners. The Japanese psyche values learning above creativity. It is not that creativity is slighted, but rather that the Japanese have never been particularly inclined to value originality.

In 1952 twenty-five members of the design community, including Tatsuzo Sasaki, JIRO KOSUGI, SORI YANAGI, and RIKI WATANABE, founded the Japan Industrial Designers Association (JIDA). That same year, under the sponsorship of the *Mainichi Shimbun* newspaper, the first New Japan Design Competition (later known as the Mainichi industrial design prize) was established. Almost all of Japan's industrial designers made their debut in this competition at that time, and during its first few years especially, it greatly influenced the direction industrial design was taking. In 1960 the World Design Conference (WoDeCo) was held in Tokyo. This marked the first time that Japanese design was truly recognized both within and from outside the country. The Ministry of International Trade and Industry (MITI) had declared 1960 as Design Year throughout Japan, and it was an occasion for Japanese architects, industrial designers, graphic designers, craftsmen, educators, and critics to sit on the same platform with their counterparts from around the world and discuss topics such as personality, individuality, regionality, universality, practicality, environment, production, communication, possibility, society, technology, philosophy, and design education. This was indisputably the dawn of Japanese design.

In 1954 Japan produced the electric rice cooker (no. 7); in 1957, the portable transistor radio (no. 32); and in 1979, the Walkman (no. 132). These are products truly created by the Japanese, with their own technical skills and their own indigenous forms, and they are designs that are still being manufactured today with little change. In my opinion, however, there is no truly original and beautiful design in these products and in the numerous reworkings that incorporated subsequent technical advances, nothing that is the result of basic research and a rethinking of the product and its design. Like the four seasons, each product has an annual style change, and many manufacturers produce large quantities each year to sell competitively for mass consumption. I believe this attitude has reached its extreme form today.

In 1973 the conference of the International Council of Societies of Industrial Design was held in Kyoto, and in 1989, in Nagoya, both sponsored by MITI. These were recognized as design years in every region of the country, and since 1990, the first of October of each year has also been designated Design Day. The congresses had great success as conferences, but although it is difficult to pinpoint exactly, the ideas and theories voiced there remained ideas and theories without yielding practical, concrete changes or results in products designed afterward.

It may be safely said that there are almost no Japanese products of industrial design that are original. Japan reproduces types of goods invented in foreign countries at cheaper prices and higher quality than anyone else. These products are not unique; they are not part of a program with their own rules and order, nor do they move us particularly. Yet they are handy, easy to use, comfortable, and through the Japanese interpretation, they are infused with a certain freshness and charm.

This is true of most Japanese products on the market today. These are products bought and used around the world. For a person in design like myself, this is a sad fact, and I do not find it satisfying. However, I believe that Japanese designers will in future produce designs that reflect the same kind of pride, confidence, and beauty of spirit in originality as is evident in the candies made by the traditional Japanese candy makers or in the tea ceremony.

Fig. 1 Soetsu Yanagi, *Bi no Kuni to Mingei* (The World of Beauty and Folk Crafts), Tokyo, 1937
Fig. 2 Raymond Loewy's proposals for the package design for Peace cigarettes, 1951–52, with the approved design in the center
Fig. 3 Japanese edition of Raymond Loewy's *Never Leave Well Enough Alone* [1951], Tokyo, 1953
Fig. 4 Japanese edition of Herbert Read's *Art and Industry* [1931], Tokyo, 1957
Fig. 5 Japanese edition of Nikolaus Pevsner's *Pioneers of Modern Design: From William Morris to Walter Gropius* [1936], Tokyo, 1957
Fig. 6 Max Bill, *Form*, Basel, 1952

The company that was at the forefront of the development of furniture in postwar Japan was TENDO MOKKO. It was established in 1940 as a cooperative of carpenters and cabinetmakers who lived in Tendo. Because it was just before World War II, materials and work were scarce, but the company was finally able to get started by obtaining military contracts. The first work for the military was making wooden boxes, mainly of cedar, for ammunition and communications equipment, according to strict specifications and schedules.

Toward the end of the war, Tendo began to experiment with methods of strengthening and forming wood so that it could be used as a substitute for such metals as steel and aluminum. It made control sticks for airplanes as well as decoy airplanes. At that time such research was conducted mainly by the Ministry of International Trade and Industry's Industrial Arts Institute (IAI), a government agency established in 1928. Until then its research had been chiefly for industry, but during the war it became a research center for the military.

The IAI was located in Sendai, about twenty-four miles from Tendo, and the company received technical directions concerning the military. From them we learned the technology of bending wood. The Tohoku area of northern Japan in which Tendo is located is rich in natural resources, and both deciduous trees and evergreens grow widely in the mountains. Its rich lumber resources include beech, oak, and ash, which are used for bentwood and molded plywood. Plywood, made by laminating lumber, was especially suited as a material for aircraft, and the technology for molding plywood quickly improved. The high-frequency radio waves used in radar were adopted to generate heat for making molded plywood. ISAMU KENMOCHI, later a furniture designer, was transferred from the IAI to the Ministry of Armaments and engaged in research on the use of wood for aircrafts. The association of Tendo and Kenmochi began at that time. The European nations and the United States were also developing techniques of using molded plywood for aircraft at the same time, and it is interesting that the American Charles Eames, who was to become internationally known as a furniture designer, was similarly engaged in the study of wooden aircraft.

After World War II, Tendo Mokko first began to make low tables and cabinets for tea utensils from surplus ammunition-box materials, and in order to develop new production techniques, a five-kilowatt high-frequency generator and heat-forming presses were installed in a full-scale effort to develop the capability for molding plywood. The IAI in Sendai assisted the company not only on technical matters but also on design, and this was the foundation of design development at Tendo. The IAI was the first design center in Japan, and it contributed greatly to the development of the industrial arts throughout the Tohoku area. It was at this time that we adapted our knowledge of Japanese furniture to Western furniture design.

In 1950 an exhibition of prototypes made of molded plywood was held at the Takashimaya department store in Tokyo and drew attention for its novel designs. These were for the most part responses to requests for furniture given to the IAI. These new designs were first used in public buildings, and when we learned that they were also in demand by architects, it became the practice for us to go to the architects directly to create designs.

Tendo opened a branch in Tokyo, and through contact with these architects, new designs and techniques to carry them out were developed simultaneously. In 1958 the beginning of the mass production of molded-plywood furniture came with the commission for 4,500 chairs for the Shizuoka Prefectural Sports Arena designed by Kenzo Tange. Other architectural firms, such as those of Kunio Maekawa and Junzo Sakakura, also began to design furniture. In those days, mainly because of the lack of good ready-made designs, most furniture was specially ordered by architects to complement their architectural designs.

In the 1950s design was generally done as a part of an architectural plan, but from the 1960s it became the work of independent designers. In 1956 SORI YANAGI, a product designer, created his Butterfly stool (no. 27); although it was a simple construction, assembled from two identical molded-plywood parts, its shape was exciting, and it gained an international reputation.

In the 1960s Scandinavian design had a great impact on Japanese furniture, and prominent department stores vied to sponsor Scandinavian design exhibitions. Plywood furniture was particularly indebted to Danish design, which was helpful not only for its design but also for its technology. Its feeling of natural wood went well with the Japanese preference for unpainted finishes.

In 1960, to commemorate its twentieth year, Tendo Mokko sponsored the first furniture-design competition in Japan (repeated five more times). REIKO (MURAI) TANABE's stool (no. 46) was the first prize winner. The 1960s were glorious years for the firm and many well known designs, which are still in production, were introduced then. The opening of new hotels also gave impetus to the work of interior designers; Isamu Kenmochi's Kashiwado chair (no. 52), KENJI FUJIMORI's *tatami* seat (no. 58), and DAISAKU CHOH's low chair (no. 49) are all products of the 1960s. Almost everything that could be designed in basic plywood construction was done at that time, so much so that there is very little new technologically that has come out since.

In the 1970s aesthetically beautiful designs were sought more than those with structural innovation (no. 99). The influence of Italian design appeared about this time, with its emphasis on aesthetics and on ideology. Imports of Italian furniture increased; brightly colored molded seats made of urethane foam had a great influence on the Japanese furniture industry and mass production became widespread. The office environment, which was said to be extremely backward in Japan, called for innovative designs. Tendo Mokko's new office furniture group, the OF series, was one proposal for what the new office should be, and presented an opportunity to give a fresh look to standardized and monotonous office furnishings.

The scope of Tendo Mokko's business grew rapidly. In the 1980s its sphere of operations expanded from government offices and office buildings to include hotels, libraries, and museums. It produces two categories of furniture: catalogue furniture and a relatively high proportion of special-order furniture, which actually has brought us back to our original business. In many instances, furniture that was designed for a specific project has been repeated several times, and then became a catalogue item.

Furniture design cannot be discussed without considering architecture. In Tendo's biennial company exhibition held in 1986, there was great response to the furniture designs submitted by young architects as indications of a new direction for furniture design in Japan. Now, when postmodernism has become a thing of the past, it is more difficult than ever to predict the direction that furniture design will take. The time is past for depending on Western examples; now we feel there is a possibility for the emergence of furniture unique to Japan itself.

While adaptation and emulation of Western models characterized the beginning of Japan's exploration of a postwar identity, by 1990 an internationally recognized "Japanese style" was evident in graphic art, product design, computer games, animated films, architecture, interior design, automobiles, dance, and fashion. In textiles, until recently, this style remained closely identified with the simple, bold, rustic folk textiles of blue-and-white and the elaborate, aristocratic robes of all colors. However, beyond this tradition-based production, the Japanese style now includes an array of innovative fabrics that support high-performance athletic wear and fireproof uniforms (no. 163), as well as avant-garde fashion designs.

For over eighteen hundred years, Japanese textile production accommodated the needs of its people by providing materials for clothing, home furnishings, and ceremonial activities. The tradition continues, but new developments, triggered by Japan's exposure to foreign influences over the past century and a half, have occurred. As long ago as the turn of the century, some centers of traditional textile production started to convert from narrow (14-inch-wide) handlooms to broad (45-to-60-inch-wide) mechanical, and later, automated looms. Japan was first widely exposed to Western influences during the Meiji period (1868–1912), and by the early twentieth century, some textiles were being produced for the export market only.

After the Meiji Restoration, Japanese government officials readily adopted Western-style clothing (fig. 1). The eminent educator Yukichi Fukuzawa published an illustrated guidebook to Western clothing, food, and homes in 1867,[1] and this and several of his other publications about the West were widely read. During the Taisho period (1912–1926), the more highly educated urban class wore Western apparel as a sign of acquaintance with Western learning or as a fashionable expression of style. Children, too, were sometimes dressed for special occasions in Western clothes and would don frilly aprons over kimonos. Many men began wearing Western business suits, which came to be known as *sebiro*, taking their name from London's Saville Row, where the world's most prestigious, well-tailored suits were made. Some occupations, such as those of train conductor, police- and postman, and the military, were identifiable by a Western-style uniform. By World War II most of the urban population had incorporated some articles of Western clothing into their wardrobe, although in rural areas the traditional national costume was still favored. During the war, women were discouraged from wearing kimonos because their restrictive nature made it difficult to fight or flee easily. Instead, *monpe* (work pants gathered at the ankles) was the garment judged to be most suitable for wartime activities, worn with a shortened version of a kimono or a blouse on top.

2

After the war the choice of everyday wear followed necessity as people met the hardships of recovering from wartime losses. Most women continued to wear *monpe* or Western-style clothing as a matter of convenience. By the 1950s, with the exception of the older generation, the majority of city dwellers had become accustomed to dressing in Western-style clothing although they often relaxed in kimonos at home, wearing a lightweight version called a *yukata* in summer, and a heavier, padded garment called a *tanzen* in winter. Formal kimonos, however, were preferred over Western clothes for special occasions, such as the observance of traditional holidays, ceremonies for school entrance and graduation, weddings, and funerals, and these needs supported the production of expensive kimonos for limited use.

By the late 1970s it was no longer expected that one would wear a kimono on special occasions; instead it became more acceptable to express one's individuality freely through a choice of clothing, and many people preferred Western styles, especially clothes with the prestige of known designers. However, in the 1980s, there was an effort to cultivate interest in the kimono as a fashion statement, especially among the younger generation. The presentation of these garments in a boutique setting, replete with dramatic lighting and sleek, contemporary fixtures, appealed to sophisticated young shoppers. This trend was part of *retoro shumi*, the popularization of a taste for old styles. The new kimono collections reflected Western taste by offering subdued, muted colors that had been traditionally reserved for older women, and by combining kimonos and *obi* sashes of the same hue, which previously would have seemed somewhat strange. During this same period, formal kimonos began to be made of improved synthetic silk to appeal to consumers who wanted to be able to wash the garments easily. Traditionally, kimonos were sent to a specialized cleaner who painstakingly removed

stains and matched dyes to restore the colors, or they were taken apart and the rectangular panels carefully washed and blocked to dry on a long board.

Although such fads as *retoro shumi* come and go, it is the exclusive market for the kimono as a work of art that sustains the kimono artist and craftsperson now. For centuries textile artisans were generally anonymous, even those who created the masterpieces that are today's national treasures. The decline in the number of craftspeople after World War II led the government to establish the designation Holder of Intangible Cultural Property (Living National Treasure) in order to recognize the artisan's creativity and mastery of skills in many different fields.

The word "kimono" comes from *kiru* (to wear) and *mono* (a thing or things), and it has come to be used as a generic term for various T-shape, straight-cut garments that are worn wrapped across the front, left over right (fig. 2). For over two thousand years different styles and types of garments evolved, all loosely termed kimonos by Westerners. The basic T-shape, as it appears when the kimono is displayed on a kimono stand, is constructed of units of narrow rectangular cloth. Its flat surface invites individual creative expression as if it were an uninterrupted canvaslike plane. Treatment of this "canvas" might involve pictorial design or an allover repeated pattern.

A master dyer such as KAKO MORIGUCHI using the paste-resist and hand-painted *yuzen* technique executes a sweeping design across an entire kimono by painstakingly matching the panels (no. 57). Moriguchi has perfected the art of *maki-nori*, a resist technique using rice-paste flakes to create an effect similar to the delicately speckled ground of lacquerware, yielding dappled areas that seem more like the brushwork of a Pointillist painting, which he combines with traditional *yuzen* dyeing. His designs have a strong affinity with Nihonga (Japanese painting), often referring to elements of nature, but he infuses them with a contemporary flavor.

The shape of the kimono has remained simple and unchanged. In contrast to the clothing of European couturiers, who project an ideal body shape by cutting, making darts, fitting materials, and applying layers of trim, Japanese garments have been of a straight cut with minimal variations in the basic shape. Embellishment has been focused on the surface of the cloth. Although some of the cloths were very simple (not all were embellished), patterning the fabric became an integral part of the weaving; hundreds of variations of plaids or stripes, for example, were created with the use of colored yarns. A outstanding example of this aesthetic is seen in the work of FUKUMI SHIMURA, another Living National Treasure, who has created simple patterns, such as stripes, plaids, or a variegated color field, by weaving with naturally dyed yarns (no. 121). Her work imparts the quiet beauty of simple folk textiles and is suffused with an elegant light.

Kimono artists represent diverse aesthetic concerns, including those of the Kyoto decorative tradition, the Edo tradition of indigo dyeing, and the *mingei* (folk-crafts) movement. KEISUKE SERIZAWA was a prominent member of the *mingei* movement, which brought attention to functional objects made for everyday use by unknown craftspeople. He was a principal exponent of *kata-zome* (paste-resist stencil dyeing), which had been commonly used in decorating *mingei* textiles (no. 60).

The Japanese view cloth as having its own life and spirit, interacting with the human body in movement and at rest, and versatile enough to fit the life-style of the wearer. This outlook can be related to the way in which Japanese garments are constructed: narrow, rectangular cloth is cut into modules to create a garment that adapts to the wearer. Compared to French haute couture, the designs of ISSEY MIYAKE and REI KAWAKUBO interpret clothing as kinetic sculpture (fig. 3) rather than embellishment or covering for the body. Sometimes the experience of wearing Miyake's and Kawakubo's clothing can be akin to a happening of the 1960s. Miyake's tricolor sweater with three arms allows the wearer to use one of the arms as a scarf to tuck in or toss over the shoulder, and Kawakubo's dress with elasticized holes invites the wearer to extend head, arm, or both arms through any of the holes. These creative fashion designers work very closely with textile artists who develop materials for each collection, a collaboration that is not often seen in Western fashion design.

Kimono culture rests upon a large foundation of textile artistry: sericulture, dyeing, weaving, patterning methods, embroidery, and various techniques of embellishment. There is a tradition of working out textile-related problems, and there are established physical and practical means to solve problems related to textile work. For example, the production of woven kimono fabric involves design, filature, pattern making, dyeing, loom preparation, weaving, and finishing. Each area is a specialization. In Japan a small niche is always explored in great depth. Whatever the craft, it is characteristic of the Japanese approach to say the most with the least, to aim for simplicity, to respect the power of a clean line, and to cooperate with the nature and limitations of the material or medium. The finishing process, which can change the texture and character of the cloth, is considered to be as important as designing and weaving.

The emphasis upon finishing is evident in the innovative contemporary fabrics created by such designers as JUNICHI ARAI, MAKIKO MINAGAWA, and EIJI MIYAMOTO.

Arai's wool yardage Korean Carrot (Ginseng) (no. 119) exemplifies the use of a complex weave structure and a finishing process that, though simple, transforms the textile in an imaginative, totally unexpected way. What was ordinary yardage, with long warp floats, becomes a cloth with short dreadlocks that resemble scraggly ginseng roots. It is sophisticated and yet rich in primal energy and playfulness. Arai also relies extensively on computers to create complex weaving structures to be utilized with dissimilar fibers, such as cotton and wool, which react differently to water, heat, and certain finishing processes. He works closely with scientists and laboratory technicians to develop the most advanced types of yarn. Ojo de Dios (no. 188) and Big Wave (no. 230), for example, are woven with what Arai calls "slit yarn," a flat, filmlike, synthetic fiber coated with an ultrathin layer of aluminum. This fabric is extremely light in weight and strong. It can be colored by the action of chemicals (tarnishing) or by coating the metallic yarn with a minute amount of protein, then dyeing it. Alternatively, by using an alkali solution, metal can be melted away from the slit yarn to create sheer, translucent areas that work effectively with the remaining opaque fabric.

Makiko Minagawa, the accomplished textile director who has collaborated with Issey Miyake for over two decades, believes that her efforts contribute to the final artistic statement created by the fashion designer, and she therefore feels no need to have her work recognized independently. Nonetheless, one should not overlook the essential role her fabrics play in the success of Miyake's clothing. Minagawa truly understands the materiality of textiles—the way color, for example, is experienced through the properties of a tangible piece of cloth, which has a specific weave structure and a surface that reflects and absorbs light in a particular way. In her collection of kasuri-type fabrics (no. 175), the usual blurry, "leggy" edges associated with the ikat dyeing technique are minimized because the weave is so fine. The clean edges of the stripes—also possible because of the fine weave—provide a wonderful complement to the soft-edged latticework and polka-dot patterns. The restrained and earthy color palette is enlivened by streaks of bronze metallic yarn. Such a fabric is the result of Minagawa's unique talent for challenging the skills and expertise of Japan's weaving mills. She is responsive to the strengths and limitations of the mills, many of which produced traditional kimono fabrics until they converted to the production of Western-style fabrics in the early part of the twentieth century. Although it is now common practice for Japan's large fashion houses to develop their own fabrics and to control the entire process from beginning to end, this was not the case two decades ago. Minagawa enters into a vibrant dialogue with such textile producers as Junichi Arai and Eiji Miyamoto, who are based in those traditional weaving centers, to create unique fabrics for Miyake. Miyake has always had specific fashion ideas, and relies on Minagawa to envision and produce the materials he needs.

Until very recently many of the uses of furnishing textiles that were common in the West were alien to Japanese culture. Traditional houses contained not curtains but shoji (sliding panels covered with translucent paper), not rugs but tatami (thick matting covered with woven reed), not chairs and couches but zabuton (square, padded, seat cushions). Fabric was used for futon (coverlets), noren (door curtains), and nobori (banners for festivals). Textiles took many forms, functioning as a furoshiki (square wrapping cloth), yutan (cloth to protect chests of drawers), tenugui (towel or head scarf), and fukusa (ceremonial cloth used in the tea ceremony or for gift presentations). In the early 1950s the Japanese rebuilt their lives around a Western model but still retained the structure of Japanese culture. When they moved into an era of relative prosperity in the 1960s, they started to buy automobiles, televisions, refrigerators, and other modern conveniences. Many traditional houses were replaced by modern, multistory, concrete apartment buildings, and sofas, chairs, coffee tables, and carpets became popular. The futon was often replaced by a bed, but the bed was still placed in a room of tatami mats.

The work of many designers of printed fabrics reflects the long tradition of graphic design in Japan, which in turn can be traced to a tradition of woodblock prints and folk textiles. FUJIWO ISHIMOTO (no. 148), HIROSHI AWATSUJI, and KATSUJI WAKISAKA are leaders in this field in Japan as well as in Europe and the United States. In Awatsuji's Bicycle fabric (no. 152), the modern image of a bicycle is treated in a way that evokes the lightheartedness and simplicity of the costumes worn in classical comic theater. Silkscreened cottons designed by Awatsuji, titled Hibiki (no. 212) and Sou (no. 213), utilize a finely patterned ground that is reminiscent of the small allover print design that was once popular in the material worn by men of the ruling warrior class. Wakisaka's printed cotton fabric Ma (no. 118) plays with light and dark with great subtlety. Each of the elements is familiar—stripes, dots, checkerboard—yet they are juxtaposed in a new way to create a strong visual rhythm. Like Wakisaka, who has lived in Finland and the United States, other Japanese designers have also worked abroad, as he did for Marimekko and Jack Lenor Larsen. By contrast, MASAKAZU KOBAYASHI, designer at the Kawashima textile mills, has not lived abroad but has moved away from the industry and become known for elegant experimental fiber works. His Space Age fabric (no. 145) shares the sophisticated understanding of line and color displayed in his three-dimensional fiber installations.

Despite the technological advancements that today make handcrafted materials a luxury, Japanese society, with its emphasis on detail and on exploring a niche deeply, gives a place for its exquisite and laborious traditions to survive. In Japan textiles have a status equal to other art forms, including painting. An exquisitely made robe that once belonged to a warlord would be treasured as much as a prized tea bowl, which a feudal lord may have spent a fortune (even the equivalent of a small castle) to obtain. People have expressed their aesthetic values through the materials they create and live in, from the most aristocratic robe to the humblest work jacket.

1 Seiyo Ishokuju; see Julia Meech-Pekarik, The World of the Meiji Print (New York, 1986), p. 65.

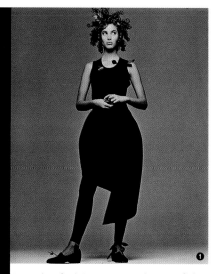

In the 1970s Japan first made a strong impact on international fashion. Japan had accepted Western clothing since the beginning of the Meiji period (1868–1912), but Westernization in clothing accelerated rapidly following World War II. In step with economic prosperity, Japanese fashion gained great strength as an industry and made great progress. Since the 1960s Japanese designers have been sending a message to the world with their unique fashion creations. HANAE MORI was the designer who paved the way. In 1965, in New York, she introduced a collection that would be identified as hers in the traditionalist Western society. She achieved this by exploring her individuality through the use of Japanese silks with distinctively elegant designs drawn from nature (no. 92). At that time, although it was in the process of change, fashion was still dominated by a hierarchy with haute couture at its apex.

The fashion atmosphere of the 1970s necessary for the appearance on the Paris scene of such Japanese designers as KENZO and ISSEY MIYAKE was the emergence of a counterculture following the 1968 student revolution. Kenzo made his debut in 1970 as a leader of ready-to-wear, which began to overshadow haute couture; he became one of the most copied designers in the world, surpassing designers like Christian Dior and Balenciaga. His clothing was in keeping with current trends—casual and unaffected—incorporating a touch of Japanese design into Western fashion. However, his Japan was not that of the exotic flowers and birds that had been the former image of its culture but of the everyday working clothes of the masses (no. 97).

Issey Miyake also began to gain a reputation in the United States and France in the 1970s. The underlying concept of Western tailoring is to drape a flat cloth on a three-dimensional human body; Japanese kimonos are also made by draping a piece of cloth on the body, but the surplus is left hanging and not cut off. It is the remnant that makes the significant difference between Western and Japanese clothes. Since the late 1980s Miyake has been greatly acclaimed for his creations comprising a series of pleats (no. 235). Pleats have been used in Japan since ancient times, but Miyake developed the flexibility of pleats a step further by creating the pleats first instead of following the usual method of pleating after the garment is cut. He invented a new type of garment by letting the material dictate its shape and then incorporating its function, thus linking together material, shape, and function. Adhering firmly to the tradition of Japanese clothing, which attached great importance to the material, his creations were firmly rooted in the technologically advanced Japanese textile industry. Through the work of Kenzo and Miyake, the ideas of layered and oversize, or one-size-fits-all, clothing, which are characteristics of Japanese apparel, became widespread during the 1970s.

The "Japanese gain a foothold in Paris" was proclaimed by Hebe Dorsey, writing on the 1981–82 fall-winter collections in the *International Herald Tribune*. REI KAWAKUBO and YOHJI YAMAMOTO proposed to bring into fashion a kind of Japanese beauty that was different from what had been shown by earlier designers. Their work was characterized by the type of traditional materials, construction, and lack of ornamentation that Bruno Taut had observed in Japanese architecture. They created an impact in Europe and America that was dubbed "Japan shock." Principally in black, their clothing was draped asymmetrically and shown on pale mannequins who were not even wearing lipstick. The shape of their dresses had no relation to the form of the body, and from the Western idea of structural beauty as dependent on the body, their garments were baggy. But the unfinished look, deliberate defects, and acceptance of shabbiness were understandable to those versed in the underlying Japanese concepts of beauty, *wabi* (simplicity) and *sabi* (patina) (fig. 1).

New concepts of Japanese fashion emerged in the 1970s. First, as clearly shown by Miyake, came the idea of making an organic garment from one piece of cloth, creating a new garment without categorizing it as Eastern or Western (fig. 2). Next came the idea that material takes precedence over shape in a flat garment. In Japanese tradition, which does not clearly differentiate arts from crafts, design begins with the choice of the material and emphasizes the importance of texture in fashion design. Third was the idea that clothing does not exist to adorn women as sex objects but as logical attire. The object is not clothing in the shape of an ideal female figure but clothing that acquires its shape when it is worn. In other words, it is people who control clothing, which nevertheless can be sensual as well as chic.

Simply stated, these designers dismantled the symbolism that had become ingrained in Western clothing during the nineteenth century. If Japanese fashion revolutionized the international fashion scene, and it continues to do so, this is not just an indication of Japan's originality; it may be a proposal for a clothing of the future that will transcend ethnic and gender differences and even the confines of an establishment called fashion.

Fig. 1 Comme des Garçons, Fall–Winter 1986–87 Collection, designed by Rei Kawakubo, 1986 (Photograph by Steven Meisel)
Fig. 2 "A Single Piece of Cloth," Spring–Summer 1977 Collection, designed by Issey Miyake, 1977 (Photograph by Akimitsu Yokosuka)

The 1950s began as a decade that was by no means prosperous, yet day by day it seemed we could feel we were moving toward a life in which there was hope. For Japan it was a period of rapid change and extreme significance. During the five years after the monochromatic drabness of the war, the bright, dazzling colors of American products and designs spread throughout Japan. Among them what most directly influenced us was probably package design, symbolizing a prosperity unimagined by the Japanese people.

Lucky Strike cigarettes, Wrigley's chewing gum, Hershey chocolate bars in their triple layers of aluminum foil, paper, and cellophane wrapping, cosmetics like Pond's Cream (filled to the brim)—all these gave us a refreshing shock in the materials that were used, in the printing, and of course, in the beauty of their designs.

Eventually Japanese manufacturers also awoke to the importance of design and began to commission package designs from America. Raymond Loewy was first, the designer of the Peace brand cigarette pack in 1952. The design fee that Loewy received was anywhere from ten to a hundred times as high as a Japanese design fee at the time, and was an even bigger topic of conversation than the design itself. After that there were innumerable package designs commissioned by Japanese clients, such as those for Asahi and Sapporo beer. Thus Japanese package design awakened under the influence of America. Of course, Japan already had a long history of design activity, but until then it had been most strongly influenced by Europe. The 1950s marked a new point of departure, when package design was created as a practical sales strategy along the American lines of rationalized marketing.

Japan had its own, long history of package design, which was directed to the institution of gift giving. The custom of packaging as part of the ceremony of gift giving developed in the 1700s as a feature of the warrior society, and its formalities became standardized. The elaborate forms of this ritual gift exchange spread among the populace in general and have been passed down to the present day.

The Japanese marketplace now includes both the package design that evolved from Western-style sales strategy (cosmetics, whiskey, foodstuffs) and the package design that is based on forms rooted in our traditional customs (Japanese-style candy, sake, and other traditional gifts). This package design reflects the two-tiered structure of our cultural heritage.

If I were to sum up Japanese package design, the custom of sending gifts in a box or in a wrapping has led to a tendency to attach great importance to the package, and has resulted in much posturing in this field. In recent years, with the worldwide economic difficulties as well as the dark clouds that hover over the Japanese marketplace and the environmental concerns that have increased consumer awareness, the trend has been to avoid past excesses of packaging. Even famous shops and department stores have now begun to cut back on their packaging, and the tendency now is to support design that is plain but has character.

Japanese package design was compelled by circumstances to start again after the war under American influences in the 1950s, and now after forty years of repeated advances and retreats, must truly do some soul-searching in the present difficult economic climate. The era is long gone when design had merely to be amusing or pretty. Both the seller and buyer have become aware of design, so those who create design must become more concerned than they had been with current issues, or we will not be able create appealing package designs that are valid for today's market. I think it is more likely that even with our rich tradition, it will become more and more difficult to create packaging that has a unique appeal.

Graphics

Yusaku Kamekura

In recent years the world has begun, rather suddenly, to sit up and take notice of graphic design produced in Japan. The phenomenon appears to be tied to the rise in Japan's economic fortunes, and is only natural since design is in essence symbolic of corporate cultural development.

The history of Japanese graphic design in the postwar era has already been documented with great detail and precision by such internationally known critics as Richard S. Thornton[1] and Alain Weill,[2] but the overviews presented by these experts and others differ considerably depending on the perspective of each writer. Their evaluations of course vary widely as well, according to their individual tastes and preferences.

To understand the achievements of Japanese graphic design in the postwar period, it is necessary to keep in mind the dramatic progress made during the prewar era. A number of commentators have offered great praise for Japan's prewar design (fig. 1), but such views are clearly tinged to a great extent by the romantic predilection toward nostalgia, which is so prevalent nowadays. In my opinion, Japanese design of the prewar period was by no means of high quality. In those days, design was seen merely as a sideline that was undertaken by artists; it enjoyed neither a high social position nor the support of a true artistic conscience.

When World War II ended in 1945, Japan's crushing defeat brought millions of its citizens to the brink of starvation as the food rationing system of the war years atrophied. The situation grew progressively worse the following year and reached its nadir in 1947. According to historical records of the period, the daily rationing of rice in those days was a mere 297 grams per person—barely enough to fill a medium-size cup. When rice was unavailable, rations of potato, sorghum, soybeans, or the like would be offered in its place. Daily vegetable rations were 75 grams, the equivalent of about half a carrot. The fish ration was one sardine every four days—when available, that is. Japan, of course, was not alone in its suffering, for the entire world was racked by soaring inflation and labor unrest.

The reason I bring up this dismal state of postwar Japan here is because it holds the key to understanding the success that Japanese design has achieved in recent years. Japan in those days was quite backward in matters of design; unlike the United States and Europe, the nation had no history of modern design whatsoever. This is not to say that Japan was lacking in modal beauty in the realm of its traditional arts; on the contrary, a fine legacy of decorative excellence was well established in such fields as painting, sculpture, architecture, and theatrical arts (Kabuki, Noh). From a Western perspective, Japan no doubt could have been adequately served simply by carrying on this native design legacy. And while I find no argument with this way of thinking, at the time we Japanese believed firmly that the only way to extricate ourselves from the devastation brought by our defeat and thereby rebuild our nation was to introduce the modern social structures of the West. We believed that unless we pursued a path toward democracy and modern capitalism, we would remain no more than a feudal and backward nation at the edge of the "Far" East. And so we pursued an aggressive and fast-paced course toward Western-style modernization.

It was in this context that Japanese corporations came to understand that they could move forward only if they broke away from the yoke of their old ways, spurring them to adopt the latest American management methods. But trends of this kind were not confined to the corporate sphere; they touched all aspects of the nation's framework. To borrow a phrase, this was an era during which Japan went through a veritable "cultural revolution." Naturally, design was no exception. Those of us in

Fig. 1 Poster for Shiseido, designed by Takeo Yamamoto, 1937. Photogravure (Courtesy of Toppan Printing Co.)

graphic design longed to catch up to the artistic level that we recognized in American design. We knew, however, that to accomplish this would require a total revolution in the thinking of all Japanese designers. To begin, we would have to undertake constructive initiatives aimed at elevating our position in society. We would have to band together in order to work toward our common goals.

Thus, in 1950, just five years after the end of the war, full-fledged preparations got under way toward the founding of a nationwide alliance of graphic designers, marking the true start of modern design history in Japan. One year later the first national organization of designers, the Japan Advertising Artists Club (JAAC), was inaugurated. The organization ultimately played a major role in the emergence and advancement of Japanese graphic design. In September 1951 the JAAC held its first joint exhibition, in Tokyo's Ginza district. Members showed original posters, each conceived freely and with no inherent restraints upon the artist. Starting with the third in what became an ongoing series of shows, nonmember designers were also invited to participate. The shift not only enabled the discovery of outstanding new talents, but also garnered wide social support for the burgeoning design movement. Newspapers and art magazines, which had consistently ignored graphic design, also began to devote substantial space to introductions and critical evaluations of new works. Designers, in turn, were inspired by this broad social response, and they took to honing their skills and artistic inventiveness with increasingly competitive drive. The result was an explosion of both talent and energy.

It has been said that in the first twenty-five years after the war, Japan carried out democratic, economic, and cultural reforms that normally have taken other countries an entire century. The same remark holds true in the realm of design. Earlier I noted why graphic designers in Japan chose to cut their ties with traditional arts and rushed headlong toward Western styles. Over the course of time, however, as the new culture reached a level of high achievement, the result was in fact a natural modernization of traditional Japanese aesthetics. During the 1950s and 1960s Japanese designers abandoned themselves to the pursuit of Western forms, but by the end of this period new designers began to appear who possessed sufficient talent, skill, and vision to develop a new Japanese style (fig. 2). One might mention here that this new Japanese style has evolved more recently into the introduction of computer graphics.

In 1952, one year after the establishment of the JAAC, the Tokyo Art Directors Club (ADC) was inaugurated. This organization was in essence a copy of its well-established American counterpart. At the time, however, art direction did not exist as a profession in Japan and the term "art director" itself was poorly understood. The system was adopted simply because it appeared to be a fashionable new way of creating advertising. Art directors in those days were not creative artists but merely personnel involved in the creation of advertisements. The Tokyo ADC was thus in reality nothing more than a collection of staff from corporate advertising departments. These dubious beginnings notwithstanding, the Tokyo ADC quickly succeeded in building up sufficient capabilities to influence the course of Japanese design.

In 1957 the organization instituted the production of an annual publication, which won increasingly high critical acclaim worldwide (fig. 3). These accolades were a genuine response to attractive and effective works that came to be produced in Japan as the art-direction system came to assume a solid position in the advertising business. Gradually an ever-larger number of innovative and creative individuals of superior talent and skill appeared within the ranks of the Tokyo ADC, enabling it to surpass the JAAC in several ways. In this manner the two organizations began to exert influences on each other, causing them to become increasingly outspoken. Today Japanese advertising design boasts both expressive uniqueness and superior quality. In my opinion, Japanese art directors today also continue to possess the driving creative spirit that has generally been lost in recent years among American art directors (fig. 4).

Over time the JAAC exhibitions, started in 1951, gradually fell into a predictable pattern. As the annual event approached the end of its second decade, critics began to castigate the designers' dependency on the exhibition format, saying they were acting precisely as if they were painters. They simultaneously began to suggest that designers should address issues that had greater social relevance. These criticisms were indeed on target. The

Fig. 2 Ten graphic designers who participated in the "Persona" exhibition in Tokyo in 1965 (left to right): SHIGEO FUKUDA, Akira Uno, TADANORI YOKOO, Tsunehisa Kimura, MITSUO KATSUI, IKKO TANAKA, KIYOSHI AWAZU, KAZUMASA NAGAI, Makoto Wada, and GAN HOSOYA.
Fig. 3 Cover of the first annual of the Tokyo Art Directors Club, 1957
Fig. 4 Tokyo Art Directors Club Trophy

1 Richard S. Thornton, *The Graphic Spirit of Japan* (New York, 1991).
2 Alain Weill, *The Poster: A Worldwide Survey and History* (Boston, 1985).

JAAC by this time had become a mecca for the nation's designers, and it wielded enormous power over them. In fact, in many quarters, this power was viewed as a kind of authoritarianism. This was around 1967, a time when the student activist movement in the United States began to spread to Japan. The movement affected both private and national universities alike, with student demonstrations resulting in widespread suspension of classes and harsh criticism of professorial staff. I can still recall quite vividly how, on August 2, 1969, dozens of helmeted students wearing masks over their faces to conceal their identities stormed into the JAAC committee meeting where judging was under way to select works by nonmembers for exhibition. The intruders represented a left-wing group intent on crushing the JAAC and its activities. Ongoing debates followed between the JAAC and the students, but no mutually satisfactory outcome was reached. Confusion reigned even within the ranks of the JAAC, and some members began to suggest that the organization would have to give in to prevailing trends. And so, on June 30, 1970, the JAAC ultimately yielded to the pressures of the times and disbanded. One newspaper likened the event to the toppling of a grand old tree.

At the time some lamented what they saw to be the end of graphic design in Japan; others, however, hailed the arrival of a new age of expressive freedom in which skill would be the only determining factor for success. What is certain, however, was that designers had lost their enthusiasm for group action. This void then continued for some eight years, until once again there arose a call for a professional organization of graphic designers. During the interim designers had come to realize that as individuals their powers were limited, and that the only way to make a social statement was through group action. In Japan there has always been a clear distinction between artistic and professional organizations; apparently, the division is unique to Japan, for it does not appear in the West. Artistic organizations bring together fellow artists of excelling quality; professional organizations are gathering points for specialists who make their living through a common profession. Since anyone engaged in the pertinent profession can join the latter, professional organizations tend to be held in low esteem. It was precisely under this demeaning gaze that JAGDA—the Japan Graphic Designers Association—was created in 1978 as a professional organization of the nation's graphic designers. Initial membership was approximately seven hundred. In its first three years, JAGDA dealt primarily with the issue of defining what a professional organization of graphic designers should be and do. Any time an event was planned, it was opposed on the grounds that such activities should be performed by artistic organizations. Any time a lecture was suggested, the undertaking came to an impasse amid charges of promoting individual "stars." As a result, complaints were frequently registered by members who saw no merit in JAGDA membership and insisted that their dues were a waste of money. JAGDA then took decisive action: it disbanded its founding committee altogether and undertook elections of directors based on free elections by all members nationwide.

Under its new leadership JAGDA succeeded in reinvigorating the organization. It then shifted its focus to enhancing the quality of design production through Japan. This led, in 1981, to publication of the first JAGDA annual, *Graphic Design in Japan*. It was, and continues to be, a magnificent work of a large scale and in beautiful full color. Since all works included in the volume are by JAGDA members, gradually the organization attracted a growing body of ever-more talented members. Today JAGDA embraces nearly two thousand designers.

Through the well-balanced program of activities conducted under the auspices of both the Tokyo ADC and JAGDA, Japan continues to improve the quality of its design work. As a result Japan is giving birth to a succession of young and innovative graphic designers, and this is one of the most appealing elements of its graphic design today. Perhaps the greatest appeal of contemporary Japanese graphic design, however, is its cross-generational diversity. Each generation is blessed with designers of supreme talent whose competitive spirit serves as a constant energizing force, which in turn, I would suggest, is why the world has recently begun to focus attention on Japanese graphic design.

Graphic design in Japan does not fall into any one uniform style. Designers follow their separate paths and stubbornly refrain from imitating others, a trend that I find extremely stimulating. At the same time, the several generations understand and have friendly feelings toward each other. They engage in competition based on recognition of each other's talents. This mutual respect, perhaps more than anything else, defines the "style" of contemporary Japanese graphic design.

Viewed against the backdrop of Japan's high-level economic development as a trading nation, the contributions made by industrial design have been very great. During this century all of Japan's economic activity has been concentrated in the Tokyo area, and because of this, the majority of the work done by Japan's design community has also been centralized in Tokyo. I, however, have been conducting my design activities in Fukui, the place of my birth, a city with a population of 25,000 located some 350 miles from Tokyo. In the prefecture of Fukui, traditional crafts, once supported primarily by the cultures of Kyoto and Nara, continue to be passed down with great enthusiasm and devotion. To introduce the field of industrial design to these traditional crafts has become one of the basic principles of my design activities.

The traditional crafts, which are an integral part of Japan's cultural system, have gradually been losing their economic base. The production of handmade objects requires the acquisition of special techniques as well as intensive labor and compensation for that labor. These factors, as well as the steady departure from rural areas of the young people to whom such skills would normally be passed on, have created problems for the industry. Under such conditions, traditional crafts are no longer viewed as products suitable for use in our everyday lives but have instead become luxury items. For these reasons I have been hoping to devise a new way of relating the traditional crafts to our everyday lives, a new way of telling their story. In such a story one would find industrial-design techniques being introduced into the arena of traditional crafts, resulting in new creations for our future and a new Japanese culture. I have taken the formulation of such a story as one of my design themes (see nos. 169–70).

In the future, moreover, Japan will need to deal with the many problems associated with the rise of a rapidly aging society. I have taken these problems as my basic point of departure along with my own very personal experience as someone forced to conduct his life from a wheelchair as the result of a traffic accident. From there I have been grappling with the problems associated with designing products for the elderly and the physically impaired (no. 219).

Until now design has been something created with the intent of providing joy and pleasure. What I hope to do is to force design to acknowledge death, that is, to confront the fact that we will grow old and die, and also to address the sadness that comes when one's body

is injured. In order to achieve concrete results in the creation of this type of design, it is especially necessary to use the latest technology and to have both substantial financial resources and the backing of the general public. Knowing this, I have come to realize that my proposals must always present designs that contain a social message.

Until now industry has operated solely within a paradigm of production and consumption. Under such a paradigm, design has been conceived of simply as a profession that provides substantial support to this kind of economic activity. It is natural, however, that we should have need for a design that goes beyond this, one that recognizes the Buddhist notion of *sho-ro-byo-shi*, the human experiences of birth, aging, sickness, and death. In my mind a design oriented solely toward a cycle of consumption and production that benefits the market has seen the last of its days. A relationship that simply connects design to the economy through international markets is no longer workable. Instead, a cycle that relates consumption and production as well as disposal must work to create a newly conceptualized world view that can normalize relations between mankind and the planet's environment. The function of design must be to establish an ideological viewpoint for creating the concrete objects necessary for this world.

In my design activities I have created such objects as wheelchairs, computerized beds, and timers for the blind. It is necessary in the creation of objects of all types to embrace a design philosophy that is concerned with the needs of the physically challenged and that directly addresses issues related to both life and death.

It may seem an exaggeration, but I feel that what will make this normalization a reality will be a design that embraces idealistic principles and a truly international spirit. If one accepts this view, then it is possible to disprove the assertion that design simply mirrors the concerns of a postmodern age. Had we not better decide that it is time for designers to wrestle with design as a means of solving the ethnic strife and religious conflicts that have erupted since the end of ideological confrontation?

The United Nations has what are referred to as peace-keeping operations and peace-keeping forces, which serve to create a new world order. (These operations and funds, it should be added, can also be thought of as working to destroy such an order.) I would like to suggest a similar approach for design. What I would like to see is the adoption of something that might be referred to as "peace-keeping through design," which would be intimately connected to the creation of a new world order and a new way of relating all mankind together.

At present a large gap and a great deal of friction exist between technology and our daily lives. Throughout the world, for example, we find a wide range of problems, from the destruction of the environment to the economic imbalances between the northern and southern hemispheres to the ethnic disputes that have erupted following the death of the conflict between the two great ideologies, capitalism and communism. If solutions to these problems cannot be found, one wonders how we will ever be able to deal with the twenty-first century.

Designs like the ones to which I have referred carry a strong message. These designs I would like to call "narratives," and to narrate the shape of things through the designer's language of form is my main objective. In order to undertake such a project, one would think I must first attempt some kind of inquiry. What, I might ask, is it that Japan's high-level industrial society has been trying to construct? And what has been the nature of the industrial design that has supported such a society? At this point, however, I am forced to conclude that such an inquiry would lead nowhere.

An industrial design that diligently participates in a nation's manufacting sector, and, in so doing, ends up producing wages that are simply channeled back into the consumption of its own goods is gradually approaching its last days both in Japan and throughout the world. Industrial design must abandon its plan to construct a utopian vision based upon this model of production and consumption. The very term "industrial design," it seems to me, must be locked away inside a museum.

Thus I am thinking of creating a new expression that would stand for a new field of design. With such an expression I would hope to drive home, with each new creation, the importance of this question: Should this really be a design for our future? What will be important to us in our everyday lives and to the planet in general? And what, in the end, will be of the greatest importance to us as human beings? I would like to call the activity that integrates responses to these questions with the creation of new forms, "intelligent design."

Designer Statement
Riki Watanabe

As the war ended, I was torn between an unbearable frustration and an exhilarating sense of freedom. For someone aspiring to a designer's career, the new things brought in by the United States Occupation Forces were the objects of interest and curiosity. Jeeps running around seemed to be prototypes of design in action. Young American women, proud and confident, walked with bold strides along the street, and I stared with wonder at their shoulder bags, bright in color and lustrous like Japanese lacquerware. The bags were made of vinyl, I learned, a synthetic material we had never seen before. Soon afterward, fluorescent lamps appeared, and completely changed our homes and streets at night: their brightness was beyond comparison with the prewar lighting of 15- or 20-watt lamps.

After the war, architects promoted the idea of separating sleeping and dining areas. In traditional Japanese rooms the floor is covered with *tatami* (straw mats). Until quite recently, in the morning the bedding was put into a closet in a corner of the room and a low table was set up for eating, while at night beds were made ready for sleeping, all on the same floor in the same room. It is a fairly basic manner of living in a minimum space, which may be difficult to visualize for people in Europe and America, where sleeping and eating areas are usually separate. Even after the war, Japan was poor, and people had to endure sparse living conditions for a long time.

When I began to explore ideas for design, I felt strongly that I wanted to make low-cost chairs (no. 2). During the prewar and even the early postwar years, the overall image we had of a chair in Japan was one that was upholstered with horsehair or some such material and had spring coils. There was really no history of chairs in Japan except for the folding stools used in military camps in the feudal period by lords and generals, which were clearly considered symbols of power and authority. My first inspiration came when I thought of the *zabuton* (cushion), which is commonly used in every home for sitting on *tatami* mats; I thought of using these cushions on rope supports within a wooden frame. For the frame I chose Japanese oak, and the readily available standardized one-inch board was just the right thickness for my chair. This material was abundant at the time, mainly in Hokkaido, abundant enough, in fact, that it was exported to Europe and America.

The Rope chair marked my debut, and with it I was recognized as a designer. Masaru Katsumie, the art critic, appreciated my work and said it reminded him of *chokibune*, the traditional horn-shape boats that were popular in the eighteenth and nineteenth centuries, and even in the 1950s could occasionally be seen on the Sumida river in Tokyo. I thought I had completely freed myself from the Japanese style and traditional ideas, but I must have reverted subconsciously to tradition in order to avoid imitating the then-popular Scandinavian style.

In early 1953 the Ministry of Foreign Affairs received an invitation to enter the X Milan Triennale. The Triennale, the olympiad of the design world, has contributed greatly to the enlightenment and advancement of design. Before the war I had seen photographs of Triennale exhibits in foreign magazines; these were stunningly fresh and exciting for me. The ministry did not respond to the invitation, and it seemed likely that it would be shelved. Eight architects and designers, including Kenzo Tange, SORI YANAGI, YUSAKU KAMEKURA, and myself, got together to discuss the matter. In the end, the invitation had to be declined mainly for financial reasons and because of a lack of time, but it served as a much-needed stimulus, and our meeting resulted in the establishment of a small group called the Japan Design Committee. Its aims were to realize Japan's entry for the XI Triennale in 1957, to be the center of international design exchange, and to encourage liaison between various fields of design.

As my Rope chair marked a new phase of my career as a designer in 1952, the year 1956 proved equally significant and precious for me. In this year the International Cooperation Administration (ICA), an affiliated organization of the United States Department of Commerce, and the Japan Productivity Center invited Japanese designers to America. I was among thirteen designers who were selected to be part of an industrial design study team. At that time traveling abroad was strictly limited, and it was impossible to get out of Japan without a specific reason. There was no direct flight to New York, and a DC-4, a propeller plane, carried us off, stopping for refueling at Wake Island and Honolulu. After many hours, we were able to get out of the plane at the Honolulu airport, and I still remember vividly the scent of aromatic air that welcomed me warmly to America. When we arrived in San Francisco, two officers of the ICA and two interpreters from the Japan Productivity Center met us at the airport and attended to us during the whole journey.

Our study tour was indeed on a grand scale: we

visited ten design offices, five art and design colleges, and eleven large manufacturers in eight weeks, starting in Los Angeles. Day after day, I was continuously bowled over by new experiences. What impressed me most strongly was the General Motors Technical Center. I was amazed at the large scale of the interior and the originality of its design. In an artificial pond under the staircase, for instance, we saw ripples produced electrically, which is nothing remarkable nowadays but was something totally new for us at the time. I learned that the chief designer of General Motors was also the vice president of the company, and its annual sales were larger than the national budget of our country. Another impressive visit was the one we made to see a mock-up of the interior of a Boeing 707 in the suburbs of New York. It was entirely like the plane itself, even to the tableware and the stewardess speaking through the loudspeaker. The interior was designed by the office of Walter Dorwin Teague, a pioneer of industrial design. Later I had a chance to meet with Mr. Teague in the large drafting room in his office in New York. Our long tour came to an end in Washington with a meeting at the Department of Commerce. It proved to be an extremely fruitful and memorable journey.

Company History: GK Design Group
Kenji Ekuan

The GK design group was founded during the 1950s. The cities of Japan had been reduced to ashes during World War II. Their appearance not only brought home strongly the emptiness of the world of man-made things, but it also made us determined to create a new and bright man-made world.

GK was initiated as a design research group by six young designers who were still students at the time. We had become aware of the importance of modern, organized design for postwar reconstruction, and we made a great effort to get training in and gain knowledge of industrial design, an area that was not sufficiently acknowledged in those days. In the early period all GK projects were primarily proposals, but it was our winning of important design competitions that brought about commissions for concrete design work (no. 35). In 1957 the GK Industrial Design Associates was incorporated. During this period, we sent our members to America and Germany to study advanced industrial-design theory and technology, carving out the foundation of our present international activities.

The subsequent growth of GK has continued in tandem with the development of Japan's industry, which began to recognize the importance of design (no. 53). Particularly during the period of industrial and economic growth in Japan during the 1960s and 1970s, we augmented our staff year by year and expanded the range of our design activities to include everything from product design to product planning, architecture, and environmental and graphic design. Since the 1980s we have aimed our activities toward becoming a knowledge-intensive, high-level, creatively oriented organization. In order to accomplish this, we have made our various departments independent—basic research, computer-systems development, technology development, and design information—and organized them together as parts of our larger present GK design group.

The Japanese word dogu (tool) has served as the key concept in the nurturing of the ideology and methodology of GK design. Dogu has a comprehensive meaning that goes beyond the range of hand tools to include all man-made objects, from machines and equipment to automobiles and electric appliances. Design is the work of giving desirable shape to dogu that are in the process of being born, and the ultimate mission of design is to inculcate a "heart" or "spirit" in such dogu.

Since its founding GK has advocated the philosophy that we call doguology, by which we maintain that in the same way as there is a human world, there is also a world of dogu, and that these two worlds must coexist and correspond to each other. Dogu must be not only useful but also of high precision. Their shapes must be not only functional but also elegant. The bestowing of "heart" or "spirit" into dogu is nothing more than the creation of a deeply intimate relationship between man and objects. The phrase "Spirit to things and the world to people" has become the slogan of GK.

GK is a unique, dynamic group that is different from the conventional free-lance design office. It is built upon three pillars—movements, projects, and learning.

Movements consists of the promulgation of design throughout the world. Without such movements targeted toward a public appeal for the importance of design, design will not gain the understanding of society. In 1973 GK was the major motivating force behind the decision to hold the congress of the International Council of Societies of Industrial Design in Kyoto, and GK has continued to play a central role in design promotion activities in Japan.

Projects consists of activities for establishing design ideology in the world: the creation of concrete things in the context of social reality and their placement in the hands of the public. For this purpose GK has taken upon itself the responsibility of contributing to postwar Japanese society through its work for the continuation and development of enterprises and a broad variety of design work.

Learning is probing deeply into the world of objects. When this foundation of knowledge is missing from design, there is often the danger of falling back onto the individual expression of the designer, which results in nothing but simple wordplay. GK began as a research group that carried out all sorts of study projects. Later a basic research department was formed in the context of which a strong attitude toward learning has been maintained.

In the midst of the diversifying social environment of the 1990s, it is impossible to satisfy the needs of society through a single area of specialty. Today the GK design group takes a comprehensive approach that encompasses all areas of design in order to respond to the various problems faced by our society. It utilizes its unique organizational structure and its breadth of specialist technology (nos. 232, 237). At the same time GK also makes great efforts toward contributing to society through a thoroughgoing philosophical approach in order to bring about harmony between the "heart and things" in the context of our high-tech, information-intensive society.

Today, when voices are being raised as we confront global crises and the demise of the period of rapid economic growth, there is a great demand for changes in life-style. GK continues its search to discover the type of life-style that should be created and the type of life-scape that shoud be carved out by design.

To effect the birth of a world of new things appropriate to the new age, we of the GK design group network with experts in a broad range of fields, for we believe that we must metamorphose our organization into an ever-increasingly complex and flexible entity. Our aims are to amalgamate entirely our movement, project, and learning activities; to develop an organic, creative body that continues to renovate and alter itself like a living organism; and to work toward the creation of a new, universal life-style.

As we approach a new global era, we must think seriously about how Japan can contribute to the world through design. For this purpose, we ourselves must create an even more open international organizational structure.

Since the achievement of modernization in Japan, we have just barely begun to succeed in effecting a reorganization of the world of things. Thus we are just on the brink of entering a true design age for Japan.

Company History: Honda

Shinya Iwakura

Since the Honda Motor Company was established in 1948, it has conducted its business activities by creating products that have global appeal on the basis of our corporate philosophy: "Maintaining an international viewpoint, we are dedicated to supplying products of the highest efficiency yet at a reasonable price for worldwide customer satisfaction."

Under the leadership of our company founder, the late Soichiro Honda, we have attempted to produce creative products for our customers, proceeding with the concept of "ambition and youthfulness," and our product lines have expanded from motorcycles (no. 84) to general power machines (such as marine engines, lawn mowers, and agricultural tillers) and then to automobiles.

Honda's philosophy of creativity is not merely to make our products different from the products of our competitors in both engineering and design, but also to meet the challenge of realizing the customers' dreams and needs in the form of a product that has a new function as it responds to social responsibility. This challenge is to create a new way of thinking, that is, to create a distinctive rather than merely a different product, and it has resulted in Honda's originality, which is recognized throughout the world.

We believe that design as manifested in a product expresses the "face" of a corporation. A good "face" on a product does not necessarily signify just external appearance, but, like the face of a human being, it also expresses the deep faith and passion of those who created the designs. The unique characteristic of Honda design may be described as a design that expresses the inner spirit of people (managers, engineers, and designers) working together to provide our customers with advanced and sound products.

As time passes the needs of customers worldwide change. To accommodate those changes, we have changed and grown from an international company producing cars for the world to a global company producing cars suitable for different market areas by thinking, creating, and doing business together with the people of those areas and by reinvesting the profits back into the areas where they are made. Honda's emphasis on design has played a major role in the company's success.

Among the many products that have contributed to the growth of Honda, the Supercub, a 50-cc motorbike, introduced in 1958, is the one that built Honda into the company it is today. The attitude of Japan's industries toward design has generally been to use it as a means of creating products that would sell for prices as high as possible. In the 1950s automobiles in Japan were like flowers at inaccessible heights; bicycles were the means of transportation and delivery. The appearance of the Supercub changed that. Merchants and consumers replaced bicycles with the Supercub as it was a much more efficient means of transportation, and the motorbike became an instant hit. It was an indispensable tool for transporting goods, contributing to the economic growth of postwar Japan.

Some of the original concepts of the Supercub motorbike were convenience exceeding that of bicycles, affordability, and the 50-cc engine—50 cc was the maximum engine size for a vehicle that could be driven with an operating permit rather than a driver's license—and it was developed to utilize new manufacturing technology for mass production. The Supercub employed large plastic components, and was relatively simple to manufacture, easy to handle, and reasonably priced. Supercubs have been in continuous production—with some minor model changes to the basic, original design—for the nearly thirty-five years since its introduction.

Honda was a latecomer to automobile manufacturing in the early 1960s, when automobile production was just beginning to increase and automobiles were still a status symbol in Japan. Since then Japan has experienced unmatched economic growth. In 1963 Honda introduced a small truck called the T360 and a small sports car, the S600, both reasonably priced. The original concept of Honda automobiles was to exploit practicality in its trucks and dreams in its sports cars. Subsequently, a mini passenger sedan, the N360, was introduced, which combined the practicality of the T360 and sportiness of the S600.

The N360 minicar became the basis for a new sub-compact sedan called the Civic, introduced in 1972; the 1975 model Civic met American exhaust emission standards (no. 113). In 1976, Honda introduced the Accord, a compact sedan, as a step-up for Civic customers, and both became best-sellers owing to their technological advances and original design concepts.

In the late 1980s, Honda introduced the NSX, a sports car that heralded a new direction for the next generation of automobiles with its aluminum alloy body, suspension, and major engine components. The aluminum high-performance engine was relatively small, allowing for lighter body weight and better fuel economy, and was made of recyclable materials.

Honda is a company that eagerly faces the challenges of creating products for the future. We wish to discover new values and create dramatic products that will allow customers, sellers, and the producer to share their dream and their joy.

Company History: Kenmochi Design Associates

Tetsuro Matsumoto

In June of 1955 ISAMU KENMOCHI resigned from his position as a government official. During the twenty-three years that he had been with the Industrial Arts Institute, Kenmochi had come under the guidance of Bruno Taut, constructed his own set of design principles, and come to define his own identity as a designer. During the war he had partial responsibility for the government policy of standardization, and he had advocated the development of furniture design as a branch of industrial design. This approach to design was given concrete form when, with Japan's defeat at the close of the war, the Industrial Arts Institute was put in charge of designing furniture for the housing of American soldiers.

This role of design became more focused as Japan, amid the confusion that followed its defeat, sought to rebuild its economy by accumulating foreign capital through exports. In 1952 Kenmochi was sent on an official trip to the United States, where he made detailed observations of the state of design. In 1953 he again went to the States, this time to attend the Aspen (Colorado) International Design Conference, and through that experience he came to recognize further the importance of design and of the designer to society. There is no doubt that from this time on, Kenmochi's desire to establish himself as an independent designer grew stronger.

The significance that Kenmochi drew from his two trips abroad encouraged him to develop in the direction of what has been referred to as "Japanese Modern" design. It should be noted that in the United States at this time, Scandinavian furniture was highly prized and viewed as essential to everyday life in an expanding industrial society. People were drawn toward a beauty that was simplified, forms that were more polished, and a functionality that possessed an emotive quality. Kenmochi's encounter with this new development inspired him to promote what he called "Modern Japanese Design."

During the period from 1952 to 1954, Kenmochi was giving serious thought to what the nature of Japanese design should be. When considering the importance of design and the designer's profession within the broader social context, he naturally found himself confronting the basic conflict that exists between the individual designer and the group. As a result he began to seek out those conditions that would permit free expression to the individual designer.

His first job was to draw up the design for a

Japanese restaurant that was to be opened in São Paulo in Brazil. He and I worked together on this project at his home. The layout we produced, with its attached perspective, elevation, and floor plan, was to be our first design job, and this project remains one I will never forget. In 1955 Kenmochi rented the attic of an old three-story wooden house in Aoyama, and the Kenmochi Design Associates was born. This was the year when the Liberal Democratic Party was formed, SONY CORPORATION was mass-producing its transistor radio, and the economy was booming.

Hearing much at this time about the consumers' unions that were active in the United States, Kenmochi believed that it was necessary for both consumers and manufacturers to be concerned with issues of design. He did not feel that good designs could be produced solely by designers. Consumers, designers, and manufacturers had to work together if design of a high quality was to be produced. From its very inception, Kenmochi Design Associates promoted the new ideals embraced by Kenmochi while working to establish the image of design as a legitimate profession. At the same time, the company sought to respond to the world's expectations for Japan's unique approach to design, and worked to keep up with the international trends in design.

In 1956, Kenmochi Design Associates moved to the second floor of a new office building and opened Living Art, a shop that sold products for the interior. This, I believe, was the first stage in the realization of the ideals embraced by Kenmochi. Here he displayed everyday goods that had been either designed or selected by him, an attempt by Kenmochi to promote design and to enlighten others about the field.

Kenmochi devised a plan for bringing a coherency to the field of design and for revising the definition of the designer as an artist well versed in the technical aspects of the field (nos. 12, 38, 52, 78). His first large-scale office design was done in 1956 for the main office of the Daiwa Securities Corporation. Jobs that followed in 1957 included work done for the Shiseido beauty salon, the captain's quarters on an American tanker, and the Decora center. In 1958 Kenmochi provided designs for Kenzo Tange's government office building in Kagawa Prefecture, Komatsu's farming equipment, and the Mitsubishi showroom in Tokyo. During that year, he also worked with Kunio Maekawa on the design for the Japanese pavilion at Expo '58 in Brussels. In 1960 he designed the Hotel New Japan in Tokyo, the largest hotel in Japan at that time. While in a sense things were moving along quite smoothly in these early days, because designers at that time did not go out and

solicit work but waited for it to come to them, there were times when Kenmochi and I were without work and our income was not sufficient. In any case, we had started down a difficult path.

This painful process, however, began to change as a result of the economic growth that Japan was experiencing during the 1960s and 1970s. The focus of our work began to shift from furniture design, which had dominated our early years, to the planning of interior spaces. This shift was further reflected in the fact that commissions began to come to us directly from clients rather than through architects. In addition, a cooperative relationship began to develop between architects and designers that put them on an equal footing. All of this pointed to the fact that interior design was coming closer to being established as a recognized profession.

During this period Kenmochi worked for the Keio department store, the Kyoto international congress hall, and the Tokyo congress hall in Kasumigaseki. In addition, when Japan Air Lines introduced its B-747, Kenmochi was put in charge of designing the interior of the plane. Kenmochi died in 1971 as his final design for the Keio Plaza Hotel in Tokyo was being completed, and I was put in charge of the firm.

Since then the range and significance of our designs have continued to expand. We have designed office interiors (Japan IBM), hotels (Hotel Nikko in Saipan), public facilities (Yamagata municipal office building), exhibition spaces (national museum of history, Sakura), signage (Tama Center), a cruise ship (Japan Steamship Corporation's Asukamalu), aircraft (JAL B-747, B-767, DC-10), trains (the Shinkansen Tsubasa for JR Eastern Japan and the Shinkansen Nozomi for JR Tokai), as well as mass-production furniture. Recently we designed the furniture for the Shinfukiage palace, which is part of the emperor's residence.

After almost four decades interior design as a legitimate profession seems to have finally been recognized in Japan.

Company History: Nippon Design Center

Kazumasa Nagai

The Nippon Design Center began operations in March 1960 as an association of art directors, designers, copywriters, and photographers assembled from all over Japan, with YUSAKU KAMEKURA, HIROMU HARA, and RYUICHI YAMASHIRO as its principals. The center was founded as a joint venture of eight companies with an overriding interest in the advancement of advertising, among them Toyota Motor, Asahi Breweries, Toshiba, and NIKON CORPORATION, and nine graphic designers. Their objective was to encourage the development of Japanese industry by producing well-designed and inventive advertising through the collaborative efforts of business and design. This concept, conceived by the present chairman of the board, Matsuo Suzuki, the economic commentator, was unique and not found elsewhere.

In the year 1960 the new United States–Japan security treaty was ratified, and the Ikeda cabinet announced a program to double the national income. Subsequently, the Japanese economy grew rapidly, and until the international oil crisis of 1973 the gross national product increased at an average rate of 10.9 percent annually (1959–73). The Japanese people no longer lacked material goods, and Japan was on the road to becoming an economic power. The production-oriented economy, which had continued since the end of the war, changed to a sales-oriented economy; with marketing activities aimed at capturing the hearts of the consumers moving briskly, the importance of advertising grew rapidly.

That the Nippon Design Center was founded in those times is an indication of foresight, and the firm received a succession of awards and drew the attention of the advertising and design communities. Previously, most advertising in Japan had been produced by art directors and copywriters whose only concern was creating advertisements. With few exceptions the quality was not very high, for their work strongly reflected the ideas of the clients but lacked the creative input of graphic designers.

The nucleus of the Nippon Design Center consisted of designers such as Hara, Kamekura, and Yamashiro together with IKKO TANAKA and myself, as well as TADANORI YOKOO, who joined a little later. These designers took the challenge of advertising seriously, which is the reason that a sense of artistic quality began to appear in advertisements, and was very refreshing (see nos. 33, 55, 75, 79). This approach was carried on, and taken up later by such artists as

MAKOTO SAITO (nos. 165, 181) and Masatoshi Toda.

This does not mean that the group adopted a single style and excluded all others; it was a cooperative artistic effort, but one in which the work of each of the strong individuals was free to shine. Rather than being merely a business enterprise, the center gave the impression of being an association of designers with like minds. However, in recent years, its radical, artistic direction has been altered and become more deeply allied to marketing in accordance with changes in the times. It began to cater to consumers, resulting in precision advertisements that lacked the pioneering spirit and that were, if anything, on the conservative side. The clients of the Nippon Design Center had become the large corporations of Japan, and this phenomenon can be said to reflect new demands from industry. What has not changed, however, is the center's emphasis on creative graphics and its endeavor to obtain the approval of both clients and consumers by meeting problems squarely without pursuing fads or exploiting sensationalism.

Hiromu Hara has passed on, and Kamekura, Saito, Tanaka, Toda, Yamashiro, and Yokoo have left the Nippon Design Center. I am still here, and I intend to maintain their tradition of quality. It is my opinion that a strong aesthetic of originality can be reflected in a single poster created through the artistry of one person or an advertising campaign produced with the involvement of an entire team. In this age of information overload, disagreeable or superfluous information must be disregarded and overlooked. This is also true of design. No matter how much a cliché is repeated, it results only in the self-satisfaction of the designer. This means that there is no longer a single mass public and that only original and targeted information will be received by an intended audience. To make that possible, I think it is necessary for those who communicate to seek out their own individuality and create powerful designs with that in mind. When the concepts of originality of business, which is the primary sender of the message; the designer, who creates the substance and form of the message; and the consumer, who is its recipient, are at the same level, it can be said that optimum communication has been achieved. With these aspirations in mind, I have endeavored to work within the framework of the Nippon Design Center to create designs that retain their originality within an organization that is becoming conservative in its direction.

Company History: Sony

Matami Yamaguchi

Sony is often referred to as a design-minded company. Sony's design has become synonymous with its products, recognized and valued the world over. We for whom product design is our daily work cannot honestly hide the delight we feel about this, but at the same time we feel rather embarrassed. The reason for this is that design results from the combined efforts of many people, including electrical and mechanical engineers, and those concerned with manufacturing technology and parts procurement, and not just the product designer. Therefore the credit for each product should be shared by all professionals in the firm. Sony designers know that what they create must be achieved by overcoming daily challenges, by never imitating others, by imposing self-discipline, by maintaining leadership, and by steadily working toward modest goals in the face of indescribably fierce competition within the industry. Most Sony products have come into being in that way. It can be said that each designer contributes individually but is also a member of a team that works toward the manifestation of collective strength, always reflecting the viewpoint of the founders.

For the designers, the greatest satisfaction comes when the collective strength is channeled into a form that is accepted by society as a product. When designers see products that they have designed displayed in stores, and purchased and used by consumers, they are gratified. And when that design becomes the source of copies and achieves value and universality, designers derive satisfaction knowing that their objects have achieved the status of symbols. However, they must not forget that the pleasure they feel comes from working as a team of many players acting as one body.

In October 1945 a handful of people arrived in Tokyo when it was a field of burnt rubble and established the Tokyo Telecommunications Research Laboratories. In May of the following year, Tokyo Telecommunications Engineering Corporation, the forerunner of Sony Corporation, was born, capitalized at 190,000 yen; it had some twenty employees. The founders were Masaru Ibuka, then thirty-eight years old, and Akio Morita, then twenty-five. The company was first located on the third floor of the Shirokiya (now Tokyu) department store in the Nihonbashi section of Tokyo. Although the building can be said to have survived the ravages of war, the room was shabby, with crumbling concrete and broken windows that let in the drafts. It was difficult just to obtain food, so even if the founders wanted to make repairs they could not get the materials. But the founders were young and fired with hope and ideals. The spirit of the new firm's prospectus, which was written by Ibuka, continues to be handed down to the designers as the Sony philosophy: 1) to build a model factory that is open, generous, and welcoming, which will bring out the best skills of devoted technicians; 2) to participate actively in the cultural advancement for the reconstruction of Japan through technology and production; 3) to eschew unwarranted profits, always placing emphasis on validity and quality, and to avoid the pursuit of expansion for expansion's sake; and 4) to give utmost effort to the selection of products without avoiding—indeed welcoming—technological difficulties.

"Regardless of quantity, the firm's goal should be high-quality technical products that contribute the most value to society. Avoid conventional classification of products into such categories as electrical or mechanical, and strive for unique combinations that other companies cannot emulate.

We cannot compete by doing the same thing as other companies. There are plenty of technical opportunities. We must do what large companies cannot do and take a part in the reconstruction of our country through the power of technology. Even if we do not have money or machinery, we have brains and technology. By using them we can do anything. Let us find ways to do what others do not do."

These words were part of a speech Ibuka gave at the inauguration of the new company. This speech may not be noteworthy today, but when one considers the conditions at that time and that it was the start of a new company, it was quite a daring statement.

By setting the goal of making appealing products different from those of other manufacturers, Ibuka was looking through the eyes of a design director from the inception of the company. He had a special interest in design. He sought the opinions of specialists, and in 1954 he sought out designers among college graduates. The founders made their own suggestions from the point of view of design, and that enthusiasm has not changed. Several times a year Sony's design center makes a presentation to the top executives, and the executives in turn make suggestions from the view of the consumer.

The TR-610 radio (no. 32) is a good example of a successful product resulting from the joint teamwork of managers, electrical and mechanical engineers, and designers. This was a small, thin radio that could fit into a pocket and be operated with one hand. It became a landmark in the history of design through the conjunction of several elements: the technology of producing a speaker just a millimeter thinner, the challenge of developing a new battery, the skillful placement of electrical parts, the concept of elevating the speaker to a position that would preserve the thinness of the body, and the banding of the speaker with a brass ring plated with real gold, thus making it the focal point of the total radio design, and offering it in an unprecedented three-color version—ivory, red, and black.

Finally placed on sale in June 1958, its price was 10,000 yen. It was quite expensive, being the approximate monthly starting salary of an officeworker at that time. In spite of that, it was well received. It was exported throughout the world, chiefly to the United States, and became a big hit, with sales of over a half-million units for the one model alone. The name "Sony" came to be used throughout the world as a prefix for transistor radios themselves. Imitations cropped up all over the world, even in design capitals like the United States and Italy. It has been said that in the United States, where the radio was highly valued, it was rumored to have been designed by Raymond Loewy.

In September 1992 the founder of Sony, Masaru Ibuka, was awarded the medal of honor. This was the first time a cultural award was given to a manufacturer; the recognition by the nation of manufacturing as a part of culture was an historical event.

In 1990 designers from various manufacturing areas, such as CANON, Nissan, NEC CORPORATION, and Sony, joined with the academic community to begin a campaign to bring design within the framework of culture and to seek a new sense of values. Based on the outcome of that research, they will make recommendations for the form that industry should take in the future.

There is no longer any doubt that design, hence production, is the language that expresses culture today. It is Sony's hope that we will send a message to the future based on our unique culture—production. It is our wish to develop new areas of design that will enable us to create unique objects that everyone can respond to and that everyone can admire.

1 Record player and radio, 1952
Designed by Sori Yanagi
Made by Nihon-Columbia, Tokyo
Prize: Mainichi industrial design competition, 1952
Wood, plastic, and metal, 16 1/8 x 19 11/16 x 12 3/16"
(41 x 50 x 31 cm)
Yanagi Product Design Institute, Tokyo

Designed in 1952 by Sori Yanagi, who opened one of the
first independent design firms in postwar Japan, this
combination record player and radio took top prize in
that year's first annual Mainichi industrial design
competition, where it was cited for the use of new
technology in its three-speed automatic record changer.
The prototype exhibited in the competition was then
modified so that it could be mass-produced and sold at
an affordable price. Typical elements of the Japanese
aesthetic reveal themselves in its low profile and
repeated horizontal lines and in the asymmetrical
placement of the company name and the control knobs.
The gently rounded tone arm previews the organic form
that would become a signature of Yanagi's designs.

1

2 Rope chair, 1952
Designed by Riki Watanabe
Made by Sokensha, Tokyo
Oak and rope, 28 3/4 x 20 7/8 x 29 1/2" (73 x 53 x 75 cm)
Museum & Library, Musashino Art University, Tokyo

Riki Watanabe was among the first to produce low-cost
furnishings for residential use in postwar Japan, and the
prototype for this wood and rope chair was shown at the
1952 exhibition of the Shinseisaku (New Works) group in
Tokyo, an art and design association that showed new
and often unconventional work. The design of the chair
accommodates the Japanese preference for floor-level
seating: the seat is quite low, and the standard square
cushions (*zabuton*) used for sitting on *tatami* mats can be
used on the seat and back. Its inexpensive materials and
ease of assembly appealed to the ideals of functional
design as well as to the need for economy.

2

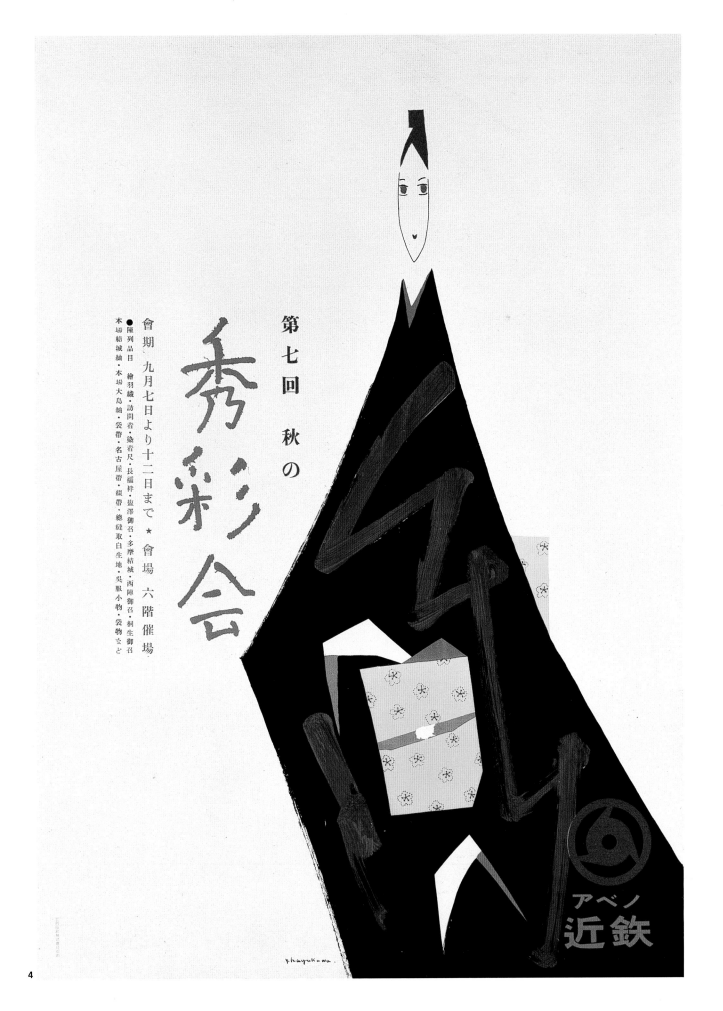

第七回　秋の　秀彩会

會期、九月七日より十二日まで★會場　六階催場

●陳列品目　繪羽織・訪問着・染着尺・長襦袢・塩澤御召・多摩結城・西陣御召・桐生御召

本場結城紬・本場大島紬・袋帯・名古屋帯・綴帯・總縫取白生地・吳服小物・袋物など

アベノ
近鉄

3 Symbol for Japan Industrial Designers Association, 1953

Designed by Yusaku Kamekura

At the urging of the critic Masaru Katsumie, Yusaku Kamekura joined the Japan Industrial Designers Association a year after it was established; he designed this symbol for the association the same year, using as his motif the Roman letter *d*. "I like this logo, it's simple and clean," wrote Kamekura. "The funny thing was that when Walter Gropius visited Japan [in 1954] and saw the logo, he said '[it] looks like some farm tool and that makes it Japanese,' a remark that became well known in the industry."[1]

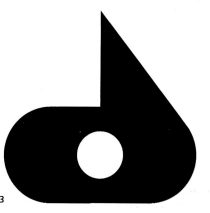

3

4 Poster for Kintetsu department store, 1951

Designed by Yoshio Hayakawa
Lithograph, 29 1/2 x 20 13/16" (75 x 53 cm)
Collection of Yoshio Hayakawa, Tokyo

This poster advertises the seventh annual autumn kimono show of the Shusaikai (Beautiful Color Society) at the Kintetsu department store in Osaka, where Yoshio Hayakawa worked as art director. Through their fashion shows the Shusaikai emphasized the materials, design, colors, and techniques of the kimono, and with this poster of a Westernized redheaded model hoped to create a revival of its use among the younger generation. In his design for the poster Hayakawa used the first character of the society's name, *Shu* (Beautiful), as the brown calligraphic element that sweeps across the black trapezoidal kimono, and combined it with a yellow decorative paper collage to suggest the *obi*, or broad sash, traditionally worn with it. The inspiration for this device may lie with a type of seventeenth-century kimono that incorporated characters and lines of poetry in striking asymmetrical patterns. Hayakawa depicted his subject in a boldly abstracted manner, aiming to create, as he said of his fashion posters, "an immediate sensation by the use of large forms and simple colors. As to their style and spirit I endeavor to interweave traditional Japanese graphic art with modern Western art, for I think that contemporary art and poster art in particular should speak a universal language."[2]

5 Poster (*Sheltered Weaklings—Japan*), 1953
Designed by Takashi Kono
Silkscreen, 74 13/16 x 98 7/16" (190 x 250 cm)
Aichi Prefectural University of Fine Arts, Aichi

Takashi Kono, a founding member of the Japan
Advertising Artists Club, designed this poster for its third
exhibition in 1953. Titled *Sheltered Weaklings—Japan*, it
illustrates graphically the political relationship that
existed between the United States (the big shark), Japan
(the school of little fish), and the Soviet Union (the sharks
at the upper right) in the early 1950s. Shown in a private
exhibition, this poster allowed the designer to express his
personal views in a way that commercial work did not.

6 Sewing machine, 1953
Designed by Jiro Kosugi
Made by Janome Sewing Machine Company, Tokyo
Prize: Mainichi industrial design competition, special
award, 1953
Enameled cast iron, 11 x 16 1/8 x 7 1/16"
(28 x 41 x 18 cm)
Janome Sewing Machine Company, Tokyo

Special-award winner of the second annual Mainichi
industrial design competition in 1953, Jiro Kosugi's
prototype of this sewing machine incorporated several
innovations to increase efficiency and to streamline its
appearance, which were included in the production
model when it appeared several years later. Compared
to traditional models, which had round bodies and black
housing, Kosugi's design featured a light-blue casing and
an interplay of round elements with an overall squared-
off shape.

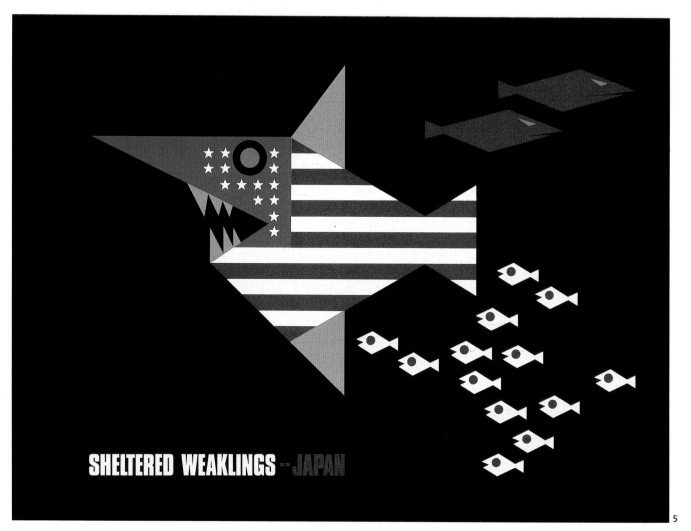

7

7 Rice cooker, 1954
Designed by Yoshiharu Iwata
Made by Tokyo Shibaura Electric Company, Tokyo
Prize: G-Mark, 1958
Aluminum and plastic, 10 1/4 x 11 13/16 x 9 7/16"
(24 x 30 x 26 cm)
Toshiba Science Institute, Tokyo

A steaming bowl of rice is still one of the essential ingredients of a Japanese meal. When this electric rice cooker was introduced in 1955, it replaced the time-consuming method of preparing rice in wooden bowls over wood or gas heat; it guaranteed a perfect pot of rice every time, even to the most inexperienced cook. This was one of the first electrical appliances to be widely used in postwar Japan, a harbinger of the technological revolution ahead. The basic design of its plain white body and aluminum lid has not changed over four decades, in part because it fulfilled the requirements so well, and lately perhaps because of sentimental attachment to the original model.

8

8 Poster for *Tanko* magazine, 1955
Designed by Takashi Kono
Silkscreen, 28 5/8 x 20 1/4" (72.7 x 51.4 cm)
The Museum of Modern Art, New York.
Gift of the designer

This poster repeats the cover design of the monthly tea-ceremony magazine *Tanko*, advertised in the typography at the bottom. In the audacious simplicity of the single persimmon, Takashi Kono captured the essence of the Zen aesthetic, which informs the tea ceremony even today. The image is an allusion to a famous thirteenth-century ink painting of a persimmon beloved by tea afficionados, and represents an elegantly successful translation of traditional imagery into the twentieth-century idiom of the poster.

9 Speed kettle, 1953
Designed by Sori Yanagi
Made by Nikkei Aluminum, Tokyo
Prize: G-Mark, 1958
Aluminum and plastic, height 9 7/16" (24 cm),
width 7 7/8" (20 cm)
Yanagi Product Design Institute, Tokyo

One of the earliest aluminum household products
produced after the war, this kettle was designed by Sori
Yanagi in conjunction with the Tokyo Gas Company to
encourage the use of gas for cooking. In the center of
the kettle is a hollow cone, which conducts the heat from
the burner upward and makes the water boil more
quickly. The large kettle with its organic molded-plastic
handle was promoted for both its speed and its
economy.

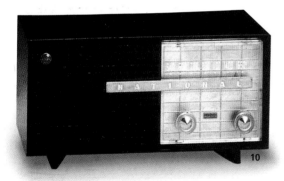

10 National radio, 1953
Designed by Zenichi Mano
Made by Matsushita Electric Industrial Company, Osaka
Prize: Mainichi industrial design competition, 1953
Plastic, 7 x 14 x 5 1/2" (18 x 35.5 x 14 cm)
Matsushita Electric Industrial Company, Osaka

In discussing the design of this National DX-350 radio for
Matsushita, Zenichi Mano explained that he "was
pondering how to make electrical appliances more
Japanese and was consciously experimenting with this
idea,"[3] here trying to suggest the lines of the best-known
icon of Japanese architecture, the Katsura Detached
Palace in Kyoto. The left half of the radio's plastic
housing echoes the wooden shutters (*shitomi*) of the
building, while the pattern on the right half is taken
from its sliding-door panels (*shoji*). Mano won an award
for this radio at the second annual Mainichi industrial
design competition in 1953.

11

12

11 Stacking stool, 1954
Designed by Sori Yanagi
Made by Kotobuki Company, Tokyo
Fiberglass-reinforced plastic,
14 9/16 x 20 1/16 x 8 5/16" (37 x 51 x 46.5 cm)
Yanagi Product Design Institute, Tokyo

Fiberglass was available commercially only after the war,
and Sori Yanagi was the first designer in Japan to adapt
it to furniture production with this stacking stool.
Strong, resistant to stains and mars, available in several
colors, and inexpensive, the stool was first produced in
1959 and remained in production for twenty years,
proving that modern industrial materials could be
successfully introduced to Japanese domestic interiors.
Following the concept of stackable furniture then
becoming popular in the West, Yanagi designed this
stool to stack, a feature that was useful in small urban
Japanese apartments.

12 Chair, 1954
Designed by Isamu Kenmochi
Made by Industrial Arts Institute, Sendai
Wood and bamboo, 32 1/2 x 16 1/2 x 15"
(82.5 x 42 x 39 cm)
Fund of Industrial Arts, Tokyo

Designed by Isamu Kenmochi and made by the
Woodwork Technical Section of the Industrial Arts
Institute (IAI), this chair was shown in one of two model
rooms created by the IAI for its "Design and Technology"
exhibition of 1954. The dining chair, which originally
had a red seat cushion, was considered competitive in
terms of its quality and cost of manufacture for both
domestic and export markets. Combining traditional
material with contemporary technology—the elements
were all machine-tooled and easy to assemble—this
furniture was geared to a new postwar life-style.

13 Sekonic light meter, 1955
Designed by KAK Design Group
Made by Seiko Electric Company, Tokyo
Plastic, 4 1/8 x 2 5/8 x 1 3/16" (10.5 x 6.7 x 3.1 cm)
Seconic Company, Ltd., Tokyo

The Sekonic light meter was the first to give readings
both for exposure setting and color filter, its compact
size enabling the photographer to use it with one hand
while adjusting camera settings with the other. Creating
a total identity program—one of the first in Japan—KAK
Design Group created the complete line of camera and
photographic equipment for Seiko's Sekonic label,
including the graphics for its logo, packaging, and
instruction brochures, for which the group won the
Mainichi industrial design prize in 1958.

13

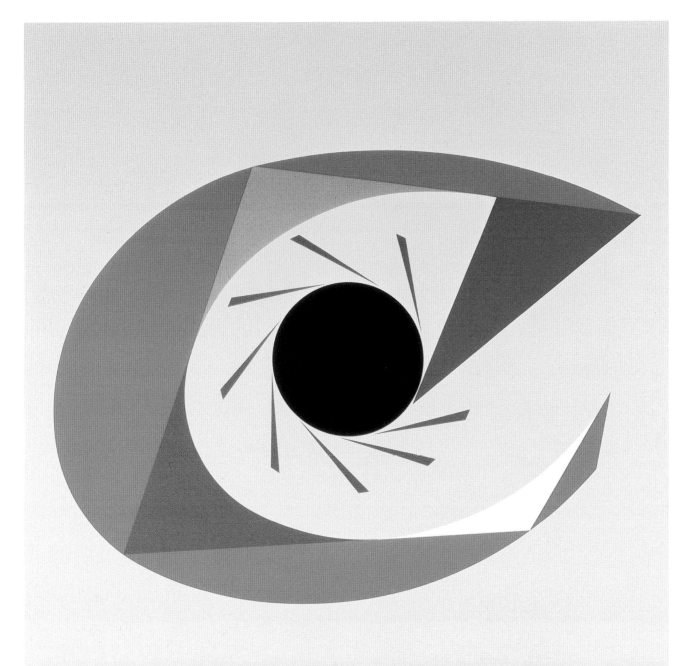

14 KEN.

15

14 Poster for Canon Camera, 1954
Designed by Kenji Itoh
Silkscreen, 40 1/2 x 28 11/16" (103 x 72.8 cm)
Collection of Kenji Itoh, Tokyo

In this poster, which was part of his complete graphic program for CANON, Kenji Itoh makes the camera lens the literal focus of his crisp and effective graphic design. The exaggerated graphic symbol reads both as an eye and as the C of the corporate symbol.

15 Poster for Daido Worsted Mills, 1954
Designed by Kenji Itoh
Offset, 40 1/2 x 28 11/16" (103 x 72.8 cm)
Collection of Kenji Itoh, Tokyo

Kenji Itoh was one of several graphic designers, among them YUSAKU KAMEKURA, commissioned by Daido Worsted Mills (now Daidoh) to create advertising posters for their Milliontex line of wool cloth in the 1950s. A pioneer in the use of photography in Japanese advertising, Itoh combined close-up photographs of the textured and striped wool fabric with typography and a hand-drawn ram, arriving at a mixed-media composition of some technical complexity and a great deal of humor.

16 Delta fan, 1955
Designed by Tetsuya Furukawa
Made by Fuji Electric Company, Yokkaichi
Prize: G-Mark, 1958
Stainless steel and plastic, 21 5/8 x 14 15/16 x 12 5/8"
(55 x 38 x 32 cm)
Fuji Electric Company, Yokkaichi

After World War II, Tetsuya Furukawa was given the
assignment to redesign the Fuji Electric Company table
fan, which had been introduced to the commercial
market in 1927. It was a particular challenge for
Furukawa because the basic technical requirements of
the product remained the same. Arriving at an elegant
solution, he replaced the standard solid base with
stainless-steel tubing, which calls to mind the triangular,
swept-back airplane wing with straight trailing edge
known as the delta wing. He added further aeronautic
allusions with the blue plastic blades and steel nose. The
deltoid shape of the base became the fan's hallmark and
gave Furukawa's model its name, Delta.

16

17 Sony radio, 1955
Designed and made by Tokyo Telecommunications
Engineering Corporation, Tokyo
Plastic and metal, 3 1/2 x 5 1/2 x 1 1/2" (8.9 x 14 x
3.8 cm)
Collection of Tom Christopher, San Jose, California

Awarded a license from the Western Electric Company to
manufacture transistors in 1953, the Tokyo
Telecommunications Engineering Corporation developed
its own transistors in 1954 and Japan's first transistor
radio, this TR-55 model, the following year. The TR-55,
with its small size and fresh styling, also marked the first
appearance of the Sony label: the name, a combination
of the Latin *sonus*, meaning "sound," with the English
"sonny," suggests a lively boy and the image of an
energetic young company. The radio's new technology,
which included the company's first use of perforated
metal for a speaker, an application borrowed from the
American automotive industry, was a source of pride for
the company. This was the firm's first export, and the
company's transistor program quickly became so
successful (see no. 32) that in 1958 the name Sony was
adopted for the firm as a whole.

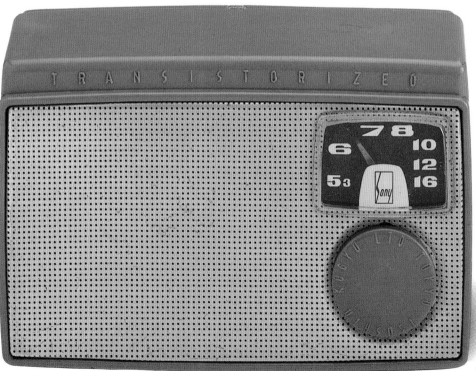

17

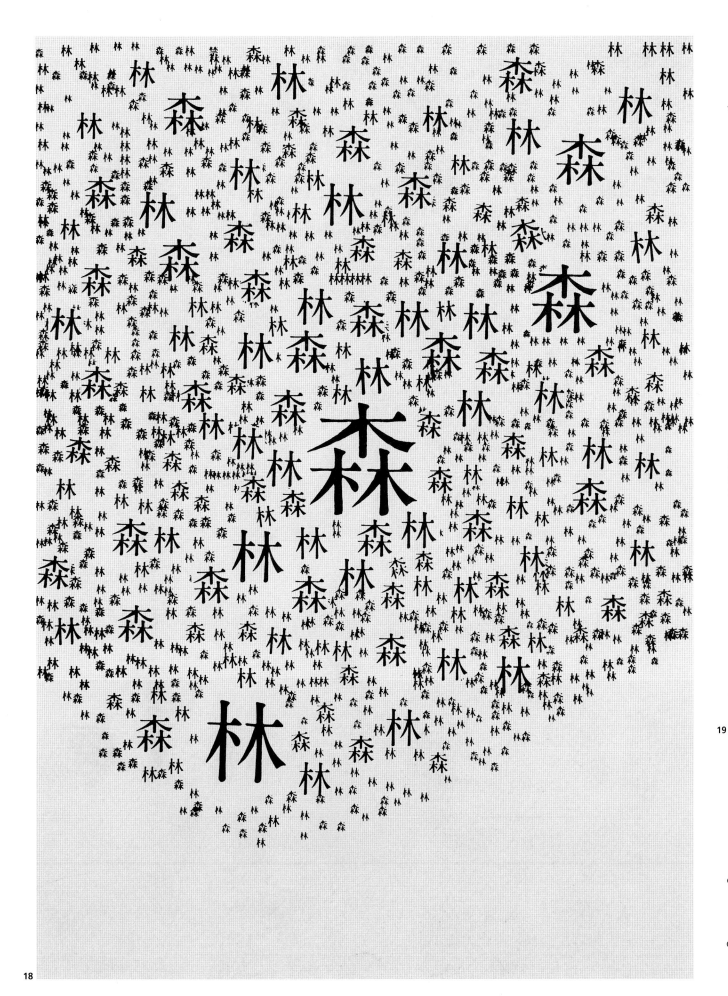

18 Poster, 1955
Designed by Ryuichi Yamashiro
Silkscreen, 40 1/2 x 28 11/16" (103 x 72.8 cm)
Museum & Library, Musashino Art University, Tokyo

Visual allusions and puns are particularly characteristic of
Japanese graphic designs where calligraphy and
typography come into play, as in this poster designed to
promote a tree-planting campaign. Ryuichi Yamashiro
expands on the linguistic structure built into the
characters for "grove" (written with two trees) and
"woods" (written with three). The black ink and the
studied use of empty space recall also the aesthetic of
Zen ink paintings.

19

19 Vase, c. 1955
Designed by Kozo Kagami
Made by Kagami Crystal Glassworks, Tsukuba
Engraved clear glass, height 7 5/8" (19.4 cm),
diameter 7 1/2" (19.2 cm)
The Toledo Museum of Art, Toledo, Ohio.
Gift of Kagami Crystal Glassworks

Kozo Kagami, who learned the techniques of glass
engraving in Germany during the late 1920s, became the
foremost practitioner of this art in Japan. He engraved
this vase with the traditional Japanese floral motif of
nadeshiko (fringed pinks) after his own calligraphic
design in brush and ink, adapting the composition to the
inverted bell shape of this free-blown vase so that the
pinks seem to grow there naturalistically.

20 Movie camera, 1956
Designed and made by Canon Camera, Tokyo
Prize: G-Mark, 1957
Metal, 5 11/16 x 2 1/16 x 3 1/2" (14.4 x 5.2 x 8.8 cm)
Canon Inc., Kawasaki

Canon's first 8-millimeter movie camera, the 8T model
was then the market's most serious model designed for
personal use. Canon approached its design from the
inside out, creating the compact external form around
the needs of its internal technology. Two lenses for
varying distances (13 and 25 millimeters) were mounted
on a double turret, which could be rotated for the
required focus. Canon's 8T movie camera was among
the first recipients of the G-Mark prize in 1957.

21

20

21 Soy-sauce container, 1956
Designed by Sori Yanagi
Made by Tajimi Porcelain Institute, Tajimi
Glazed porcelain, height 4 1/8" (10.6 cm),
diameter 2 11/16" (6.7 cm)
Yanagi Product Design Institute, Tokyo

The most radical of the many soy-sauce containers
designed in the 1950s, Sori Yanagi's server was
distinguished by the boldly curved spout and body,
which gave it a distinctive visual profile and made it
comfortable to hold and use. "In industry the straight
line is used most frequently, but it gives a hard
impression," Yanagi explained. "I try to avoid that as
much as possible and use molds that eliminate sharp
corners from the forms."[4]

22 Goblets and tumbler, 1955
22 Goblets and tumbler, 1955
Designed by Masakichi Awashima
Made by Awashima Glass Design Institute, Tokyo
Glass, height of goblets 3 1/2" (9 cm) and 4 7/8" (12.5 cm),
diameters 3 3/8" (8.7 cm) and 2 1/2" (6.5 cm); height of
tumbler 4 1/16" (10.5 cm), diameter 2 3/4" (7 cm)
Philadelphia Museum of Art.
Gift of Mrs. Masakichi Awashima

Four years after founding the Awashima Glass Design
Institute in 1950, Masakichi Awashima patented the
process for his *shizuku* (dripping water) glass, its
irregular, mold-blown surface appearing as if it were
covered with water like the rain droplets on a window,
which gives the process its name. Awashima used the
process for more than one hundred fifty glassware forms,
including the drinking glasses seen here. The straight-
sided tumbler evokes the shape and feel of Western
tableware while the footed goblets, which are used for
sake, refer to Japanese prototypes in ceramic.
Awashima's early *shizuku* glass was hand-blown into
fired ceramic molds; later it was mass-produced, using
more advanced techniques and metal molds.

22

23 Poster (*Peaceful Use of Atomic Energy*), 1956
Designed by Yusaku Kamekura
Silkscreen, 40 1/2 x 28 11/16" (103 x 72.8 cm)
Museum & Library, Musashino Art University, Tokyo

Designed for an exhibition held by the Japan Advertising
Artists Club, of which Yusaku Kamekura was a founding
member, this poster displays an image of atomic energy
that is beautiful rather than terrifying. Known nowhere
better than in Japan as a destructive weapon of war,
atomic energy became in 1955 a major tool of peace
with the First International Conference on the Peaceful
Uses of Atomic Energy held in Geneva and the initiation
of the United States' Atoms for Peace program, which
supplied uranium and technology to Japan and other
fuel-short countries for the development of nuclear
power. Despite the strict geometry of Kamekura's
composition, his abstract, light-radiating cosmic image is
poetically evocative.

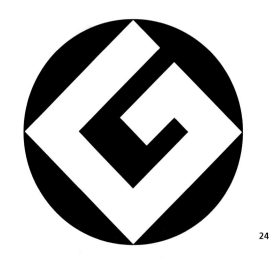

24

24 Symbol for the Good Design Selection System, 1957
Designed by Yusaku Kamekura

This mark, which is still being used, was created under
the auspices of the Good Design Selection System,
operated in its first year by the Japanese patent office
and afterward by the Ministry of International Trade and
Industry to recognize outstanding design in Japanese
products. A stylized Roman letter *G*, it is known as the G-
Mark, and was first awarded in 1957 to products selected
by a committee that included the mark's designer,
Yusaku Kamekura. "Stickers of this mark were pasted on
goods selected by this committee," he explained. "The
work of the committee was widely publicized and
screening of goods was quite strict."[5]

23

25 Surging Waves basket, 1956
Designed and made by Shounsai Shono
Bamboo, height 20 7/8" (53 cm),
width 28 15/16" (73.5 cm)
The National Museum of Modern Art, Tokyo

The calligraphic lines of this elegant formal (*shin*) basket
made to hold a flower arrangement for the tea
ceremony almost belie the fact that the piece is made of
woven bamboo rather than brushed in ink. Shounsai
Shono invigorated the conservative tradition of precision
bamboo basketry with the liveliness and movement of
this abstract modern shape, achieved here in the swirling
motion of the bamboo strips, which mimic surging
waves (*doto*).

25

26 Torii stool, 1956
Designed by Riki Watanabe
Made by Yamakawa Rattan Company, Tokyo
Prize: Triennale, Milan, gold medal, 1957
Rattan, 18 1/8 x 18 7/8 x 13 3/4" (46 x 48 x 35 cm)
YMK Company, Ltd., Tokyo

The name of this stool is taken from the *torii* gateways at
Shinto shrines, whose form is echoed in its architectonic
lines. The rattan lends it a light, airy look, and the color
and texture of the material harmonize well with the
tatami mats used as flooring in Japanese interiors. The
stool belongs to the first series of rattan furniture
produced by Yamakawa Rattan Company (now YMK),
which altered the direction of its production from
traditional crafts such as baskets and trays to
contemporary interior furnishings.

26

27 Butterfly stool, 1956
Designed by Sori Yanagi
Made by Tendo Mokko, Tendo
Plywood and metal, 15 11/16 x 16 9/16 x 12 3/16"
(39 x 42 x 31 cm)
Philadelphia Museum of Art. Gift of Tendo Company, Ltd.

Sori Yanagi employed modern materials and
manufacturing processes while exploiting vernacular
forms and traditional styles. His best-known work, this
Butterfly stool, blurred the strict distinctions between
Western and Japanese furnishings that had existed in
well-to-do Japanese houses before the war: while the
stool as a type belonged to Western interiors, Yanagi's
form suggests the exuberant curves of Japanese
architecture—the buoyant upward sweep of the legs and
overhang of the seat—giving the stool the appearance of
floating free like the butterfly that inspired its name.
Non-Western is the stool's calligraphic aptness, which de-
emphasizes its structure and three-dimensional form.
Made of two molded-plywood shells, the stool could be
assembled at home.

28

28 Teapot, 1956
Designed by Sori Yanagi
Made by Tajimi Porcelain Institute, Tajimi
Glazed porcelain and bamboo, height 8" (20.5 cm),
diameter 5 1/2" (14 cm)
Yanagi Product Design Institute, Tokyo

Among the earliest products Sori Yanagi designed after
World War II were his white porcelain tea and coffee
sets. Although they subsequently were shown at the
Triennale in Milan in 1957, where Yanagi was awarded a
gold medal for his works, these wares were rejected as
too starkly functional by the Mitsukoshi department
store when Yanagi first tried to market them: "I was told
they were unfinished, since they had no decoration," he
recalled. "That got me riled. So I took them to a French-
style coffee shop on one of the back streets of Ginza,
where they put them into use. Then they suddenly took
off."[6] Now variants of Yanagi's design are the standard
service used in almost all coffee shops in Japan.

27

29

30

31

29 Stationery box, 1957
Designed and made by Tatsuaki Kuroda
Lacquered wood, 7 1/4 x 12 1/4 x 6 1/4"
(18.5 x 31.2 x 16 cm)
The National Museum of Modern Art, Tokyo

Introducing modern abstract shapes to the traditional art of lacquerwork, Tatsuaki Kuroda carved this stationery box from a solid block of wood by gouging it out with hand tools, afterward applying many coats of red lacquer to color the surface. Its strong sculptural presence recalls the folk-craft idiom Kuroda promoted as co-founder of the Kamigamo Folk Art Association in Kyoto.

30 Angler Fish basket, 1957
Designed and made by Rokansai Iizuka
Bamboo, height 8 5/8" (22 cm),
diameter 9 1/16" (23 cm)
The National Museum of Modern Art, Tokyo

Rokansai Iizuka considered flower baskets for the tea room his most important creations. Following the three traditional styles of tea-ceremony utensils, he classified his baskets as *shin* (formal), *gyo* (semi-formal), and *so* (informal). This basket, made for the 1957 Traditional Japanese Crafts exhibition, is in the formal style, precise, tight, and symmetrical. Iizuka believed that "the line, which is the basic element of Eastern painting or calligraphy, must also come alive when weaving a piece in bamboo."[7] The Angler Fish basket demonstrates his concept perfectly: the upper half, an intricately woven pattern representing the fisherman's basket, gradually opens up into the looser plaiting of the waves below.

31 Tiger Cage basket, 1959
Designed and made by Shounsai Shono
Bamboo, 19 11/16 x 43 5/16 x 25 3/16" (50 x 110 x 64 cm)
The National Museum of Modern Art, Tokyo

In contrast to the precision basketry of his works in the formal (*shin*) mode (no. 25), Shounsai Shono employed the informal (*so*) style in the sculptural form of this large, unusually expressive Tiger Cage (Koken) basket. He emphasized the purely tactile qualities of his material, using long strips of bamboo in a loose, open weave to suggest the tiger cage that inspired it.

32 Sony radio, 1957
Designed and made by Tokyo Telecommunications
Engineering Corporation, Tokyo
Plastic, 4 3/16 x 2 1/2 x 1" (10.6 x 6.3 x 2.5 cm)
Collection of Roger Handy, Los Angeles

The Sony TR-610 model was the most popular of the
company's early pocket-size transistor radios; introduced
in 1958, it became Sony's first major export product.
Although its price at the time was about 10,000 yen, or
roughly the monthly starting salary of the average
worker, the radio became an instant hit in Japan, where
it ranked second only to Toshiba's electric rice cooker
(no. 7) in a 1959 *Kogei Nyusu* (Industrial Art News) survey
of the best Japanese products made since the war. The
TR-610 was thinner and smaller than earlier Sony
transistor models, and the housing was more efficiently
designed, with the round speaker set neatly into the
front panel and tuning and volume controls located at
the side, where they could be manipulated with one
hand. The addition of a folding stand on the back that
could also be used as a handle made the radio easy
to carry.

32

33 Poster for Nippon Kogaku, 1957
Designed by Yusaku Kamekura
Silkscreen, 40 1/2 x 28 11/16" (103 x 72.8 cm)
Museum & Library, Musashino Art University, Tokyo

Part of a complete graphic program Yusaku Kamekura
initiated with the Nikon line of cameras in 1954, this
poster was the first to demonstrate the style of
geometric forms and the precision printing techniques
for which the designer later became known and which
were so perfectly suited to his client's products.
Kamekura suggested the high and wide range of shutter
speeds of the new 35-millimeter camera with an
abstracted image of the shutter itself (a kind of curtain
with various open slits across it), using glowing colors
and white letter forms to indicate its brilliant and
clear results.

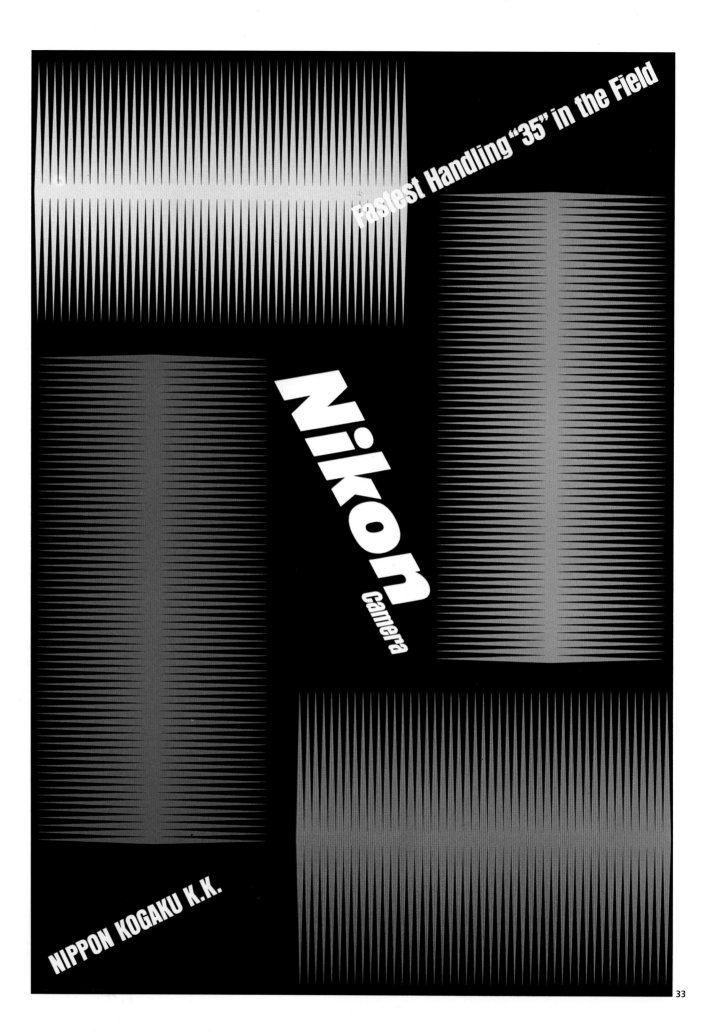

34

34 Nikon camera, 1957
Designed and made by Nippon Kogaku, Tokyo
Metal, 3 3/16 x 5 5/8 x 1 5/8" (8.1 x 14.3 x 4.1 cm)
Nikon Inc., Tokyo

Nippon Kogaku began producing cameras for the
Japanese military during World War II and developed its
first commercial camera in 1948. Like other Japanese
photographic firms, Nippon Kogaku based its early
models on German prototypes, but the Nikon
SP 35, a 35-millimeter single-lens reflex, represented a
deliberate attempt to match the firm's reputation for
high-quality lenses with an equally original body design,
typified here by the new, wide, horizontal viewfinder.
This was one of the first Japanese cameras to gain an
international reputation and was used by professional
photographers worldwide.

35

35 Motorcycle, 1956
Designed by Shinji Iwasaki and GK
Made by Yamaha Motor Company, Shizuoka
Enameled steel, aluminum, and synthetic rubber,
36 13/16 x 76 3/16 x 27 3/4" (93.5 x 193.5 x 70.5 cm)
Collection of Shinji Iwasaki, Tokyo

The first motorcycle designed by Shinji Iwasaki, then with
GK Industrial Design Associates, the Yamaha YD-1 was
hailed as a Japanese original. In designing it, Iwasaki
discarded the European models used until then and
created a machine that was distinguished by its sporty,
compact, low-slung, narrow, slate-gray body. Fitted with
a 250-cc engine, the YD-1 cycle, one of Yamaha's first
products, set the standard for the sport motorcycle
market, which Japan would later dominate.

36 Table lamp, 1957
Designed by Mosuke Yoshitake
Made by Industrial Arts Institute, Sendai
Metal and glass, height 24 5/16" (21 cm)
Fund of Industrial Arts, Tokyo

Mosuke Yoshitake's table lamp with metal supports and
a white glass shade was exhibited in the Japanese display
at the international trade fair held in New York in 1957.
Intended for export, the lamp was designed after the
form of a traditional Japanese lantern but was made
larger and in modern materials to suit the American
market. Wasp-waisted, with a light bulb set inside the
waist, the lamp casts a very bright glow from its center,
which is diffused as it travels toward the top and bottom
of the shade.

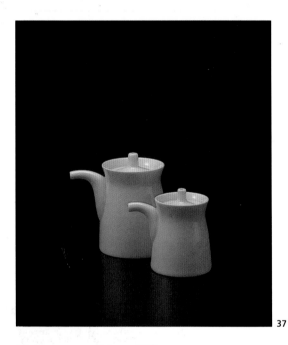

37

37 Soy-sauce containers, 1958
Designed by Masahiro Mori
Made by Hakusan Toki, Hasami-machi
Prizes: Good Design prize, 1960; G-Mark, 1961; G-Mark
long life award, 1977
Glazed porcelain, 2 7/8 x 3 1/16 x 2 1/4" (7.3 x 7.8 x
5.7 cm); 3 5/8 x 3 11/16 x 2 11/16" (9.2 x 9.3 x 6.9 cm)
Hakusan Toki K.K., Hasami-machi

When the manufacturer Hakusan Toki hired Masahiro
Mori in 1956 as a product designer for its postwar
venture into mass-production porcelain, it focused on the
soy-sauce container as one of the most common and
indispensable components of the Japanese table. Mori's
server became the touchstone of the revived ceramic
industry, an example of the new industrial role of the
designer in creating high-quality, low-cost tableware for
everyday use. This server won two significant firsts: the
Good Design prize in 1960 and the G-Mark award for
ceramics in 1961; it still remains in production several
decades later.

36

38

38 Chair, 1958
Designed by Isamu Kenmochi
Made by Yamakawa Rattan Company, Tokyo
Prizes: G-Mark, 1966; G-Mark long life award, 1982
Rattan, 28 3/8 x 36 5/8 x 33 7/8" (72 x 93 x 86 cm)
Museum & Library, Musashino Art University, Tokyo

Like RIKI WATANABE (see no. 26), Isamu Kenmochi was commissioned by the Yamakawa Rattan Company to update the firm's rattan craft products. Originally designed as part of a bar-lounge series for the Hotel New Japan in Tokyo, this chair makes bold use of the flexibility of the rattan material to create round contours and a sense of volume. Its light weight and solid, simplified shape suited both commercial and domestic interiors, and the chair has remained in continuous production with only slight modifications since its introduction in 1960. The acquisition of this model by the Museum of Modern Art in New York in 1964 was considered one of the milestones for the recognition of contemporary furniture design in Japan.

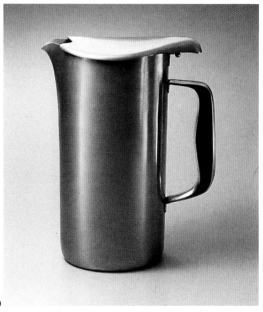

39

39 Pitcher, 1958
Designed by Sori Yanagi
Made by Uehan Shoji Company, Tokyo
Stainless steel and Bakelite, height 8 7/8" (22.5 cm),
diameter 5 1/8" (13 cm)
Yanagi Product Design Institute, Tokyo

This water pitcher was one of a series of stainless-steel tableware that Sori Yanagi designed for Uehan Shoji in 1957–58. Yanagi wanted to use modern industrial materials but sought to apply them in forms that evoked the human qualities and aesthetic he associated with traditional craft: "Handicrafts," he said, "should seek a hand-made beauty, and machine products should design a beauty of its own kind. But as the beauty in both grows out of the involvement of man's life, the basic ground whence it emerges is the same."[8]

40 Radicon robot, 1955
Designed and made by Masudaya Corporation, Tokyo
Tin, 14 15/16 x 8 11/16 x 5 7/8" (38 x 22 x 15 cm)
Collection of Mark Montifiore, New York

The first in a series of five battery-powered robots sold
by Masudaya from 1958, the Radicon robot was greeted
with great acclaim in Japan and abroad. The earliest
radio remote-controlled robot, it was on the cover of the
December 1958 issue of the American magazine *Popular
Electronics*. Operated from a bulky box with several
switches and an aerial, the robot moves forward and
backward on small wheels set into its base, its head
turns, and its eyes and a gauge on its chest light up.

40

41 Packaging for Arakiya, 1958
Designed by Kunibo Wada
Made by Daiichi-Shiko Company, Kyoto
Printed paper, 13 3/16 x 12 9/16 x 4 1/2"
(33.5 x 32 x 11.5 cm)
Meibutsu Kamado, Kyoto

When in 1951 Saburo Araki, founder of Arakiya, established an annual tea ceremony (*daichakai*) to promote his confectionery and the tea cakes it made, Kunibo Wada was commissioned to design fans as gifts for the guests. Their success prompted Araki to hire Wada to create a complete graphic program, from wrappers for individual cakes and candies to boxes, paper, shopping bags, and the black-and-white logo that decorates all Arakiya stores. A former cartoonist, Wada adapted a humorous long-nosed goblin (*tengu*) and other legendary creatures from traditional Japanese folk art for his designs, identifying the popular affection for such images with Arakiya sweets. The strength and simplicity of these designs (originally executed in the woodcut medium) had enormous popular appeal, even enabling Wada to overcome the Japanese prejudice against black—commonly associated with sadness and grief—which he used prominently in his designs and in each of the some fifty thousand boxes produced daily for the firm's products.

42 Mazda truck, 1958
Designed by Jiro Kosugi
Made by Toyo Kogyo Company, Hiroshima
Enameled metal, 56 5/16 x 50 3/8 x 117 1/8"
(143 x 128 x 297.5 cm)
Mazda Motor Corporation, Hiroshima

Jiro Kosugi began designing small, three-wheeled trucks for Toyo Kogyo in 1948, and continued to turn out designs for them over the next twelve years. Suited to narrow Japanese streets (and particularly useful for making deliveries), the trucks gained instant popularity in the early postwar years, owing not only to their size but also to their relatively low cost. In this Mazda 360, the largest and most evolved of Kosugi's models for Toyo, the engine was moved from the front to the immediate rear of the driver's seat, allowing for more space in the cab.

43 Teapot, 1958
Designed by Sori Yanagi
Made by Kyoto Gojozaka Kiln, Kyoto
Prize: G-Mark, 1960
Glazed porcelain and bamboo, height 8 1/2" (21.5 cm),
diameter 5 1/2" (14 cm)
Yanagi Product Design Institute, Tokyo

Sori Yanagi's teapot won the G-Mark prize in 1960 for
the innovative way in which it was designed for mass
production. Unlike the construction of most teapots, the
spout and handle base are made in the same mold as the
body, which results in more efficient manufacture and
makes the protruding elements less vulnerable to
breakage. The bamboo handle is replaceable.

44 Television, 1959
Designed and made by Sony Corporation, Tokyo
Prize: Triennale, Milan, gold medal, 1960
Metal, 8 x 8 1/4 x 9 1/2" (20.3 x 21 x 24.1 cm)
Collection of Roger Handy, Los Angeles

"TVs used to be heavy and were not meant to be carried
around, but now they can be moved from your living
room to a bedroom and you can even enjoy a sports
program in a car," *Kogei Nyusu* (Industrial Art News)
reported enthusiastically in a feature article in 1960.[9]
The article was referring to the world's first all-transistor
television, this small black-and-white model (TV 8-301)
with an eight-inch screen, which operated on either
household current or batteries and weighed only 13 1/4
pounds—light enough to be easily portable. Sony's staff
capitalized on its earlier experience in applying transistor
technology to radios (nos. 17, 32), designing and
engineering circuitry and miniaturized parts for very
small spaces. This television was created from the inside
out, with the internal components laid out first and the
exterior designed around them. Fitting snugly around
the picture tube, the chassis, with its irregularly spaced
knobs and grill slots, and the projecting hood signaled
that Japanese design was equal to its technology.
Overhanging the screen like a traditional Japanese roof
or a sun visor, the hood was intended to reduce glare
and make viewing easier.

45 Poster for Japan Advertising Artists Club, 1959
Designed by Hiromu Hara
Silkscreen, 40 1/2 x 28 11/16" (103 x 72.8 cm)
Museum & Library, Musashino Art University, Tokyo

The character *sho* (small, few), which appears on this poster, is commonly used to teach the four basic brushstrokes of Japanese calligraphy. Here, Hiromu Hara played with the character, using a standard Japanese typeface, Mincho (Ming Dynasty), to create a bold new look for its printed form by mimicking the variations in thickness of line normally achieved in writing with brush and ink. Hara added a humorous touch to the resulting "portrait" with the beauty-mark asterisk. The poster was designed for the 1959 typography exhibition sponsored by the Japan Advertising Artists Club, of which Hara was a founding member.

46 Stool, 1960
Designed by Reiko (Murai) Tanabe
Made by Tendo Mokko, Tendo
Prize: Tendo prize, 1960
Teak plywood, 14 3/16 x 17 11/16 x 17 1/16"
(36 x 45 x 43.3 cm)
Tendo Company, Ltd., Tendo

A winner of the first furniture design competition held
by the Tendo company in 1960, Reiko (Murai) Tanabe's
lightweight stool was put into production the following
year. One of several postwar Japanese products to
explore the possibilities of molded plywood, it is made
of three identical panels glued together to produce a
deceptively simple object that works equally well as
seat, table, or step stool. "I focus on spirituality in my
design," Tanabe has said. "This may result in a certain
ascetic quality, but I am aiming for timelessness."[10]   Her
success is proven by the fact that her design is still in
production more than thirty years after its first
appearance.

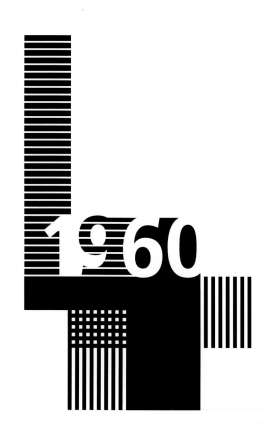

46

Among the Japanese-style furniture produced by Tendo
during the 1960s, this long bench was designed by Riki
Watanabe specifically for use in public spaces. Echoing
the unadorned austerity of the great, curving roofs of
Shinto architecture and in keeping with the traditional

47

Japanese preference for using natural materials, this
bench has been in continuous production and is used in
a variety of interiors, among them the National Museum
of Modern Art in Tokyo.

48 Table, 1960
Designed and made by Tendo Mokko, Tendo
Prizes: G-Mark, 1964; G-Mark long life award, 1981
Plywood with zelkova veneer, 13 1/4 x 47 3/4 x 29 1/2"
(33.5 x 121 x 75.5 cm)
Tendo Company, Ltd., Tendo

The low table for use with floor seating was and
remains the traditional focus of the *tatami* room. These
tables were originally made of solid wood and were of

48

fairly heavy, uniform vernacular design until Tendo
Mokko developed this model, with its elegantly slim
profile and rounded corners, and other plywood
versions for domestic and commercial use in the 1960s.
Plywood made the table lighter and more mobile;
however, a veneer was added to give the table a sense
of traditional materials and craftsmanship.

49 Chair, 1960
Designed by Daisaku Choh
Made by Tendo Mokko, Tendo
Plywood with Japanese oak veneer and upholstery,
25 1/2 x 21 3/4 x 26 1/4" (65 x 55 x 66.5 cm)
Tendo Company, Ltd., Tendo

In the late 1950s and early 1960s Tendo Mokko launched
a new line of chairs, which included this elegant low
model designed by Daisaku Choh. Choh literally took
the concept of the floor-level seat one step up, raising it
about 11 1/2 inches off the floor and integrating it with
the upholstered back and seat common to Western
chairs. The two canted *J*-shape supports of wood
laminate combine comfort with a stylish linear profile,
and the runners prevent damage to the *tatami* mats by
distributing the weight of the seat. This chair was
exhibited at the Triennale in Milan in 1960.

49

world
design
conference
1960 世界デザイン会議

50

The World Design Conference (WoDeCo) in Tokyo in
May 1960 was the first international professional design
conference held in Japan. More than 150 delegates
from twenty-four countries met to discuss the role of
design and designers in "the coming age." A team of
Japanese designers created a complete graphic program

51

for the conference, ranging from invitations and
programs to the publication of the proceedings. It was
one of the first coordinated graphic-design efforts
sponsored by the Japanese government and was a
precursor of the 1964 Tokyo Olympic designs (nos. 65,
67–69). The World Design Conference logo—the
blocklike Roman letters *cd*, with the white spaces
suggesting the interstices of a *W*—designed by TAKASHI
KONO, appeared on all the conference's printed
materials, including the poster (by Ikko Tanaka) and the
cover of the proceedings (by Kiyoshi Awazu and Hiromu
Hara), where, printed in bright colors, it provides
decoration as well as identity.

52 Kashiwado chair, 1961
Designed by Isamu Kenmochi
Made by Tendo Mokko, Tendo
Prize: G-Mark, 1966
Lacquered Japan cedar, 24 3/4 x 33 1/2 x 30 1/4"
(63 x 85 x 77 cm)
Tendo Company, Ltd., Tendo

This chair was one of the first joint projects between the
designer Isamu Kenmochi and the Tendo company,
which began to employ outside designers on royalty
contracts from the 1950s. "As a manufacturer,"
Hisayoshi Sakurai, head of Tendo's Tokyo office, later
recalled, "we did not know the first thing about design,
so we had to respond to the requirements of the
designers. . . . We took the position of trying to meet
the demands of the designers even if it was for
something we had not done before."[11] The chair is built
up of blocks of Japan cedar in a unique fashion; the
blocks form a *U*-shape profile with their wood grains
arranged in a decorative pattern. The stout, imposing
shape of this armchair inspired its name, Kashiwado,
after a famous sumo wrestler of the period.

52

53 Soy-sauce container, 1960
Designed by GK
Made by Kikkoman Shoyu Company, Noda
Prize: G-Mark long life award, 1986
Glass and plastic, height 5 7/8" (15 cm),
diameter 2 5/8" (6.5 cm)
Private collection

This compact, disposable soy-sauce container created by
the GK design group has become the symbol of its
contents. Designed to replace small table-size
containers that had to be repeatedly filled from large
soy bottles, it was first introduced by the Kikkoman
company in 1961 and is still in production. The shape is
reminiscent of the bottles traditionally used to serve
sake, the other essential accompaniment to a Japanese
meal. With its red plastic drip-proof pourer and the
corporate logo—a new "applied ceramic label"
intended to adhere even after repeated use—the design
was bright and discreet enough to make it acceptable
for use straight off the supermarket shelf, at home as
well as at restaurants.

53

54

54 Oil and vinegar bottles, 1960
Designed by Saburo Funakoshi
Made by Hoya Glass Works, Tokyo
Prizes: G-Mark, 1960; G-Mark long life award, 1980
Blown glass and wood, height of oil bottle 8 11/16"
(22 cm); height of vinegar bottle 7 1/4" (18.5 cm)
Philadelphia Museum of Art. Gift of Hoya Corporation

Saburo Funakoshi was one of the first professional
designers to enter the Japanese tableware industry after
the war, joining the Hoya Glass Works in 1957. The
forms of these sleek, architectural oil and vinegar
bottles were revised by Funakoshi from a set of crystal
sake containers he had designed in 1958, and were
intended to encourage the then newly introduced
Western practice of eating vegetables in salads with oil
and vinegar dressing. "Traditionally, Japanese ate
pickled vegetables," Funakoshi said, "but I wanted to
create an instrument that would help make fresh
vegetables tasty and enjoyable for people who don't
like pickled vegetables."[12] The bottles were designed
not for kitchen use but for the dining table where the
wooden stopper, fastened to a glass ball with epoxy
resin, provided a suitably formal accent.

55 Poster for the Museum of Modern Art, Tokyo, 1960
Designed by Ryuichi Yamashiro
Silkscreen, 40 1/2 x 28 11/16" (103 x 72.8 cm)
Museum & Library, Musashino Art University, Tokyo

Some images in Japanese graphics have become so pervasive that designers take great risks in trying to give them fresh appeal. One of the most successful reinterpretations of a famous nineteenth-century woodblock print, *Under the Waves at Kanagawa* (often called "The Great Wave"), by Hokusai, was rendered in this poster by Ryuichi Yamashiro for the Second International Biennial of Prints at the Museum of Modern Art in Tokyo in 1960. To the image of Mount Fuji and the huge wave rolling in at the left of the sheet, Yamashiro has added a large diamond shape as a backdrop for the crest of the wave. This shape may be meant to suggest the *mimasu* (three sake measures) crest used in Hokusai's time by the actor Danjuro Ichikawa, which is represented in Edo period prints of Kabuki actors, or it may just be an abstract graphic device. The ambiguity itself makes it more interesting for the fan of Japanese prints, both traditional and contemporary.

57

56 Rippled Sand kimono, 1961
Designed and made by Tokio Hata
Resist-dyed silk, 65 x 55 1/8" (165 x 140 cm)
Collection of Tokio Hata, Kyoto

A master *yuzen* dyer, Tokio Hata was inspired to create this kimono by rippled patterns on sand dunes. In this early example of what would become one of his favorite subjects, straight lines combine with curvilinear forms to suggest a sensation of wind-whipped sand. Hata's

56

technical virtuosity is such that this undulating design is composed entirely of horizontal lines. Pairs of fine lines drawn with paste resist, applied with a double-tipped cone (*renkan*), fill the entire surface. The expansiveness, depth of field, and sense of movement achieved in this piece are difficult to accomplish in this traditional art form.

57 Swift Stream kimono, 1961
Designed and made by Kako Moriguchi
Resist-dyed silk, 62 5/8 x 48 13/16" (159 x 124 cm)
The National Museum of Modern Art, Tokyo

Kako Moriguchi is known as a master of *maki-nori yuzen-zome*, a combination of two resist techniques: *yuzen-zome*, the application of rice-paste resist followed by hand painting, and *maki-nori*, the use of dried rice-paste flakes scattered over the surface of the fabric to create a mottled effect. The classical Japanese motif of a flowing stream is dramatically interpreted here by large-scale, high-spirited lines that resemble great brushstrokes. The hand-applied paste resist dividing the streams of color is particularly suited to such an abstract flowing design.

58 Chair, 1961
Designed by Kenji Fujimori
Made by Tendo Mokko, Tendo
Prize: G-Mark long life award, 1981
Plywood, 15 3/4 x 13 x 19 1/4" (40 x 33 x 49 cm)
Tendo Company, Ltd., Tendo

This legless seat, with its light plywood shell designed to
be stacked and stored economically in the small spaces
that typically define Japanese interiors, gave new formal
expression to the Japanese custom of dining at floor
level on *tatami* mats. Utilizing factory technologies
developed during the war for the plywood-molding
process, Kenji Fujimori aimed to create chairs of high
quality that could be mass-produced inexpensively. In
continuous production since 1961, this chair contributed
a new furnishing form to the postwar period.

58

59 Spoke chair, 1963
Designed by Katsuhei Toyoguchi
Made by Tendo Mokko, Tendo
Oak and upholstery, 32 3/4 x 31 1/2 x 26 1/4"
(83 x 80 x 67 cm)
Tendo Company, Ltd., Tendo

Katsuhei Toyoguchi has experimented with chair seating
for Japanese homes throughout his career. His ample
Spoke chair has a seat at mid-height, low enough to
allow for comfortable conversation with someone
sitting on the floor, and a long, fan-shaped back with
vertical supports, which give the chair its name.

59

60 Kimono, 1961
Designed and made by Keisuke Serizawa
Stencil-dyed cotton, 57 7/16 x 48" (151 x 122 cm)
The National Museum of Modern Art, Tokyo

Like the ABC's, the Japanese syllabary is often arranged
into a poem, beginning with *i-ro-ha*, memorized by
every schoolchild and used as a system for ordering all
manner of things. Here it appears as an almost abstract
pattern on a kimono designed by Keisuke Serizawa,
applied with a stencil-dyeing (*kata-zome*) technique
practiced in Japan for more than four hundred years.
Cut-paper stencils were laid over the cloth and rice paste
was applied as a resist to the fabric in the pattern areas.
After the paste dried, the fabric was dyed with a large
brush, and the sharply defined lettering revealed where
the dye could not take because of the resist.

60

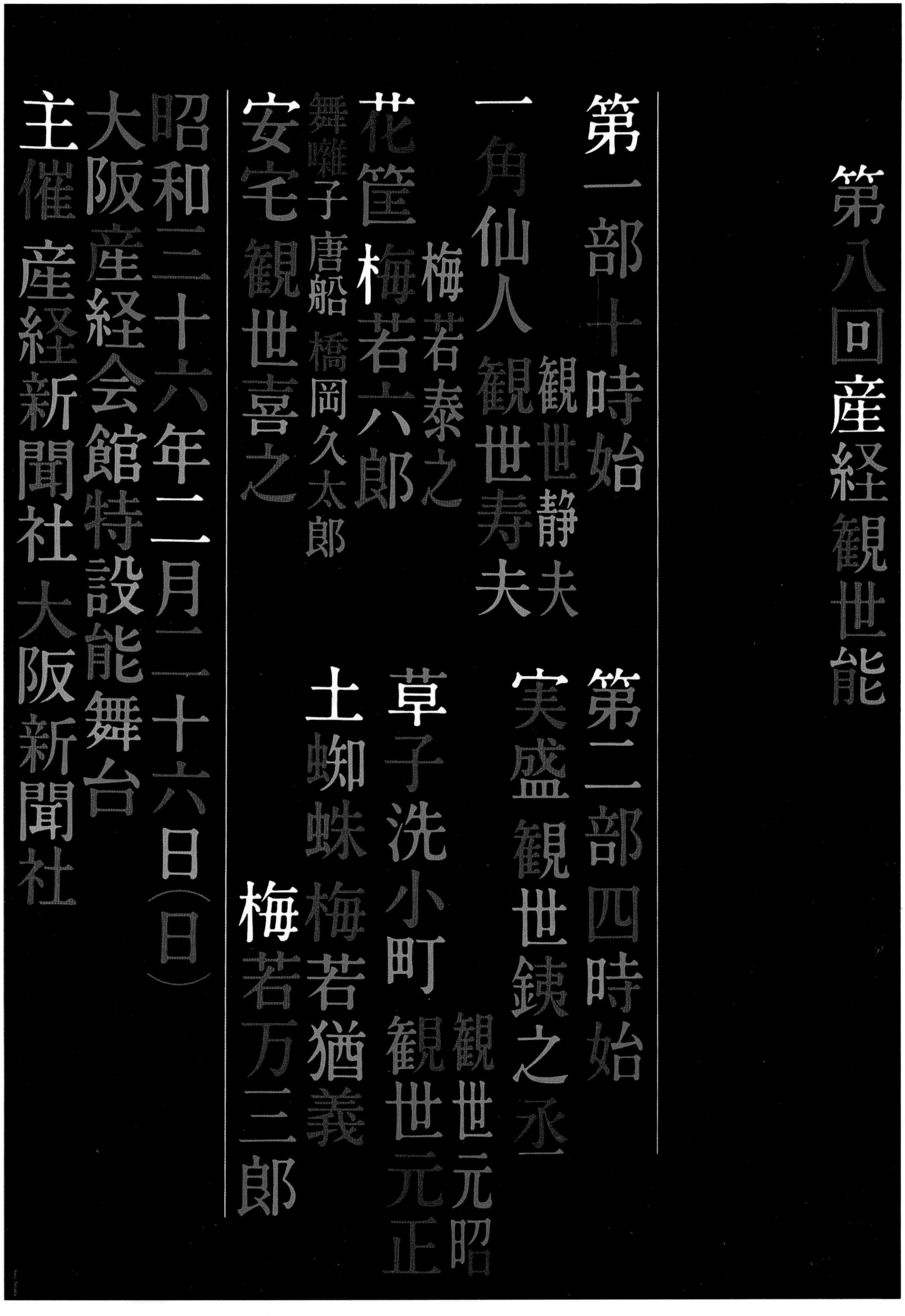

第八回産経観世能

第一部十時始
一角仙人　観世静夫
　　　　　観世寿夫
花筐　梅若六郎
　　　梅若泰之
舞囃子唐船　橋岡久太郎
安宅　観世喜之
　　　観世喜之

第二部四時始
実盛　観世銕之丞
草子洗小町　観世元昭
　　　　　　観世元正
土蜘蛛　梅若猶義
　　　　梅若万三郎
梅

昭和三十六年二月二十六日（日）
大阪産経会館特設能舞台
主催　産経新聞社　大阪新聞社

61 Poster for Sankei Shimbun, 1961
Designed by Ikko Tanaka, 1961
Prize: Tokyo Art Directors Club, silver medal, 1961
Silkscreen, 40 1/2 x 28 11/16" (103 x 72.8 cm)
Museum & Library, Musashino Art University, Tokyo

From 1953 Ikko Tanaka was commissioned annually by the Sankei Shimbun newspaper company to design posters to advertise performances of the Kanze School of Noh Theater in Osaka. This, the eighth in the series, is one of his most famous, relying on typography and color rather than representation for its effect. Including only Chinese-style characters, the oldest of the three writing systems used in Japan, Tanaka suggests the antiquity and ritual of the performance. The lettering is printed in strong, vibrant colors against a dark ground, evoking the bright costumes worn in Noh theater.

62 Sharp Compet calculator, 1964
Designed and made by Hayakawa Metal Works, Osaka
Metal, 9 13/16 x 16 9/16 x 17 5/16" (25 x 42 x 44 cm)
Sharp Corporation, Osaka

Introduced in 1964, the Sharp Compet CS-10A calculator
was the world's first all-transistor-diode electronic
calculator. Called a desktop calculator yet half the size
of an average desk top, weighing 55 pounds and having
over 100 buttons and dials (and over 4,000 parts), it was
nevertheless received with tremendous excitement as it
did calculations at great speed, and very quietly. This
calculator inspired fierce competition among electronic
firms around the world, all of whom were eager to
enter the lucrative electronic calculator market.

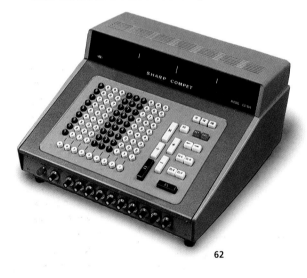

62

63 Poster for Yamaha Motor, 1961
Designed by Gan Hosoya
Offset, 40 1/2 x 28 11/16" (103 x 72.8 cm)
Museum & Library, Musashino Art University, Tokyo

In the late 1950s the newly established motorcycle
manufacturer YAMAHA MOTOR COMPANY hired Gan
Hosoya to create their advertisements. Hosoya's use of a
photographic image in this poster for Yamaha's new
250-cc YDS2 sports motorcycle is witty and sophisticated.
By standing the photograph on end, Hosoya conveyed a
sense of speed and adventure, giving the illusion that
the riders, shot on a flat road surface, are plunging
down a steep hill on their Yamaha motorcycle.

63

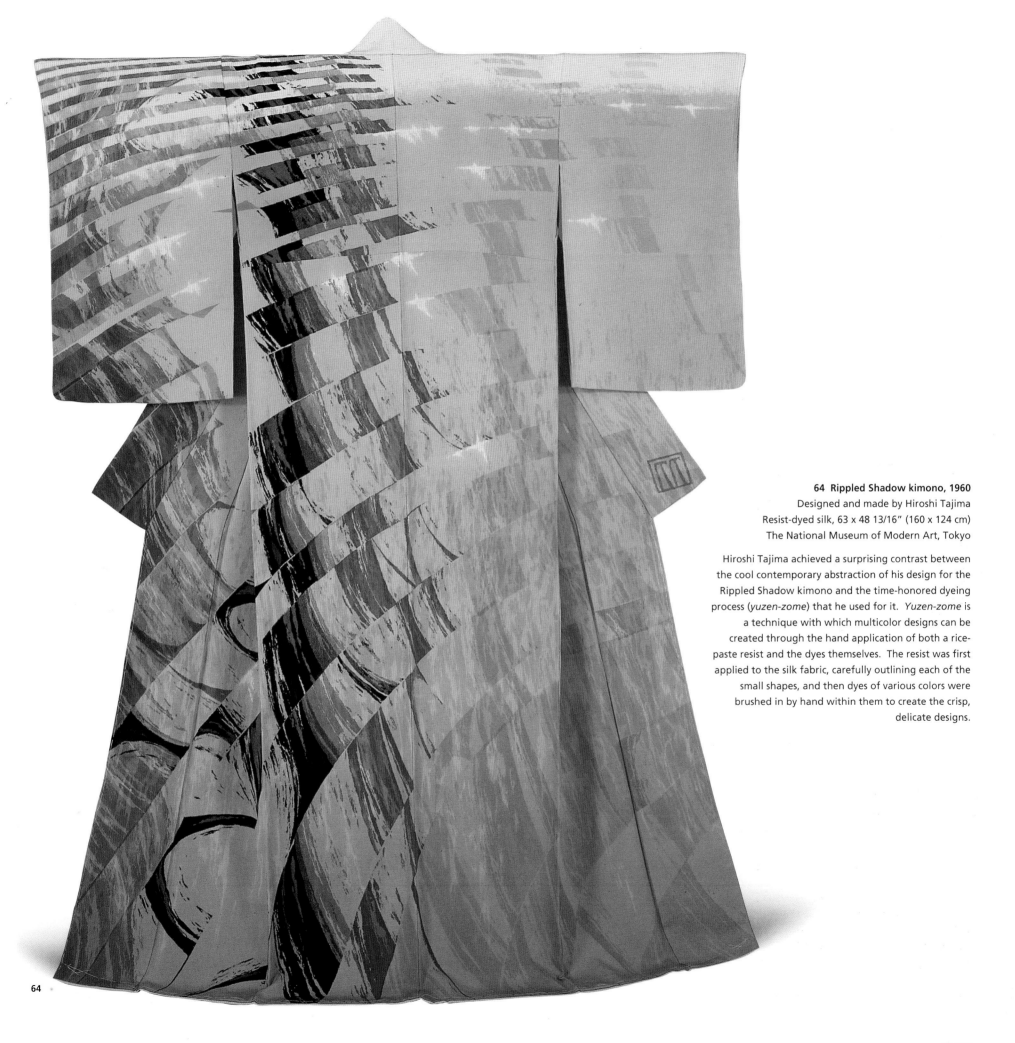

64 Rippled Shadow kimono, 1960
Designed and made by Hiroshi Tajima
Resist-dyed silk, 63 x 48 13/16" (160 x 124 cm)
The National Museum of Modern Art, Tokyo

Hiroshi Tajima achieved a surprising contrast between the cool contemporary abstraction of his design for the Rippled Shadow kimono and the time-honored dyeing process (*yuzen-zome*) that he used for it. *Yuzen-zome* is a technique with which multicolor designs can be created through the hand application of both a rice-paste resist and the dyes themselves. The resist was first applied to the silk fabric, carefully outlining each of the small shapes, and then dyes of various colors were brushed in by hand within them to create the crisp, delicate designs.

64

65 Poster for Eighteenth Olympic Games, Tokyo, 1961
Designed by Yusaku Kamekura
Prizes: Tokyo Art Directors Club, gold medal, 1961;
Mainichi industrial design prize, 1963; International
Poster Biennale, Warsaw, artistic graphics edition medal,
1966
Gravure, 40 1/2 x 21 9/16" (103 x 54.8 cm)
Museum & Library, Musashino Art University, Tokyo

In 1961 the Tokyo Olympics Organizing Committee
invited six Japanese graphic designers to compete for
the creation of the official image of the Olympic Games
to be held in Tokyo in 1964: Ikuichiro Inagaki, Yusaku
Kamekura, TAKASHI KONO, KAZUMASA NAGAI, KOHEI
SUGIURA, and IKKO TANAKA. Kamekura's design was
chosen unanimously for its power to communicate with
three simple elements: the red rising sun of the
Japanese flag, the Olympic rings, and the words "Tokyo
1964." These elements are combined vertically in a
bannerlike arrangement that borrows the use of
simplified emblems and their direct frontal presentation
from both traditional Japanese crests and Western
advertising. The success of the design brought
Kamekura additional commissions for posters from the
Tokyo Olympics Organizing Committee (nos. 67–68), and
set a precedent for the unified graphic programs now
standard in Olympic design.

**66 Torch for Eighteenth Olympic Games,
Tokyo, 1964**
Designed by Sori Yanagi
Oxidized aluminum, height 7 11/16"
(19.5 cm), diameter 6 11/16" (17 cm)
Yanagi Product Design Institute, Tokyo

The flame burning at every Olympic Games is carried
from Olympia in Greece, where the first games were
held, to the host country by a relay of thousands of
runners. For the 1964 games, the flame was flown to
Japan and then divided into four flames, which were
carried from the farthest corners of the country to
Tokyo and were united into one outside the Imperial
Palace. For the opening ceremonies, a young man born
near Hiroshima on the day of the atomic bomb blast
brought the flame to the stadium to light the torch that
would burn throughout the games. The torches used by
the thousands of Japanese runners were designed by
Sori Yanagi. Made of aluminum so that they were very
light and portable, they had short handles and flared
upward, with "XVIII Olympiad Tokyo 1964" inscribed
around the top. The aluminum was oxidized to give the
torches the appearance of being more weighty.

66

65

67

67 Poster for Eighteenth Olympic Games, Tokyo, 1962
Designed by Yusaku Kamekura
Prizes: Tokyo Art Directors Club, gold medal, 1962;
Mainichi industrial design prize, 1963
Gravure, 40 1/2 x 28 11/16" (103 x 72.8 cm)
Museum & Library, Musashino Art University, Tokyo

68 Poster for Eighteenth Olympic Games, Tokyo, 1963
Designed by Yusaku Kamekura
Prizes: Tokyo Art Directors Club, silver medal, 1963;
Mainichi industrial design prize, 1963
Gravure, 40 1/2 x 28 11/16" (103 x 72.8 cm)
Museum & Library, Musashino Art University, Tokyo

For his second Olympic commission, which was to advertise the athletic competitions, Yusaku Kamekura studied previous Olympic posters and discovered that none had ever used photographs. "I decided that I would be the first to challenge this convention," Kamekura later wrote. Kamekura meticulously planned every detail of the posters, providing, for example, sketches to illustrate the effects he wanted for the beginning of a foot race in one poster and for a swimmer approaching head-on at full speed in another. "The most difficult part of the shooting was to get a perfect start of a running race. The dynamic tension of

68

the work was to be in this sprint from the starting blocks. A very long telephoto lens was used to compress the depth of field so that the runners would appear to be almost directly on top of each other. A shutter speed of 1/1000 second was used to freeze the action. Out of 40 photos I found one that was perfect. For the swimming poster . . . I set the camera under the water so that the swimmers could swim over the camera without having to lower their speed."[13]

69 Pictograms for Eighteenth Olympic Games, Tokyo, 1962

Designed by Yoshiro Yamashita

As host for the first Olympic Games held in Asia, the Japanese organized a committee headed by the design critic Masaru Katsumie to coordinate the graphic program. Because few participants or visitors would be fluent in Japanese, a program of pictograms was devised to designate the events and services, and it was essential that its visual vocabulary be easily comprehensible. Yoshiro Yamashita, a young graphic designer at the Nippon Design Center, produced these pictograms, the first used for an Olympic meet, featuring stylized silhouettes of athletes in action. The angular forms and use of blank space to read as clothing distinguished the pictograms and gave them a fresh, contemporary look.

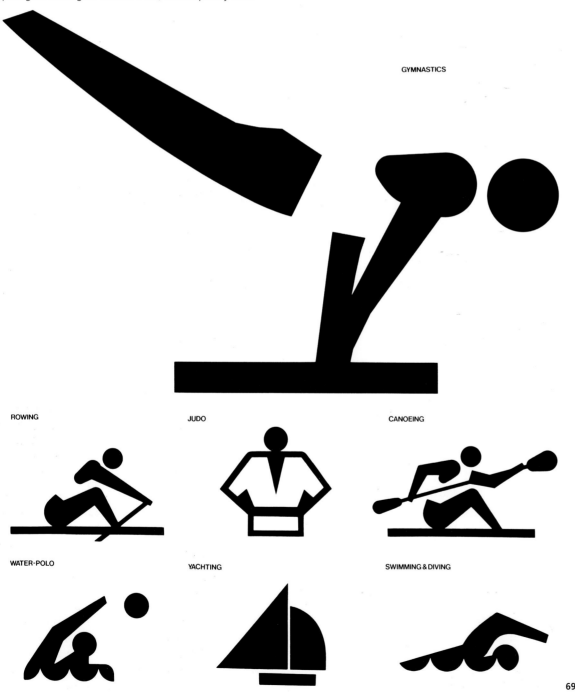

GYMNASTICS

ROWING   JUDO   CANOEING

WATER-POLO   YACHTING   SWIMMING & DIVING

69

70

71

70 Poster for Eleventh Winter Olympic Games, Sapporo,
1966
Designed by Kazumasa Nagai
Offset, 40 1/2 x 14 1/16" (103 x 35.7 cm)
Collection of Kazumasa Nagai, Tokyo

The Japanese flag, a snowflake, and the Olympic symbol of five interlocking rings are combined in this poster for the 1972 Sapporo Winter Olympics. The official mark for the games, this poster works equally effectively with the three figures vertically or horizontally, and was produced both ways. With this poster, Kazumasa Nagai won a 1966 competition to design promotional materials for the Sapporo games.

71 TV robot, 1963
Designed and made by Alps Shoji Company, Tokyo
Tin and plastic, 14 15/16 x 6 11/16 x 3 15/16"
(38 x 17 x 10 cm)
Zip's Toys To Go, Ardmore, Pennsylvania

This robot was made by the Alps Shoji toy company for export under the name Television Spaceman. The robot has a television in its chest, which shows a picture of a space station when the toy is turned on, and the robot walks forward and its coglike eyes rotate when the antenna is depressed. The robot was inspired both by NASA's Apollo project and by the increased popularity of television in Japan (where the first color television broadcasts were made in 1960). Alps made several versions of this popular robot between 1964 and 1969. The early ones were primarily tin plate with metal antennas and plastic heads and chests; later versions also had plastic legs and radar shields.

72 Ashtrays, 1962
Designed by Mosuke Yoshitake
Made by Yamasho Chuzo Company, Yamagata
Prize: G-Mark, 1964
Cast iron, height of hailstones 2 1/16" (5.3 cm), diameter
6 5/16" (16.1 cm); height of rings 1 3/8" (3.5 cm),
diameter 4 3/8" (11.2 cm)
Museum & Library, Musashino Art University, Tokyo

Mosuke Yoshitake was commissioned by the Yamasho
Chuzo foundries in his native Yamagata Prefecture to
broaden their product line of kettles for the tea
ceremony with designs for mass-produced housewares
such as these ashtrays. Adapting the classical *arare*
(hailstone) and ring patterns from the kettles, where
they aided in retaining heat, Yoshitake used them both
as decoration and as a means of holding cigarettes on
the rim.

73 Tray, 1962
Designed and made by Mashiki Masumura
Lacquer, 1 3/4 x 14 1/4 x 14 1/4" (4.5 x 36.2 x 36.2 cm)
The National Museum of Modern Art, Tokyo

Mashiki Masumura was a leader in the field of *kanshitsu*
(dry lacquer) technique, but unlike many craftsmen
using traditional techniques, he pioneered modern
vessel forms and abstract decorative effects that were
often intended to express symbolic or referential
content. This footed tray is decorated in black lacquer
with a pattern of wavy, parallel red lines that suggests
water flowing in a stream between black banks. By
varying the space between the lines, Masumura
represented fast and slow currents in the stream. The
lines were made by applying red lacquer to the edge of
a circular plate and rolling it over the black-lacquered
surface of the tray. The application of red over black
lacquer follows the traditional *negoro* technique named
after the Negoro-dera temple in Wakayama Prefecture,
where this type of lacquer was first produced.

74 Classic casserole, 1961
Designed by Yusuke Aida
Made by Bennington Potters, Bennington, Vermont
Glazed stoneware, 5 1/2 x 10 5/8 x 8 7/8"
(14 x 27 x 22.5 cm)
Collection of Chris Gauthier, Bennington, Vermont

Yusuke Aida designed the Classic oven-to-table service
when he was chief designer of Bennington Potters in

74

Vermont (1962–64). This two-quart casserole dish, part
of a service that included bowls, a gravy boat, sugar and
creamer, and various serving vessels, was created from a
single mold that incorporated its two handles, giving
the piece a sleek and continuous profile. According to
Aida, "the design was important for me as a Japanese
designer because it was embraced by the average
American consumer—not just those interested in things
Asian."[14]

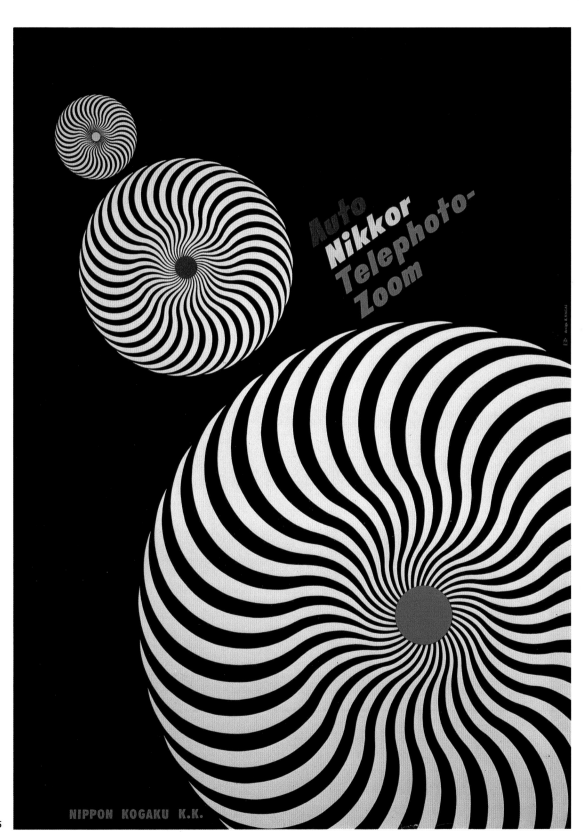

75

75 Poster for Nippon Kogaku, 1962
Designed by Kazumasa Nagai
Silkscreen, 40 1/2 x 28 11/16" (103 x 72.8 cm)
Collection of Kazumasa Nagai, Tokyo

Reflecting the influence of Western Op art in its swirling
forms and juxtaposition of colors, this poster graphically
demonstrates the zoom effect of the lens it is
advertising. Kazumasa Nagai repeated the swirling circle
three times, enlarging it from a tiny, distant object in
the upper left to one so large at the bottom right that it
cannot fit completely within the frame of the poster.

76 Poster for the 2nd World Religionists Conference for Peace, 1964

Designed by Kiyoshi Awazu
Offset, 40 1/2 x 28 11/16" (103 cm x 72.8 cm)
Collection of Kiyoshi Awazu, Tokyo

This arresting monochromatic illustration was originally created by Kiyoshi Awazu in 1961 when he was experimenting with expressions of identity, including heads composed from fingerprints and this hand containing dozens of seals of family names. As the poster for a meeting of individuals representing the world's religions in Tokyo in 1964, it was particularly appropriate as an image of unity from diversity. The raised hand itself recalls the symbolic gesture (*mudra*) used in Buddhist iconography to mean "fear not."

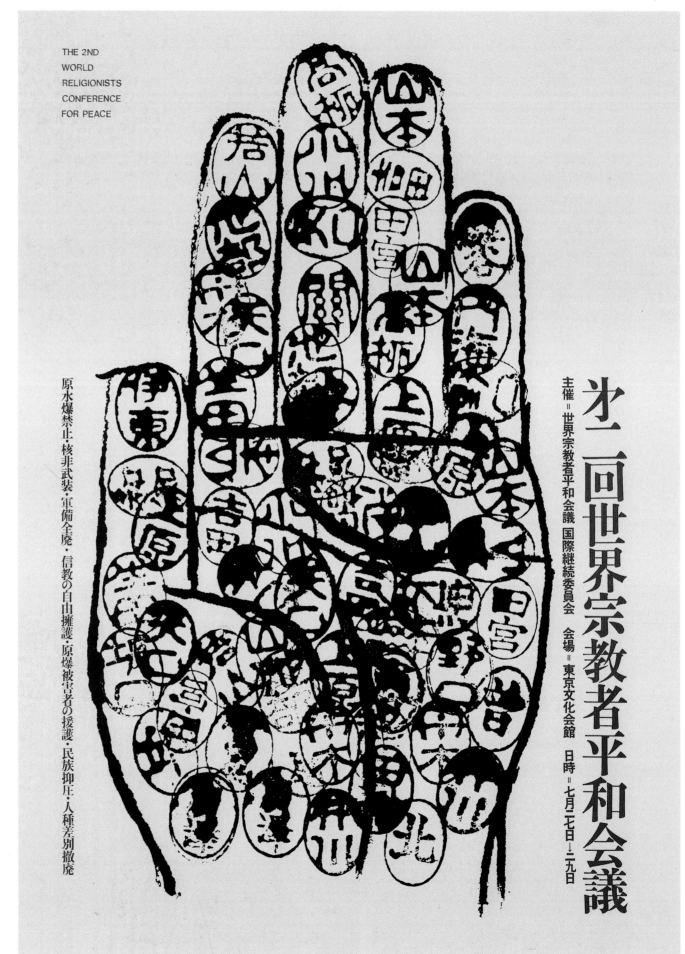

THE 2ND
WORLD
RELIGIONISTS
CONFERENCE
FOR PEACE

第二回世界宗教者平和会議

主催＝世界宗教者平和会議 国際継続委員会　会場＝東京文化会館　日時＝七月二七日→二九日

原水爆禁止・核非武装・軍備全廃・信教の自由擁護・原爆被害者の援護・民族抑圧・人種差別撤廃

76

77 Bird Tree puzzle, 1965
Designed by Shigeo Fukuda
Made by Tanaka Craft, Sahara
Prize: Mainichi industrial design prize, 1970
Cherry wood, 10 1/2 x 8 x 3 1/2" (26.5 x 20.2 x 8 cm)
Private collection

Looking for toys for his young daughter and finding none that he thought were really fun, Shigeo Fukuda decided to design some for her. The Bird Tree was the first of a series of wood puzzle toys, which also included a Squirrel Tree, a Rainbow Tree, and a Skyscraper Tree. One of his goals in designing these toys was to make a game of cleaning up. "When I was a child," he recalled, "my weak spot was putting away my wood blocks after playing with them, and I remember always being scolded on that score."[15] The toys presaged Fukuda's whimsical designs in both two and three dimensions, among which the Bird Tree remains a popular best-seller today.

78 Tower ashtrays, 1964
Designed by Kenmochi Design Associates
Made by Sato Shoji Company, Tokyo
Prize: G-Mark long life award, 1982
Chrome-plated steel and melamine,
height 1 1/4" (3.2 cm), diameter 2 7/8" (7.4 cm)
Philadelphia Museum of Art.
Gift of Kenmochi Design Associates

Kenmochi Design Associates, then comprising ISAMU KENMOCHI, Tetsuo Matsumoto, and Eiji Fujimoto as principal designers, was commissioned in the early 1960s by the Sato Shoji company to find a use for the scrap metal generated in the production of the firm's other products. The result, these chrome stacking ashtrays, became one of Kenmochi's best-known designs, and they remain ubiquitous elements in coffee shops and offices. The size of the ashtrays was made deliberately small and deep to prevent the ashes from being blown by the wind, while the quarter-inch cut in the side acts as an automatic extinguisher for the burning cigarette in case of a forgetful smoker. The concept of stacking to save space is typically Japanese, used by generations of delivery men who can still be seen riding their bicycles piled high with stacked lacquer trays of sushi. The removable plastic liners come in bright colors to contrast with the polished chrome finish of the ashtrays.

79 Poster for Asahi Breweries, 1965
Designed by Kazumasa Nagai
Prize: International Poster Biennale, Warsaw,
gold medal, 1966
Offset, 40 1/2 x 28 11/16" (103 x 72.8 cm)
The Museum of Modern Art, Toyama

Although he is best known for his hand-drawn
silkscreened posters (no. 75), Kazumasa Nagai also
worked with photography. He was the art director for
this eye-catching poster advertising the Asahi company's
Stiny beer, photographed by Muneo Maeda. According
to the text on the poster, three million bottles of the
beer were sold in Japan in 1965, and the hundreds of
bottle caps in the photograph graphically depict the
beer's immense popularity.

80 Zuiko Pearlcorder tape recorder, 1969
Designed by Yoichi Sumita and Olympus Optical
Company
Made by Olympus Optical Company, Tokyo
Aluminum, 2 1/4 x 8 1/4 x 3/4" (5.8 x 20.8 x 2.2 cm)
Museum & Library, Musashino Art University, Tokyo

Inspired by the success of miniature cameras, the
engineers at Olympus applied new precision technology
to producing the world's smallest tape recorder, its
Zuiko Pearlcorder, which used microcassette tapes.
Despite its small size—the unit fits easily into the palm
of the hand—the model has a two-hour recording
capacity. The sleek aluminum housing, designed by
Yoichi Sumita of Toyoguchi Design Associates, is without
any decoration except for the white and red of the
control buttons and the brand name. The compact body
is made in three parts, recorder, speaker unit, and
power pack; in later models a radio component was
added as well.

発売1年3億本 マイペースで飲もう アサヒスタイニー

79

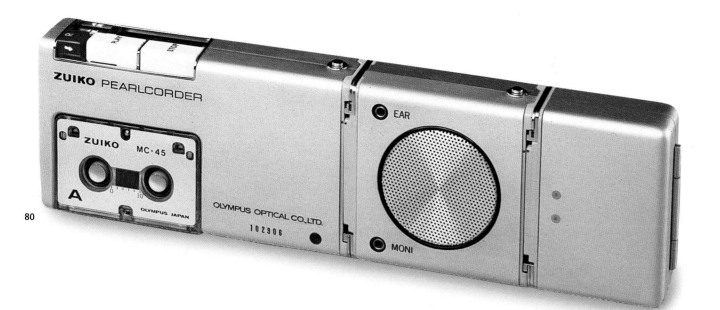

80

81 Canola calculator, 1964
Designed and made by Canon Camera, Tokyo
Metal, 10 1/4 x 15 3/8 x 20 1/16" (26 x 39 x 51 cm)
Canon Inc., Kawasaki

Canon expanded into the field of electronic calculators in 1964 with its Canola 130 model, joining over thirty manufacturers in an intense competition to put the most advanced product on the market. Capitalizing on the transistor-diode technology of Sharp's Compet (no. 62) of the same year, Canon was the first to simplify its product substantially and reduce the number of buttons and mechanisms, resulting in the now familiar ten-key numeric pad. The elimination of as many as 100 buttons found on earlier calculators, which were arranged one to nine in decimal columns, allowed for a smaller, more compact unit. However, the Canola 130 was still a substantial machine, weighing nearly 40 pounds and requiring over 300 square inches of desktop space.

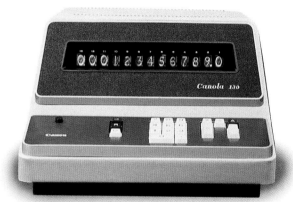

81

82

82 Cine Canonet movie camera, 1963
Designed and made by Canon Camera, Tokyo
Prize: G-Mark, 1964
Metal, 3 7/16 x 1 5/16 x 7 13/16" (8.7 x 3.3 x 19.9 cm)
Canon Inc., Kawasaki

Canon accelerated its postwar pursuit of the amateur photographer with the compact, reasonably priced Cine Canonet 8 movie camera of 1963. Smaller, thinner, lighter, and more portable than previous movie cameras (no. 20), it incorporated technical improvements such as a reflex view finder and zoom attachment. The visual clarity of its design was intended to communicate the simplicity, precision, and efficiency of the apparatus, factors that came to account for Canon's continuing success in a major export industry.

83 National vacuum cleaner, 1965
Designed by International Industrial Design
Made by Matsushita Electric Industrial Company, Osaka
Prize: G-Mark, 1966
ABS plastic, glass, rubber, and vinyl, 9 3/4 x 21 5/16 x
7 13/16" (24.8 x 54.1 x 19.8 cm)
Museum & Library, Musashino Art University, Tokyo

The first Japanese vacuum cleaner with an all-plastic
body, National's MC-1000C model was lighter and more
powerful than older metal vacuums and had better
maneuverability around small rooms, thanks to its
plastic construction and its large wheels, which were
safe for use on *tatami*-mat surfaces as well. Among its
many features were a window that showed when the
canister inside needed emptying, a magnet to pick up
pins and needles, and attachments for dusting, ironing,
and repairing *shoji* screens. The sleek styling of the
shiny red plastic body, with its inset handle, and the
golden hose added a sense of modernity to the
Japanese housewife's cleaning chores. The vacuum
became instantly popular, and in the two years it was in
production 630,000 were sold in Japan.

84 Dax-Honda minimotorcycle, 1969
Designed and made by Honda Motor Company,
Hamamatsu
Metal, 37 13/16 x 59 7/16 x 22 13/16" (96 x 151 x 58 cm)
Honda Motor Company, Ltd., Tokyo

Its name a playful reference to the small, low-slung
dachshund, the Dax-Honda minimotorcycle was
especially popular for its manageable size and
convenience. It was small enough to fit in the back of a
car or to store inside one's house, with handlebars that
could be conveniently folded down. The Dax-Honda
was a pioneer of the current recreational vehicle,
conceived by Honda for leisure-time activities as well as
for use in crowded streets.

85 Poster for "Persona" exhibition, 1965
Designed by Tadanori Yokoo
Silkscreen, 40 1/2 x 28 11/16" (103 x 72.8 cm)
Museum & Library, Musashino Art University, Tokyo

The "Persona" exhibition held at the Matsuya
department store in Tokyo in 1965 featured
autobiographical posters by sixteen graphic designers.
This poster, designed for the show by Tadanori Yokoo, is
an early example of his experimentation with the
"persona" genre, in which the artist is the main theme
of the work. The self-referential images, including a
reproduction of a photograph of Yokoo at age one and
a half, and a hanging corpse apparently symbolizing the
death of the artist's more conventional graphic style of
the late 1950s, are enhanced by the bold English text:
"Having reached a climax at the age of 29, I was dead."
Yokoo utilized Pop art aesthetics to signal his renewal of
interest in prewar Japanese commercial imagery,
incorporating the stylized rays of the rising sun of Japan
as it appeared on beer labels, matchboxes, and the logo
of his family's textile company.

83

84

86 Poster for the Jokyo Gekijo troupe, 1966
Designed by Tadanori Yokoo
Silkscreen, 40 1/2 x 28 11/16" (103 x 72.8 cm)
Museum & Library, Musashino Art University, Tokyo

This poster was designed to announce the first performance of the underground Jokyo theater troupe led by the dramatist, producer, and actor Juro Kara. Performed in the open in Tokyo in 1966, the play *Koshimaki-Osen* (Osen in Petticoats) was deliberately controversial in its rejection of the orthodoxies of contemporary Japanese theater, and the local authorities attempted to cancel the performances. Equally controversial, Yokoo's poster outraged some Japanese critics who decried the vulgarity of its imagery and rejection of the order and logic of prevailing rationalist design theories. Expressing the visual agitation of the performance it promoted, the poster combines garishly colored images and techniques from mass media and popular culture in its densely packed space, including the peach and rose pattern from a matchbox cover, rising sun, awning from a tenement house, and a string of nude comic-book superwomen flying through the air. Through posters such as this, Yokoo became a popular symbol of protest in the 1960s, revered for his disregard for convention, Pop aesthetic, and use of silkscreen and other printing techniques in collages of photographic elements and illustration.

87 Condiment set, 1969
Designed by Saburo Funakoshi
Made by Hoya Glass Corporation, Tokyo
Prizes: G-Mark, 1970; Osaka Design House prize, 1970;
G-Mark long life award, 1986
Blown glass, height of soy-sauce container 4 3/4"
(12.1 cm); height of salt container 3 1/4" (8.3 cm); height
of mustard pot 2 1/16" (5.2 cm)
Philadelphia Museum of Art. Gift of Hoya Corporation

This condiment set was first exhibited in Saburo
Funakoshi's one-man exhibition of glass designs at the
Matsuya department store in Tokyo in 1969, and again
in 1970 at the Japan Design House in Tokyo, where it
served as an example of the high quality of Japanese
exports. The shapes of these glass containers for soy
sauce, salt, and mustard were derived by Funakoshi from
traditional pottery and lacquer examples, which he
studied and considerably simplified to improve the way
the liquids poured; earlier containers, he noted, often

87

had to be accompanied by a tray to catch the spill. By
reinterpreting traditional shapes in a modern material,
Funakoshi hoped to make crystal glass a common item
on the Japanese dining table.

88 Plate, 1965
Designed by Kyoichiro Kawakami
Made by Hoya Glass Corporation, Tokyo
Clear molded glass, height 1 1/2" (3.8 cm),
diameter 11" (28 cm)
Collection of Kyoichiro Kawakami, Tokyo

During the 1960s as Western foods became popular in
Japan, Hoya Glass introduced glass tablewares from
which the new types of food could be served. Since
glass was then considered strictly for Western table use,
Hoya's designer, Kyoichiro Kawakami, elected to
decorate this plate with rings of the familiar *arare*
(hailstone) pattern borrowed from Japanese ironwork.
Radiating from the center of the plate, the rings become
larger toward the edge, making a rhythmical pattern of
texture and reflection. "If this had been designed by a
foreigner," Kawakami said, "the pattern would have
reached the edge of the plate, but I purposely left a
blank border to leave the rhythm echoing, like a
lingering taste."[16]

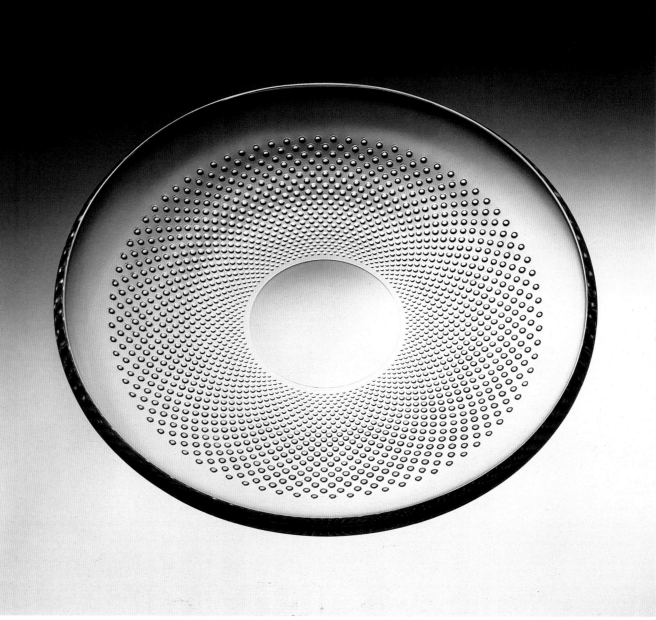

88

89

89 Tray, 1964
Designed and made by Yusai Akaji
Lacquered wood, height 2 1/2" (6.5 cm),
diameter 18" (45.6 cm)
The National Museum of Modern Art, Tokyo

Yusai Akaji's meticulous craftsmanship is displayed in
the *magewa* (bent ring) construction technique he
developed for his wooden wares to prevent distortion of
the wood. This tray is built up from multiple layers of
narrow wood strips bent into hoops and layered in tiers
to form the gently rising slope of its circular shape. The
tray has bands of black, gold, and cinnabar lacquer
around the rim and a warm, fluid surface.

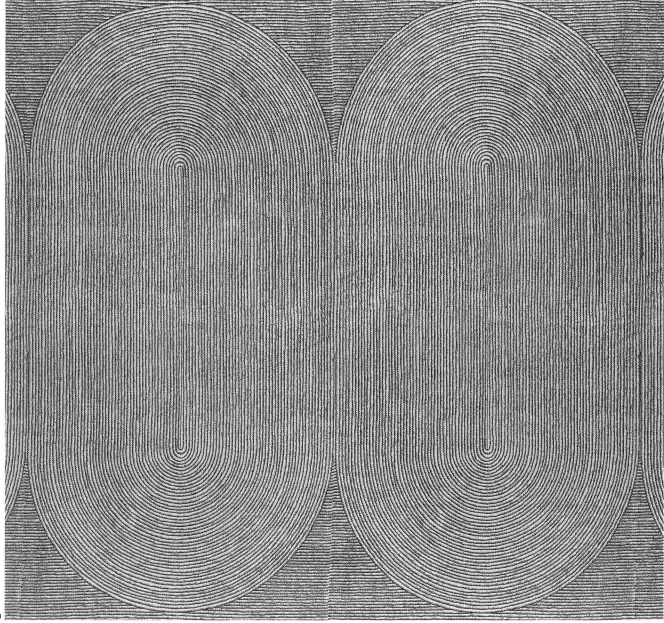

90

90 Sand Garden fabric, 1967
Designed by Hiroshi Awatsuji
Made by Hiroshi Awatsuji Design Studio, Tokyo
Screen-printed cotton, width 43 5/16" (110 cm),
repeat 19 1/16" (49 cm)
Collection of Hiroshi Awatsuji, Tokyo

Hiroshi Awatsuji based the abstract design of concentric
ovals of this fabric on the raked sand gardens of
Japanese Zen Buddhist temples. Just as Zen gardens are
intended to be quiet places for thought and
contemplation, Awatsuji created this fabric for use in a
"quiet and calm space."[17]

91 Poster for the Jokyo Gekijo troupe, 1967
Designed by Tadanori Yokoo
Silkscreen, 40 1/2 x 28 11/16" (103 x 72.8 cm)
The Museum of Modern Art, New York. Gift of the designer

This is one of a pair of posters created by Tadanori Yokoo to advertise the Jokyo theater troupe's production of *John Silver*. Borrowing the format of traditional Edo period Kabuki theater woodblock handbills, Yokoo displays the names of four of the play's sponsors—three bars and the boutique of the fashion designer Junko Koshino—in cartouches in the corners as well as a portrait of the play's producer, Juro Kara, framed in a hand mirror. The performance was held in a theater in Shinjuku, one of Tokyo's red-light districts and the area where the bars and boutique were also located. Enclosed in a frame of Japanese playing cards (*hanafuda*) is the silhouette of a nude woman with a Japanese coiffure in a *ukiyo-e* woodblock style. Deliberately provocative, Yokoo intended to draw every possible contrast to what he considered was the cultural affectation of Tokyo's official theater.

91

**92 Chrysanthemum pajama suit,
Fall–Winter 1966–67**
Designed by Hanae Mori
Made by Hanae Mori Tokyo
Silk
Hanae Mori Tokyo
(photograph by Richard Avedon)

The first Japanese fashion designer to gain international
recognition, Hanae Mori showed her collection abroad
for the first time in New York in January 1965, and her
clothes appeared regularly thereafter in such American
fashion magazines as *Vogue*. This Chrysanthemum
pajama suit appeared in the November 1966 *Vogue* in
an article entitled "Luxury: The Personal View," which
featured elegant lounge wear created by several
designers and worn by international personalities
photographed by the American fashion photographer
Richard Avedon. This pajama suit is named for the
Japanese-style chrysanthemums printed both on the
crepe de chine used for the body of the suit, with its
floating panels and harem trousers, and for the light,
flowing chiffon scarf.

93 In-line roller skates, 1969
Designed by Yoshisada Horiuchi
Made by Japan New Roller Skate Company
Leather, metal, and plastic, 9 5/8 x 2 3/4 x 14 3/4"
(24.4 x 7 x 37.5 cm)
Collection of Yoshisada Horiuchi, Koshoku

The most revolutionary change in roller skates since
their invention in 1873 was these in-line straight skates
designed by Yoshisada Horiuchi to approximate the
movements used for ice skating. The new skate, a
leather boot on double steel blades resembling those of
ice skates, has four plastic wheels set in line between the
blades. Designed by Horiuchi for ice skaters to practice
in summer, these skates were soon being used by
Japanese athletes training for the 1972 Sapporo Winter
Olympics.

94 Torch for Eleventh Winter Olympic Games, Sapporo, 1970

Designed by Sori Yanagi
Oxidized aluminum, 8 11/16 x 5 1/8 x 4 3/4"
(22 x 13 x 12 cm)
Yanagi Product Design Institute, Tokyo

The eleventh Winter Olympic Games were held in Sapporo in February 1972. Sori Yanagi, who had designed the runners' torches for the 1964 Tokyo Summer Olympics (no. 66), was commissioned in 1970 to design all holders for the Olympic flame at Sapporo: the case in which it traveled from Greece to Japan, the footed vessel to which it was transferred upon arrival and from which the relay runners' torches were lit, the runners' torches, and the stadium torch, which remained lit throughout the games. Yanagi was intent on making the torches for the runners as light as possible, and as in 1964, he designed them to be of aluminum, those in 1970 weighing scarcely half as much as those in 1964. The stadium torch, a large version of the runners' torches, was made from blued copper and stood almost ten feet tall. The lips of both the large and the small torches are asymmetrically curved, suggesting the form of a ski or skate blade.

94

95

95 Furniture in Irregular Forms: Side 2 chest of drawers, 1970
Designed by Shiro Kuramata
Made by Cappellini, Arosio, Italy
Lacquered wood, 66 15/16 x 24 13/16 x 19 11/16"
(170 x 63 x 50 cm)
Philadelphia Museum of Art. Gift of Cappellini

This chest of drawers, which curves irregularly in its width, and a companion piece (Side 1), which curves irregularly along its depth, belong to a group of experimental furniture with drawers by Shiro Kuramata that originated in the late 1960s. The chests were made by Fujiko in Tokyo for a brief period and rereleased by the Italian manufacturer Cappellini in 1986, when such seemingly irrational and individual forms of expression were widely accepted as postmodern. Leaping into a future that he would help to define, Kuramata was already a rising star of furniture and interior design known for his conceptual, even poetic, approach when Furniture in Irregular Forms was first produced. Kuramata returned to his childhood experiences for his fundamental influences. "I've loved drawers ever since I was a child," he reflected. "Mine used to be full of toys and spinning tops and colored cards: they were my hidden treasures. I loved putting my hand into those untidy drawers and rummaging about. Now as an adult I think that maybe I'm still looking for something in the drawers that isn't there."[18]

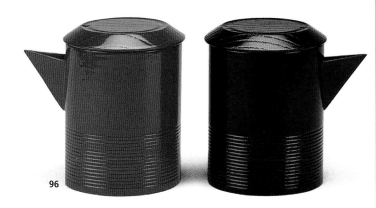

96

96 Soy-sauce containers, 1971
Designed and made by Tadashi Nakanishi
Lacquered wood, height 2 7/8" (7.3 cm),
diameter 2 3/16" (5.5 cm)
Philadelphia Museum of Art. Gift of the designer

Tadashi Nakanishi sees his lacquerworks as traditional in their technique but new in their feeling or sensibility, as exemplified by these soy-sauce containers. The red and black lacquer finishes were applied to the cylindrical wood forms using the time-honored method of building up and polishing successive layers of lacquer over a period of days. The carved bands give a firm grip to the container and add a contemporary look, as does the sharply pointed spout, which prevents spilling of excess soy sauce.

97 Dress and sweater ensemble, Fall–Winter 1971–72
Designed by Kenzo
Made by Kenzo, Paris
Wool and cashmere
Kenzo, Paris

In April 1970 Kenzo became the first Japanese fashion
designer to show his work in Paris, and the success of his
fresh, young collection encouraged others to follow.
This ensemble introduced the "schoolgirl" look for
which Kenzo first became known: its dress with wide
collar, mini length, lollipop colors, and accompanying
oversize beret were widely copied in Parisian ready-to-
wear. Although here he was working with Western
formal components, Kenzo declared his cultural
independence, adapting the loose sleeves and
unstructured silhouette from traditional Japanese dress.
For the fall–winter 1971–72 collection, Kenzo explained,
"I decided to use the straight cut with a square shape. I
called it the anti-couture cut, because it was in
opposition to the close-cut technique traditional in Paris.
I tried deliberately to create a nonstructured,
nondefined form and to introduce a new, different
fullness using the technique of the kimono."[19]

98 Outrigger ski, 1972
Designed by Yoshitaka Tanaka and Yamaha Corporation
Made by Yamaha Corporation, Hamamatsu
Aluminum, nylon-coated steel, and plastic, length of
shaft 34 1/4" (87 cm) (extended)
Yamaha Corporation, Hamamatsu

The concept of the Outrigger ski was developed by
Yoshitaka Tanaka, a designer and activist for the
disabled, who took his idea to Yamaha to be realized.
Tanaka was particularly interested in creating an
attractive product because of his strong belief that
sporting goods designed to meet the needs of the
physically challenged should also be considered from the
viewpoint of aesthetics. Replacing the ski pole, the
Outrigger has a small ski attached at the bottom of its
shaft, which provides support and balance for amputees
or skiers with one-foot paralysis. The handle is slanted
slightly and equipped with an arm cuff to add to its
comfort and stability; a lever on the handle allows the
ski to be converted to a crutch for aid in walking.
Because the Outrigger must support a considerable
amount of weight, its shaft was constructed of a
special aluminum alloy, which although strong and
somewhat thicker than usual, is still light in weight. In
development for about seven years, during which it was
subjected to vigorous trials and modifications, the
Outrigger ski was first made available to the public
in 1979.

98

97

99 Marilyn chair, 1972
Designed by Arata Isozaki
Made by Tendo Mokko, Tendo
Stained wood with polyurethane upholstery, 55 1/8 x
21 1/4 x 21 7/16" (140 x 54 x 54.5 cm)
Tendo Company, Ltd., Tendo

Representation and historic reference are recurrent
themes in the buildings of Arata Isozaki, which the
architect applied to his Marilyn chair, the best known of
his furniture designs. Isozaki has compared the creation
of this chair, which draws on two disparate sources, to
the writing of Japanese *honka-dori* poetry, which adapts
existing poetic texts: The "Marilyn chair's black finish
and narrow back which emphasizes its vertical line recall
the image of Mackintosh's [high-back] chair. The other
image is the body line of Marilyn Monroe. By
combining various curves of her body taken from the
famous nude photographs, I had a special 'French curve'
made for [use] in my atelier. The curve of the 'Marilyn'
chair's back is drawn with this 'French curve.' Thus, the
images of these two originals, superimposed, invade
each other. This is seen, for example, in the contrastive
juxtaposition of the curve of the back and the rectilinear
form of the seat."[20] The chair was made first by ICF in
Italy and then Sunar in Canada during the 1970s; since
1981 it has been produced by Tendo in Japan.

99

100 Packaging for Hyobando, 1972
Designed by Akemi Kashima and Yoshimasa Kawakami
Offset on cardboard, 3 15/16 x 6 11/16 x 3 15/16"
(10 x 17 x 10 cm) (each)
Collection of Akemi Kashima, Tokyo

These packages for traditional rice crackers (*arare*) were
designed by Akemi Kashima and Yoshimasa Kawakami
in a pie shape, emphasized by the bold semicircular
pattern on the boxes. The packages, which fold down
into a single flat unit, come in four colors and fit
together into a playful variety of groupings. The simple,
effective design, interrupted discreetly by the Hyobando
name in a vertical cartouche at one side of the box,
makes the packages memorable and easy to spot on a
store shelf.

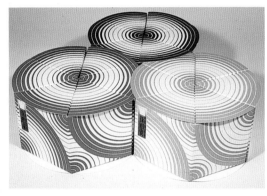

100

101 Cattail chair, 1972
Designed by Katsuo Matsumura
Made by Johoku Wood Manufacturing Company,
Matsumoto
Larch and cattail, 33 1/4 x 18 1/2 x 32 5/8"
(84.5 x 47.5 x 83 cm)
Collection of Miyoko Ono, Tokyo

Katsuo Matsumura designed this low-slung chair as part
of a series of mass-produced items of furniture in larch,
winning both the Mainichi industrial design prize and
the Japan Interior Designers prize for the line. Finding
paulownia and other high-quality woods expensive and
in great demand for architectural and interior use,
Matsumura began to experiment with alternate choices
such as larch. Although readily available, larch had been
impractical as a furniture wood until a low-cost
technology for removing its resin was invented in the
late 1960s. Here Matsumura combined a simple larch
frame with a seat and back woven from cattail,
providing a pleasing contrast of texture and pattern.

101

102 *Enku*, 1973
Designed by Hiromu Hara
Published by Kyuryudo, Tokyo
Printed book, 9 5/8 x 7 5/8 x 1 1/8" (24.5 x 19.4 x 2.9 cm)
Museum & Library, Musashino Art University, Tokyo

This book, a deluxe presentation of the works of Enku, an itinerant seventeenth-century carver of rough-hewn Buddhist statues, was the creation of Hiromu Hara, the most productive book designer of postwar Japan, with over one thousand volumes to his credit. The two calligraphic characters that almost fly off the red binding form the name of the sculptor, and the vitality of this usage announces at once Enku's bold and prolific oeuvre. Designed and supervised by Hara, this book demonstrates his loving attention to detail, as well as his uncanny instinct for capturing just the right tone for each subject, the quality for which his work is much admired.

102

103 Rocking chair, 1972
Designed by Takeshi Nii
Made by Ny Furniture Company, Tokushima
Steel, wood, and canvas, 37 1/16 x 23 5/8 x 28 3/8"
(94.2 x 60 x 72 cm)
Museum & Library, Musashino Art University, Tokyo

When his family's business was temporarily prevented from producing equipment for the traditional martial art of *kendo*, Takeshi Nii began experimenting with the surplus canvas and bamboo staves to create designs for folding chairs, making his first model in 1956. He started using steel tubing in 1960, and has been refining these designs ever since. This version of his Ny X series from 1972 is a rocking chair with a high-back canvas seat and wooden arm rests. It is sturdy and comfortable, and compact for easy storage when not in use, features that have made Nii's folding chairs best-sellers in Japan and abroad. "Ny" in the series title is a play on the Danish word *ny*, meaning "new," and the designer's name Nii.

103

104 Allex scissors and letter opener, 1973
Designed by Tokuji Watanabe
Made by Hayashi Cutlery Company, Iseki
Prize: G-Mark, 1974
Stainless steel and plastic, length of letter opener
6 11/16" (17 cm); length of scissors 6 1/2" (16.5 cm)
Hayashi Knife Co., Ltd., Iseki

Tokuji Watanabe was inspired by the form of the child's
scissors in his son's schoolbag to create equally user-
friendly, high-quality scissors for adults. The handles are
designed to be comfortable for both right- and left-
handed users. The plastic rings on the inside of the
handles cushion the grip and add comfort as well as a
dash of color to the organic stainless-steel shape of the
scissors and the letter opener.

105 Cobra desk lamp, 1973
Designed by Masayuki Kurokawa
Made by Yamagiwa Corporation, Tokyo, and Fuso Gomu
Industry Company, Tokyo
Rubber and steel, height 23" (58.3 cm) (extended)
Collection of Masayuki Kurokawa, Tokyo

Exploring the plastic possibilities of rubber, Masayuki
Kurokawa created this electric lamp with a ribbed
rubber neck and hooded shade that transformed it into
a swaying cobra poised as if to strike. For Kurokawa,
the form of the lamp was of primary importance:
"When a lighting fixture is turned on," he wrote, "the
fixture itself cannot be seen. The viewer just sees the
light. When the light is extinguished, only the form is
seen. It would be good if this dual aspect could be
shown more dramatically. . . . In my case, I first think of
a shape that I would like."[21]

104

105

106 Yukaida animals, 1973
Designed by Shigeo Fukuda
Printed paper, 14 9/16 x 10 1/4" (37 x 26 cm) (flat)
Private collection

The Yukaida animal series, novelty items commissioned by a soft-drink manufacturer as a promotional giveaway, were based on the idea of paper airplanes. They were distributed as flat sheets and punched out and assembled by the recipients. The name is a combination of the Japanese word *yukai* (happiness, delight) and the English word "glider." Shigeo Fukuda's twenty-one different animals were chosen from species not normally thought of in flight, such as pigs, lions, and turtles. He recalls that the turtle was his biggest worry before "flight testing" because of its rounded form and weight, but "against my expectations, it turned out to be the one that flew the best. I kept making it fly higher and higher until my arm started to hurt."[22]

106

107

107 Wireless microphone, 1972
Designed by Chubu Design Research Center
Made by Hoshiden Company, Yao
ABS plastic, 1 1/2 x 5 7/8 x 1" (3.8 x 15 x 2.5 cm)
Chubu Design Research Center Inc., Nagoya

Designed primarily for professional use but a forerunner of the personal microphones used in *karaoke* singing, this sleek and compact wireless model allowed its user considerable freedom to move about while performing. Its angled easy-to-hold design provided an alternative to conventional microphones, which had round heads, and was intended to provide the purest sound possible.

108 Packaging for Ehime Nori, 1974
Designed by Shozo Kakutani
Printed paper, 11 x 8 9/16 x 3 15/16" (28 x 21.7 x 10 cm)
Collection of Iwao Suzuki

Part of a series of package designs for Ehime brand seaweed (*nori*), this box features a sea-dark background with stylized waves. The curves of the waves echo the shape of the first syllable *no*, with the second syllable *ri* suspended below like two droplets of the water in which the seaweed is grown. The logo of the maker, Ehime Nori, appears below at center, and at lower left is the logo of Daimaru department stores, which sell the seaweed. The understated design is varied through the six-color scheme, lending the package a traditional yet sophisticated air.

108

109 Snail boxes, 1974
Designed and made by Yoshiko Kitagawa
Paper, width of large box 20 1/16" (51 cm)
Collection of Yoshiko Kitagawa, Tokyo

Taking her cue from the flexible modular system used in arranging traditional *tatami*-mat flooring in Japanese houses, Yoshiko Kitagawa translated the concept into package design in her Snail (Katatsumuri) series of folding boxes. The modules are flat, oblong sections of black or silver paper, which when combined in three to seven units, form triangular, square, or in this case, hexagonal boxes, fastened at the side in an overlapping snail-shell pattern that gives the series its name. The box was originally designed as a hatbox, but the interlocking pieces can be limitlessly reconfigured into boxes of different shapes and sizes.

109

新音楽媒体
NEW MUSIC MEDIA

新魔法媒体
NEW MAGIC MEDIA

110

110 Poster for New Music Media Committee, 1974
Designed by Koichi Sato
Offset, 40 1/2 x 28 11/16" (103 x 72.8 cm)
Collection of Koichi Sato, Tokyo

Advertising with this poster the "new magic" of the
summer 1974 contemporary music festival held in the
resort town of Karuizawa, Koichi Sato irradiated the
image of a carp in a glowing pool of light. Its quiet,
poetic aura acts as a visual metaphor for the "new
wave" music. Borrowing his subject of a carp swimming
in a box from traditional Japanese painting, Sato
infused it with futuristic meaning, wielding the airbrush
as earlier painters wielded their paintbrushes and, in the
process, adding unexpected expression. Having
established himself with this poster as a new creative
talent in graphic design in Japan, Sato continued to
explore luminosity as an expressive device in his later
work (no. 203).

111

111 Packaging for Eitaro-Sohonpo, 1975
Designed by Eitaro-Sohonpo
Printed paper, 5 1/8 x 3 1/8 x 2 3/4" (13 x 8 x 7 cm) (each)
Eitaro-Sohonpo Company, Ltd., Tokyo

Eschewing an overall graphic consistency, the Eitaro-
Sohonpo confectionery firm has instead produced
different packaging for each of its different products.
Here New Year's candies are whimsically boxed as lions,
referring to the traditional New Year's festival dance
(*shishimai*) in which the celebrants, wearing brightly
colored lion costumes, appear and drive away bad spirits
for the coming year.

112 Exhibition poster, 1975
Designed by Shigeo Fukuda
Silkscreen, 40 1/2 x 28 11/16" (103 x 72.8 cm)
Collection of Shigeo Fukuda, Tokyo

Shigeo Fukuda created this humorous chorus-line poster
for an exhibition of his work at the Keio department
store in Tokyo in 1975. Here Fukuda explores optical
illusions, playing with reversible black and white images,
which can be read both as a row of men's legs on a
white ground and a row of women's legs on a black
ground.

113 Civic automobile, 1975
Designed and made by Honda Motor Company,
Hamamatsu
Metal, 147 13/16 x 59 1/4 x 52 1/8"
(375.4 x 150.5 x 132.4 cm)
American Honda Motor Company, Inc., Torrance,
California

Using a revolutionary new fuel-efficient CVCC
(Compound Vortex Controlled Combustion) engine that
burned a leaner mix of gasoline and would pass the
strict emissions tests set in the United States in 1975, the
Honda Civic became an international best-seller when it
was introduced in 1972, quickly capturing the American
auto market at its low end, the first Japanese car to do
so. The world oil crisis beginning in 1973 caused exports
of compact, fuel-efficient, moderately priced Japanese
cars like the Civic to rise sharply, and by 1980, the
production of cars and trucks in Japan was the highest
in the world. Originally a motorcycle manufacturer (see
no. 84), Honda had produced cars and trucks only since
the 1960s. The Civic was based on the Honda N360
minicar of 1967, which had the squared-off chassis with
front-mounted engine and front-wheel drive that came
to typify Japanese auto body design, just as the CVCC
engine represented the technological achievements that
made Japanese cars so competitive on the global market.

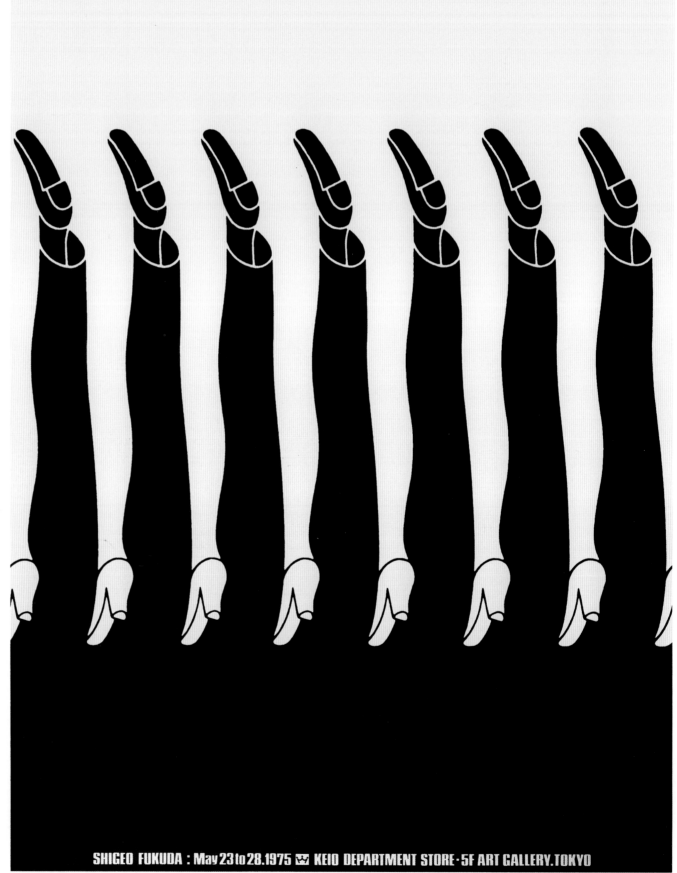

SHIGEO FUKUDA : May 23 to 28.1975 ▨ KEIO DEPARTMENT STORE · 5F ART GALLERY.TOKYO

112

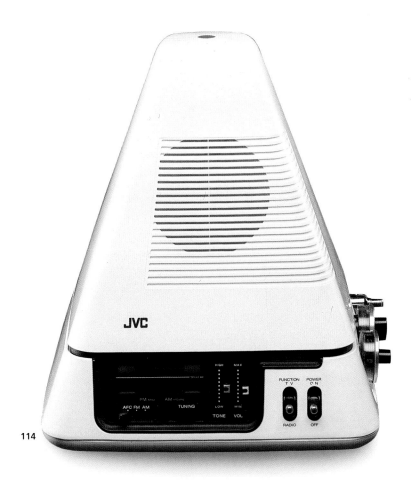

114

114 Television and radio, 1972
Designed by International Industrial Design
Made by Japan Victor Company, Tokyo
Styrol resin, 12 3/16 x 11 1/2 x 11 5/16"
(31 x 29.2 x 28.7 cm)
Collection of Daniel Ostroff, Los Angeles

This portable television, weighing slightly over ten pounds, is a model of compactness, emphasized by its light cream color and minimal control panel. In the early 1970s Japan Victor experimented with producing appliances in unusual shapes: round television sets that resembled space helmets or pyramids like this one, which echoed the form of the robots they were testing at the time. Its pyramidal case, which houses the speaker, opens up to reveal a seven-inch screen.

113

115 Poster (*Victory*), 1975–76
Designed by Shigeo Fukuda
Prize: Thirtieth Anniversary of Victory in War
International Poster Competition, Warsaw,
grand prize, 1976
Silkscreen, 40 1/2 x 28 11/16" (103 x 72.8 cm)
Museum & Library, Musashino Art University, Tokyo

This image, created with a clarity and wit typical of
Shigeo Fukuda's work, was the winner of the poster
competition held in Warsaw to commemorate the
thirtieth anniversary of the victory of the Allies in World
War II. By turning the bullet back into the barrel of the
gun, Fukuda makes the point that ultimate victory
comes only with the end of war altogether. The poster
was subsequently printed and sold in Poland, and the
proceeds went to further the peace movement.

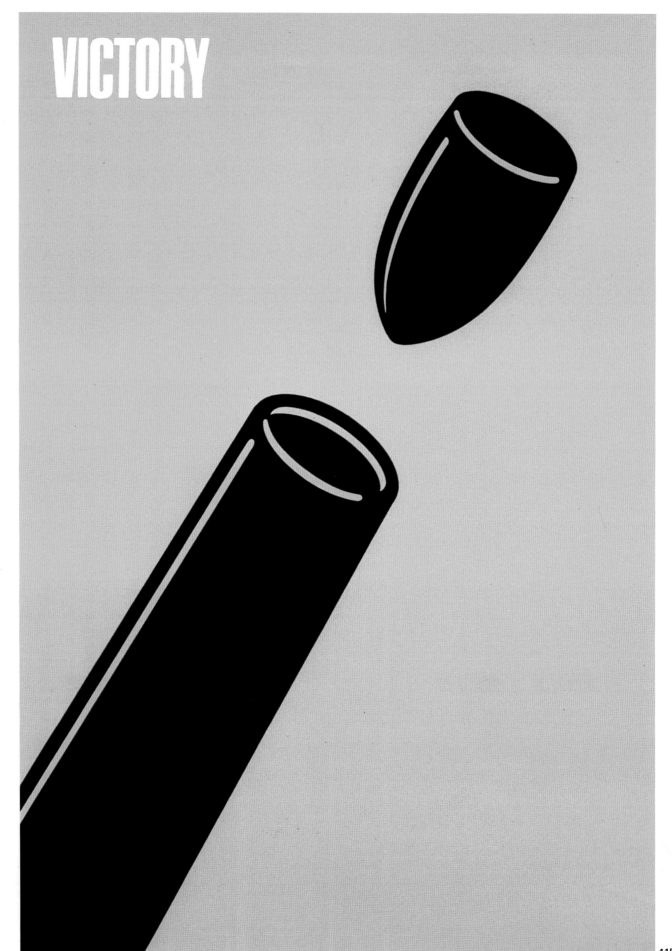

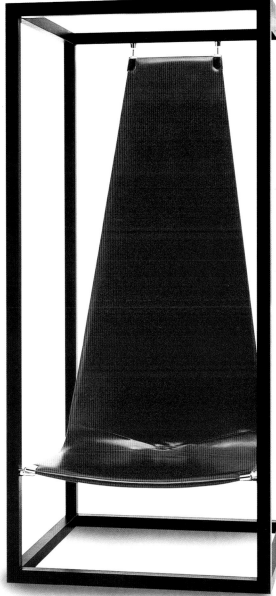

116

116 Open=Close or Plus=Minus Space chair, 1976
Designed by Shutaro Mukai
Made by Aoshima Shoten Manufacturing Company,
Tokyo
Metal with leather upholstery, 74 9/16 x
35 5/16 x 35 5/16" (190 x 90 x 90 cm)
Private collection

Shutaro Mukai was one of the few postwar Japanese
designers to study in Germany, where at the Hochschule
für Gestaltung in Ulm he absorbed the advanced theories
then being propounded by its faculty, as well as other
avant-garde Western ideas, particularly the studies of
semiology and linguistics that inform this Open=Close or
Plus=Minus Space sling seat. "I have seen the essence of
the chair not in its simple function of supporting the
human body but rather as an ambivalent or ambiguous
space enveloping the seated person," Mukai said. "As
the seated person crouches in the chair and reads a book,
the mere fact of there being a space limited and
confined by four supporting columns suggests a personal
universe closed within itself on a psychological plane."[23]
A poem written by Mukai accompanies the chair:

Chair as space
Space as chair
Space in space
Space outside of space
Space around space
Space in time
Time in space
Open in closed
Closed in open
Many in one
One in many·
Alone in common
Useless in usefulness
Useful in uselessness
Identity, structure / significance, change
In time and space for man and nature.

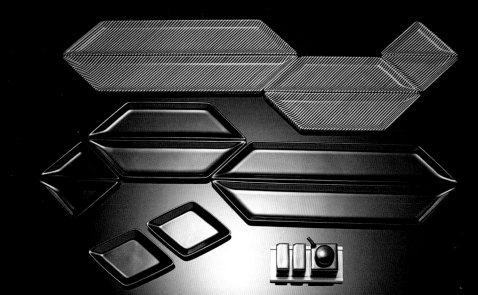

117

117 Trays, 1976
Designed by Masahiro Mori
Made by Hakusan Toki, Hasami-machi
Prize: G-Mark, 1979
Glazed porcelain, length of longest section 18 3/4"
(47.5 cm)
Hakusan Toki K.K., Hasami-machi

This set of party trays represented a new type of
tableware for Japan, accommodating the postwar
custom of entertaining in the home. The set consists of
a modular system of porcelain trays in various sizes that
nest in a square configuration but can be used in various
combinations to suit the size of the party and the
allotted space. The pattern of blue stripes adds a pleasing
accent to the diamond- and trapezoid-shaped trays.

118 Ma fabric, 1977
Designed by Katsuji Wakisaka
Made by Wacoal Interior Fabrics, Kyoto
Printed cotton, width 54 3/8" (138 cm)
Philadelphia Museum of Art. Gift of Wacoal Interior
Fabrics

Like a number of other Japanese designers, Katsuji
Wakisaka gained professional experience in the West,
working between 1968 and 1976 for the Finnish textile
firm of Marimekko. On his return, Wakisaka reasserted
his Japanese identity, borrowing for his Ma fabric
elements of traditional kimono design: the small,
geometric stripes, checks, and pinpoints and the
principles of light and dark (*notan*); however, he
translated them into the clear, bright colors for which
Marimekko is known. These small-scale patterns are
arranged in geometric units or patches in a dense,
allover asymmetrical combination divided only by the
clear intervals (*ma*), or borders, between them.

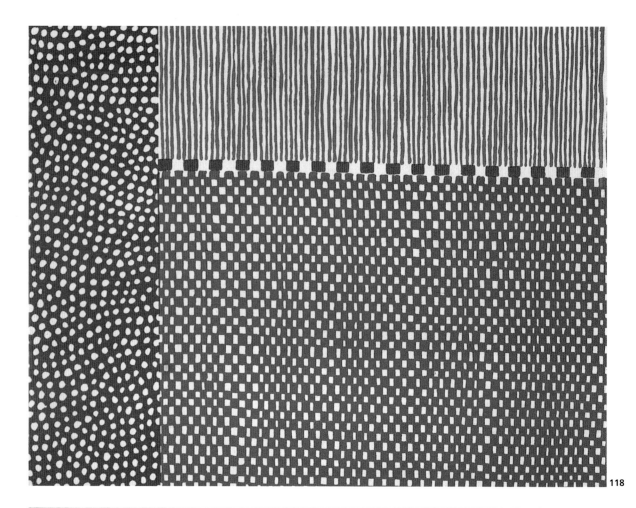

118

119 Korean Carrot (Ginseng) fabric, 1979
Designed by Junichi Arai
Made by Anthologie, Tokyo
Wool, width 51" (129.5 cm)
Philadelphia Museum of Art

Junichi Arai gained recognition in the 1980s for his
imaginative use of computers to create intricate
weaving structures. This piece combines a complex
weave and a finishing process that, although simple,
transforms the textile in a totally unexpected way. Arai
finished the cloth by cutting loose, long yarns from one
of the weft structures and felting the whole piece in a
washer and a tumbler. What was ordinary yardage with
long floats became a cloth with short, pointed
dreadlocks resembling scraggly ginseng roots. His
signature style—sophisticated technology— was used
here with primal energy and humor.

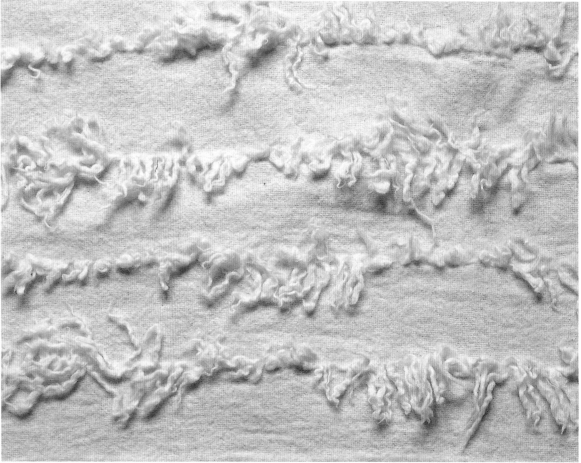

119

120 Packaging for Kibun, 1977
Designed by Shin Matsunaga
Printed paper, height of large can 4" (10.1 cm);
height of small can 3 3/16" (8.1 cm)
Collection of Shin Matsunaga, Tokyo

In a witty contradiction of the package as a protection and enclosure for its contents, Shin Matsunaga turned this concept inside out with his design for these canned-vegetable labels. The contents, from tomatoes to lotus roots, are "exposed" in life-size photographs on the printed labels, an expression of Matsunaga's design philosophy that "it is very important that appearance and content of the merchandise have a good fit. Only in this way will it appeal to consumers."[24] In a daring marketing move, the client, Kibun foods, allowed its name and all accompanying printed matter to be moved to the back of the cans, so that the product, through Matsunaga's design, could stand on its own with unadorned honesty on the supermarket shelf.

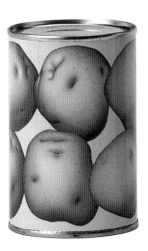

120

121 Gradation kimono, 1978
Designed by Fukumi Shimura
Dyed silk, 65 3/4 x 51 15/16" (167 x 132 cm)
The National Museum of Modern Art, Tokyo

Working with silk for nearly forty years, Fukumi Shimura
has dedicated herself to exploring the essential elements
of cloth: yarns, dyes, and plain weaves. As in Western
color-field painting, Shimura's canvas, the kimono,
speaks of her impressions, her feelings, the seasons, and
the cycle of life. As painters use brushes and pigments,
she uses the loom and yarns to weave a subtly textured
surface in a multitude of graduated colors derived from
natural dyes. While restricting her language to plain
weaving, she has created a body of work that has great
depth and sophistication.

121

122 Cigarette package for Japan Monopoly Corporation, 1976
Designed by Takashi Kanome
Printed paper, 3 3/4 x 2 x 3/4" (9.5 x 5.1 x 1.9 cm)
Japan Monopoly Corporation, Tokyo

Commissioned by Japan's tobacco monopoly to create packaging for its Elegance cigarettes, Takashi Kanome fulfilled the promise of the product's name with this unusually beautiful wrapper, which has the aura of a

122

gift rather than a package destined to be wadded up and thrown away. Superimposed on a dark background of the traditional textile *komon* (small pattern) designs are the elegant calligraphic characters for *miyabi* (elegance) printed in unexpected and brilliant colors. The reference to traditional pattern design was deliberate, for according to the designer, the monopoly wanted the package "to project the elegance of Japanese tradition and the ancient capital of Kyoto."[25]

123 Poster for Shiseido, 1978
Designed by Makoto Nakamura
Offset, 40 1/2 x 28 1/16" (103 x 72.8 cm)
Shiseido Company, Ltd., Tokyo

Shiseido traditionally used images of beautiful women to advertise its cosmetics, but it was Makoto Nakamura, working with the photographer Noriaki Yokosuka, who made photography as distinguished a form of expression as illustration for the firm's posters. One of a series featuring the well-known model Sayoko Yamaguchi, Nakamura's advertisement for Shiseido's Mai perfume abandoned the conventional image of the smiling beauty and focused solely on Yamaguchi's eye and part of a gold fan. Nakamura cropped the original photograph to create this astonishing detail and worked with the printer to retouch the model's eyelashes and skin for enlargement. Posed behind her fan like one of Utamaro's "beauties" in a *ukiyo-e* print, Yamaguchi as conceived by the designer represented a new ideal of Japanese feminine beauty (enhanced, of course, by Shiseido's perfume) that could reveal itself with a single compelling glance.

ほのかな香りほど悩ましい。資生堂練香水 ［舞］

2,500円

123

124 Capsela toy system, 1975
Designed by Nido Industrial Design Office
Made for Mitsubishi Pencil Company, Tokyo, by Play
Tech, Hong Kong
Plastic and metal, length of capsule 2 1/2" (6.4 cm)
Private collection

Capsela is a complex toy and an educational aid for
young children, its design partially based on the physics
curriculum taught in Japanese elementary schools. It
was developed, according to Yoshiyasu Ishii, one of the
Nido Industrial Design Office team responsible for its
design, "as a very unique formation toy to help children
systematically recognize the general principles of
mechanisms based on electricity, magnetism and
dynamics . . . and to allow them to know a new world
that has never been experienced by them."[26] Looking
like a space-age version of the Tinker Toy, Capsela
consists of a series of transparent spherical body
capsules, each with a specific mechanical or electrical
function, and equipped with six octagonal extensions
that connect to other units. Bright-yellow floats,
brilliant-red wheels, blue axles, and green propellers can
be combined to create a wide array of colorful machines
such as cranes, motorboats, airplanes, and elevators.

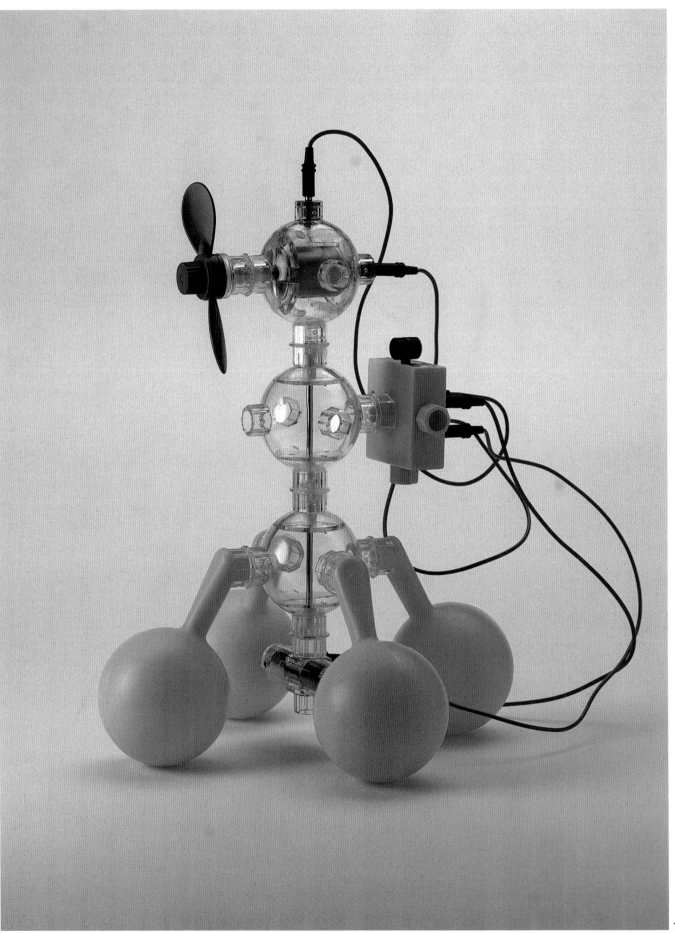

124

125 Pocket camera, 1978
Designed by Yoshihisa Maitani
Made by Olympus Optical Company, Tokyo
Prize: G-Mark, 1981
Plastic and metal, 4 x 2 9/16 x 1 9/16" (10.2 x 6.5 x 4 cm)
Olympus America Inc., Woodbury, New York

This pocket camera, ready to be taken anywhere, was the first miniature camera to use full-frame 35-millimeter film, and became an instant classic. Yoshihisa Maitani's innovative housing of molded plastic provides protection for the lens and eliminates the need for lens cap or camera case. The camera's compact size, sleek black casing, and ease of operation immediately made it equally popular with women as with men, winning a new market segment for Olympus.

125

126

126 Calculator, 1979
Designed and made by Sharp Corporation, Osaka
Metal and plastic, 3 3/4 x 2 1/8 x 1/16" (9.6 x 5.4 x .16 cm)
Sharp Corporation, Osaka

By 1973 Sharp had used the liquid-crystal display (LCD) in calculators for the first time, incorporating all its circuitry on a tiny glass panel. Replacing larger and less efficient light-emitting diodes (LED), LCDs enabled Sharp to produce smaller calculators than ever before. During the 1970s the technology was refined further, and calculators continued to shrink. This calculator, the thickness of a credit card (about 1/16 inch), was the thinnest and smallest calculator that had yet been produced, although six years later, after another technological breakthrough, Sharp replaced it with one only half as thick.

127 Whisky bottle, 1978
Designed by Iwataro Koike
Made by Ozeki Brewing Company, Nishinomiya
Glass and printed paper, 7 1/2 x 3 9/16 x 4 5/16"
(19 x 9 x 11 cm)
Collection of Yoko Koike Johanning, Haverford,
Pennsylvania

Iwataro Koike brought his humor and originality to a
design field not always known for either. Creating
whisky bottles for mass production within the strictures
of the manufacturer's budgetary limitations was a
challenge that Koike met with a conspiratorial nod to
the consumer. Instead of following the conventional
format for liquor bottles, which usually emphasized
their manly weightiness, Koike gave his bottle for Ozeki
brewers a light-hearted touch: by putting the label on
the narrower side and making that its front, he created
a bottle that was easier to take off the shelf. The
rocking aspect makes it memorable and also adds a
soothing element to its external appearance, meant
perhaps to parallel the internal effects of its contents.

127

128 Motocross helmet, 1978
Designed by Michio Arai
Made by Arai Helmet, Ohmiya
Fiberglass-reinforced resin laminate, expanded
polystyrene, and nylon-rubber thermoplastic,
depth 34 1/4" (87 cm)
Arai Helmet (America) Ltd., Daytona, Florida

Designed for off-road motorcycle riders, this futuristic
helmet with its projecting peak and rock guard was
meticulously engineered to help prevent injuries and
wind buffeting at high speeds. Produced by the firm
that made Japan's first modern fiberglass motorcycle
helmet in 1950, the Motocross MX helmet exploits the
properties of its plastic materials: reinforced fiberglass
for strength and lightness in the shell and an array of
expanded polystyrene foams for impact absorption and
comfort in the interior lining. The interior of the helmet
is assembled and fit into place by hand, a process
ensuring conformity to facial contours. The extremely
high quality and attention to detail shown in this
helmet are characteristic of the firm known
internationally for its safety products.

128

西洋は東洋を着こなせるか

129 Poster for Parco, 1979
Designed by Eiko Ishioka
Offset, 40 1/2 x 28 11/16" (103 x 72.8 cm)
Private collection

Can West Wear East is the best known of the three
advertising campaigns Eiko Ishioka created in 1979 for
the Parco retail group featuring the American actress
Faye Dunaway. The last campaign included this poster
(produced in two sizes) and two television commercials,
which were shot by the photographer and
cinematographer Kazumi Kurigami. Working with an
American actress gave Eiko the opportunity to
communicate Parco's philosophy on an international
level. "Japan has learned from the West," Eiko wrote of
this campaign, "but the situation is now changing.
Today's trends show that the West is starting to look
East. Our rather bold question, 'Can West Wear East?' is
being asked by the New Japan, which looks forward to
the future—and to a time when East and West become
one."[27] Making Parco's message literal and visually
comprehensible, Eiko transformed the American actress
into the Japanese Buddhist deity Kannon, with a
majestic satin robe and headpiece created for her by
ISSEY MIYAKE. Completing the image are Eiko's two
nieces, who play the *doji*, or children who serve Buddha
in the temple.

130 Blitz folding chair, 1977

Designed by Motomi Kawakami
Made by Skipper International, Milan
Prizes: International Furniture Design Competition,
San Diego Chapter, American Institute of Architects,
first prize, 1977; Japan Interior Designers prize for
excellence, 1984
Steel and polyurethane, 29 5/16 x 20 7/16 x 18 7/8"
(74.5 x 52 x 48 cm)
Collection of Motomi Kawakami, Tokyo

Motomi Kawakami was one of a number of Japanese
furniture designers who studied and worked in Italy,
where manufacturers were willing and had the ability to
produce new and sometimes unconventional forms in
synthetic materials. Using Western formal components—
here the folding leg frame—and industrial technology,
Kawakami still retained a Japanese aesthetic sensibility,
which is evident in the calligraphic elegance of the
chair's profile. The production of this chair, like that of
other plastic furniture, was affected by the world energy
crisis of the 1970s and it took five years before it was
merchandised. "As it was right after the oil crisis, when
the price of petrochemical products surged," Kawakami
related, "the development of a comfortable yet slim
chair minimizing the use of material was sought."[28]
Designed for use internationally, the chair is marketed
with two different seat heights for different body types.

131

131 Packaging for Futatsuido Tsunose, 1978

Designed by Takashi Kanome
Enameled tin and printed paper,
tin box 1 9/16 x 4 11/16 x 9 1/16" (4 x 12 x 23 cm);
paper package 1 x 6 1/8 x 5 1/2" (2.5 x 15.5 x 14 cm)
Collection of Takashi Kanome, Tokyo

For the 225-year-old confectionery firm of Futatsuido
Tsunose in Osaka, Takashi Kanome designed these two
packages to suggest that the Naniwazoshi rice and
millet cakes inside were being offered as a traditional
celebratory gift. "I gave this old product a double
image of tradition and modernism," Kanome wrote.
Traditional are the shapes of the boxes and the sparing
color scheme with its use of "antique purple and a
feeling of lacquer";[29] modern are the boldly abstracted
characters for naniwazoshi, which serve as the chief
decorative element on their plain backgrounds.

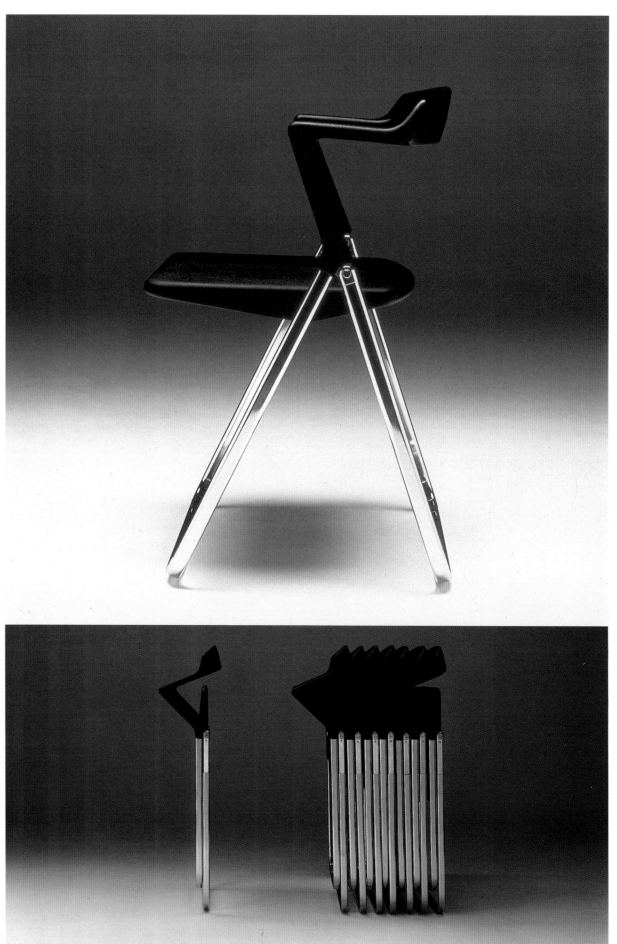

130

132 Walkman personal cassette player, 1979
Designed and made by Sony Corporation, Tokyo
Anodized aluminum, 5 1/2 x 3 9/16 x 1 1/8"
(14 x 9 x 2.8 cm)
Sony Corporation, Tokyo

Sony's Walkman evolved from the pocket cassette recorders developed for dictation by the company in the late 1960s. Converting from recording to audio functions, Sony engineers refitted their Pressman recorder to make the Walkman by removing the recording circuit, speaker, and microphone and replacing them with a stereo amplifier, and adding miniature headphones. The Walkman was originally designed for the youth market: it was colored a metallic "blue jean" blue and the dual headphone jacks were labeled "guys" and "dolls" (soon changed to read simply "A" and "B"). Subsequent models were made lighter and smaller, reduced as closely as possible to the size of the cassette-tape case they contained. Within a decade and a half the phenomenally successful Walkman had spawned some forty-three varieties and descendants, from a micro television (Watchman, 1982) to a compact disc player (Discman, 1985).

133

133 Opus sewing machine, 1979
Designed by Takuo Hirano
Made by Brother Industries, Nagoya
Prize: G-Mark, 1979
Aluminum, 12 1/4 x 17 1/2 x 7 3/16" (31.2 x 44.5 x 18.3 cm)
Brother Industries, Ltd., Nagoya

Since its founding in 1934 Brother Industries had been a leader in introducing technological innovations into home sewing machines. In the late 1970s the company commissioned Takuo Hirano to design this Opus 8 model, the first to use a microcomputer, which controls its five speeds and twenty-six types of stitches and can be preset to change automatically. The stitching choices are shown in a band of pictograms at the top, giving the otherwise undecorated cream-color body an advanced, computerized look. The compact case houses the attachments and electric cord, as well as a small emergency repair and cleaning kit.

132

134 Symbol for Hanae Mori, 1978
Designed by Ikko Tanaka
Offset, 40 1/2 x 28 11/16" (103 x 72.8 cm)
Ikko Tanaka Design Studio, Tokyo

Ikko Tanaka designed this symbol as a present for the fashion designer HANAE MORI when she opened her building, which was designed by Kenzo Tange, on the Omotesando in Tokyo in 1978. He chose a butterfly because Mori loves them and uses them often to ornament the clothes she designs. Tanaka's symbol is boldly graphic, its dark, angular form accented by bright circles and stripes, with the two straight lines forming the butterfly's antennas bracketing Hanae Mori's name. The symbol was used in color throughout the building on banners, doors, and elevators, and has appeared in both black-and-white and color on Mori's shopping bags.

135

135 Record jackets, 1976
Designed by Yoshiaki Kubota
Printed paper, 14 9/16 x 14 9/16" (37 x 37 cm) (each)
Collection of Yoshiaki Kubota, Tokyo

In an attempt to make Western classical music more accessible to children, Yoshiaki Kubota designed a set of five record jackets with appealing images. Each cover sported a "portrait" of the composer, Bach, Mendelssohn, Chopin, Beethoven, and Verdi, made recognizable by their caricatured features and their hair styles, done in broad swaths of color. Each jacket has a cutout along the edge for easy removal of the inner sleeve, which is color coordinated with the front cover.

136 Dress and scarf-coat, Spring–Summer 1977
Designed by Issey Miyake
Made by Miyake Design Studio, Tokyo, and printed by
Rainbow Studio, Milan
Screen-printed silk
Miyake Design Studio, Tokyo

Since establishing his own studio in 1970, Issey Miyake
has led avant-garde fashion design through his use of
unprecedented constructions and unique fabrics. One
of Miyake's quests during the mid-1970s was to create a
garment that would seem to have been made from a
single piece of cloth, with no visible seams or fastenings.
He was fascinated with the relationship between the
two-dimensional fabric when it lay flat and the garment
made from it, which would assume three-dimensional
shape. When worn over the straight-cut, ankle-length
dress, this large, rectangular "scarf" with slits for arms
becomes a dynamically draped coat. Believing that
"designing clothes isn't solitary work, it requires
associations with many people,"[30] Miyake has had
numerous collaborators whose skills lie beyond the field
of fashion, among them artists who provide designs for
his printed fabrics. "These prints, created by artists'
hands, refuse to be only the last [decorative] phase of
making clothes; instead they wish to be a chosen
message."[31] The image of Adam and Eve printed on the
back of the scarf-coat and entitled "Paradise Lost" was
designed by TADANORI YOKOO. With its companion
scarf-coat, "Paradise," it was first shown in Miyake's
1977 spring–summer collection.

136

137 Box and Cox, from 1979
Designed by Katsu Kimura
Made by Zonart Company, Tokyo
Printed paper, banana 1 15/16 x 2 3/16 x 11 5/8"
(4.9 x 5.5 x 29.5 cm); cheese 4 x 4 x 6 1/4" (10.2 x 10.2 x
15.9 cm); egg 3 3/4 x 2 1/2 x 2 1/2" (9.5 x 6.3 x 6.3 cm)
Katsu Kimura and Reinhold-Brown Gallery, New York

In his Box and Cox series of 1979 produced by Zonart,
Katsu Kimura developed ingenious boxes in the shape of
food and other objects. Although most of the boxes can
be used to store small items, they are primarily

137

ornamental, made to display Kimura's bravura paper
engineering skills: the sides of a banana-shape box peel
back to expose an inner box in the form of a peeled
banana, a rectangular egg-shape box opens along a
crack to reveal a square-yolk box inside, and a wedge-
shape cheese box is studded with random holes. Also
included in the series are cigarette boxes filled with
cigarettes and sneakers complete with tube socks.
Kimura says that these boxes visually express his "eternal
theme" of "making a package playing the parts of a box
and an object."[32] Kimura enjoys creating these packages,
which have proven
immensely popular, and
continues to design and
redesign them. There are now
over twenty boxes in the series.

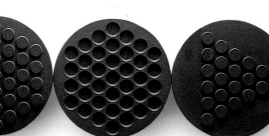

138 Gom office accessories, 1973–83
Designed by Masayuki Kurokawa
Made by Fuso Gomu Industry Company, Tokyo
Rubber, stainless steel, and glass, height of ashtray 3 1/8"
(8 cm), diameter 3 1/8" (8 cm); height of coaster 3/16"
(.5 cm), diameter 3 1/8" (8 cm); square lighter 1 5/8 x
2 3/4 x 2 3/4" (4.2 x 7 x 7 cm); height of round lighter
1 5/8" (4.2 cm), diameter 2 3/4" (7 cm); diameter of clock
11 7/16" (29 cm)
Collection of Masayuki Kurokawa, Tokyo

Confounding expectations that desk accessories, clocks,
and door knobs and handles should be made of hard
metal or plastic, Masayuki Kurokawa developed a series
of soft versions of these objects for the Fuso Gomu
company made of black rubber. With their soft matte
finish, they became pleasantly tactile, well suited for
being held and used frequently. Designed over a
decade, the Gom series gave Kurokawa the opportunity
to create new applications for a material then commonly
associated not only with tires, as he said, but also with
machine parts, packing, and shock absorbers (no. 105).

138

139 Bustier, Fall–Winter 1980
Designed by Issey Miyake
Made by Issey Miyake Inc., Tokyo
Plastic, 15 x 12 3/4 x 8 1/2" (38.1 x 32.4 x 21.6 cm)
Philadelphia Museum of Art

This plastic bustier, which brought Issey Miyake's study of the relationship between the body's form and the garment directly back to the body, was a centerpiece of his "Bodyworks" exhibitions shown in museums and galleries in Tokyo, Los Angeles, San Francisco, and London between 1983 and 1985. Molded on a human form, the bustier was made in collaboration with Nanasai, a mannequin manufacturer. Reversing the idea that clothing clads or conceals the body, the bustier replicates the body—or at least part of one—and then exposes it in a second, plastic skin. Miyake showed the bustier with a long pants-skirt, thereby transforming the corsetlike apparel from underwear to daring outerwear; at the same time, the bustier was widely published by itself, like a piece of sculpture, confirming Miyake's philosophy that clothing has an independent form and functionality.

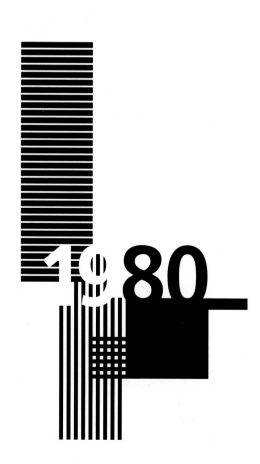

1980

139

140 Wrapping paper, 1981
Designed and made by Keiko Hirohashi
Printed paper, 25 3/8 x 25 3/8" (60 x 60 cm)
Collection of Keiko Hirohashi, Tokyo

Keiko Hirohashi's package designs originate in the simple rituals of everyday life: folding a kimono into a flat rectangle for storage, or wrapping an *obi* sash around one's waist. Hirohashi based her design for this wrapping paper on the concept used for cutting kimono patterns from a single piece of cloth. She calls it a "multiple-

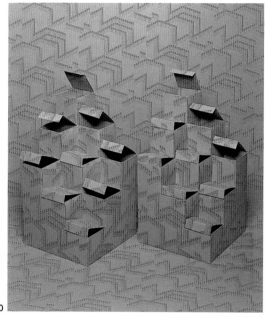

package" because each sheet of paper can be cut and folded to construct containers of many different shapes by following the graphic instruction sheets she provides. Like a kimono, the paper can be folded flat for use again on another occasion. "The functional beauty found in the unique style of clothing and cloth wrappers of Japan," says Hirohashi, "have the wisdom of the Japanese wrapped up inside of them. I have utilized this functional beauty in the creation of [this] wrapping paper."[33]

141 *Katsu Kimura: Package Direction*, 1980
Designed by Katsu Kimura
Published by Rikuyosha, Tokyo
Prize: Tokyo Art Directors Club prize, 1981
Printed book and slipcase, 11 3/4 x 9 3/8 x 1 3/8"
(30.2 x 23.8 x 3.5 cm)
Collection of Katsu Kimura, Tokyo

This volume surveys Katsu Kimura's packaging design from 1967 to 1980 (see no. 137). Kimura based its Pop-inspired slipcase on the Lucky Strike cigarette package created by the America designer Raymond Loewy in 1940, and like a cigarette package, it opens from the top. The title replaces the words "Lucky Strike" in the red target on the side of the box, and "Made in Japan" and the number of plates in the book are shown in place of the seal at the top. Continuing the cigarette theme within, the book's cover is decorated with a lit cigarette in the form of one of Kimura's boxes, a stylized plume of smoke rising from its tip, repeated in the wavy lines on the endpapers.

142 Poster for UCLA Asian Performing Arts Institute, 1981
Designed by Ikko Tanaka
Offset, 40 1/2 x 28 11/16" (103 x 72.8 cm)
Ikko Tanaka Design Studio, Tokyo

Ikko Tanaka was one of twelve Japanese designers who created posters publicizing the United States tour of a Japanese performing troupe for the Asian Performing Arts Institute at the University of California at Los Angeles in 1981. For this poster advertising the classical dance performance, Tanaka arranged geometric shapes to create a Japanese woman's head, using three black squares and two dark-gray triangles to show her elaborate traditional coiffure and two brightly colored triangles and a red square to create her kimono. Tanaka considers this poster his best work as its geometric composition graphically demonstrates what he calls "Japanese sentiment."[34]

143 Shopping bag for Citizen Watch, 1981
Designed by Takenobu Igarashi
Made by Citizen Watch Company, Tokyo
Paper, 16 1/2 x 17 1/4 x 5 11/16" (41.9 x 43.8 x 14.5 cm)
Takeo Company Limited, Tokyo

143

The second of two shopping bags designed by Takenobu Igarashi for Citizen Watch, this, like the first one of 1976, relies for its logo on the axonometric alphabets that brought Igarashi recognition and international clients after he opened his design office in 1970. Here, however, the letter forms are superimposed on a background of gridded planes that join to create dimensional shapes. For Igarashi, the x and y axes of the grid on which the computer plotted the clear, delicate lines, bring mathematical order as well as a sense of boundless space to his work. Igarashi's use of the grid and axonometric shapes in lively juxtaposition provides not only measure and clarity but also variety and ambiguity perfectly suited to the image of a company whose products are directed to younger consumers.

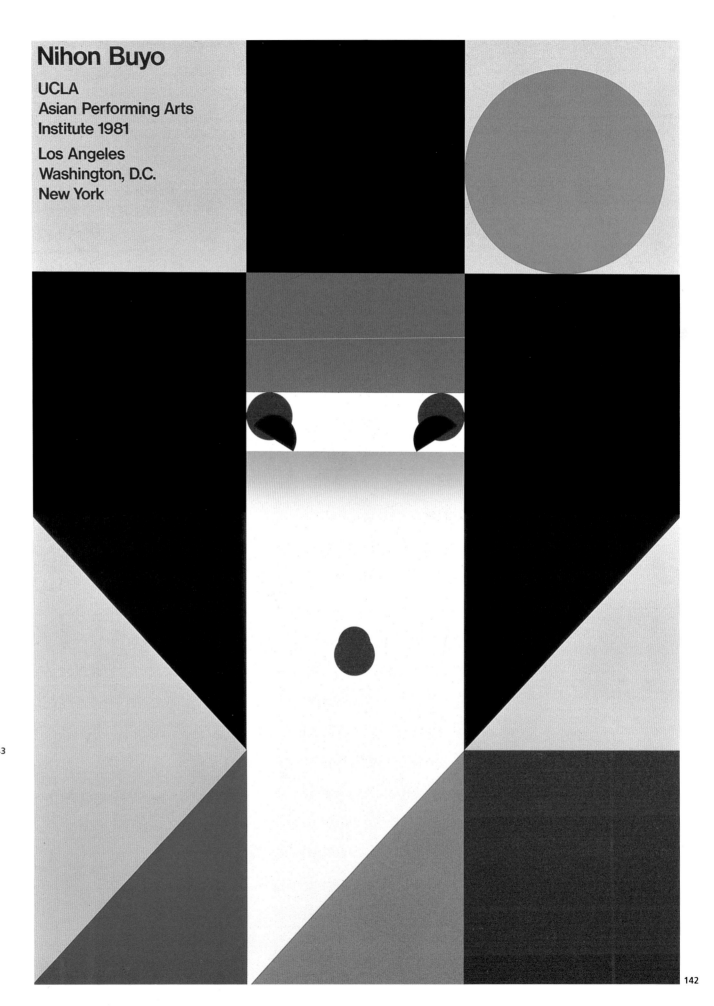

144

144 Ginza Robot cabinet, 1982
Designed by Masanori Umeda
Made by Memphis, Milan
Plastic-laminated wood, 68 7/8 x 21 5/8 x 16 1/2"
(175 x 55 x 42 cm)
Philadelphia Museum of Art. Gift of Collab
in memory of Hava J. Krasniansky Gelblum

Masanori Umeda was one of three Japanese designers
(along with ARATA ISOZAKI and SHIRO KURAMATA)
invited to design furniture for Memphis, a radical design
group based in Milan that gave postmodernism its most
characteristic expression in the early 1980s. Like the most
avant-garde of his Italian colleagues, Umeda came to
value the references to the past and the humor, symbol,
color, ornamented surfaces, and irregular shapes for which
Memphis became known. Umeda's references, however,
were specifically Japanese, drawn from Japan's popular
culture. The form of this storage unit, from Memphis's
second collection in 1982, refers to the mechanical men of
Japan's science fiction and the popular toys they inspired.
With two drawers and open shelves in bright colors, and a
decorous case of striped laminate posed on two green
legs, the Ginza Robot cabinet is named ironically after
Tokyo's most fashionable shopping district.

145 Space Age fabric, 1982
Designed by Masakazu Kobayashi
Made by Sangetsu Company, Kyoto
Screen-printed cotton, width 47 1/4" (120 cm)
Philadelphia Museum of Art. Gift of the designer

This lively printed fabric, one of two designed by
Masakazu Kobayashi for Sangetsu under the title Space
Age, was intended to represent a path through a
harvested rice field in the fall. Although the pattern of
small short strokes divided by thin colored stripes suggests
the traditional textile designs called *komon*, the vibration
of the white strokes with their brightly colored and
feathered edges against the black ground relates to
modernist painting.

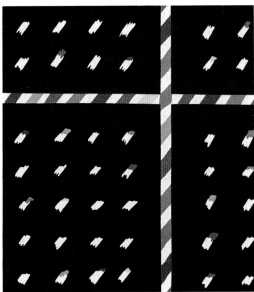

145

146 Wink chair, 1980
Designed by Toshiyuki Kita
Made by Cassina, Meda, Italy
Steel with Dacron-covered polyurethane-foam upholstery,
height 15 3/4" (40 cm), length 78 3/4" (200 cm) (extended)
Philadelphia Museum of Art. Gift of Collab

The adjustable Wink chair brought Toshiyuki Kita his first
international recognition. Made in Italy, where Kita had
worked since 1969, Wink demonstrated the designer's
ability to combine Japanese and Western formal elements

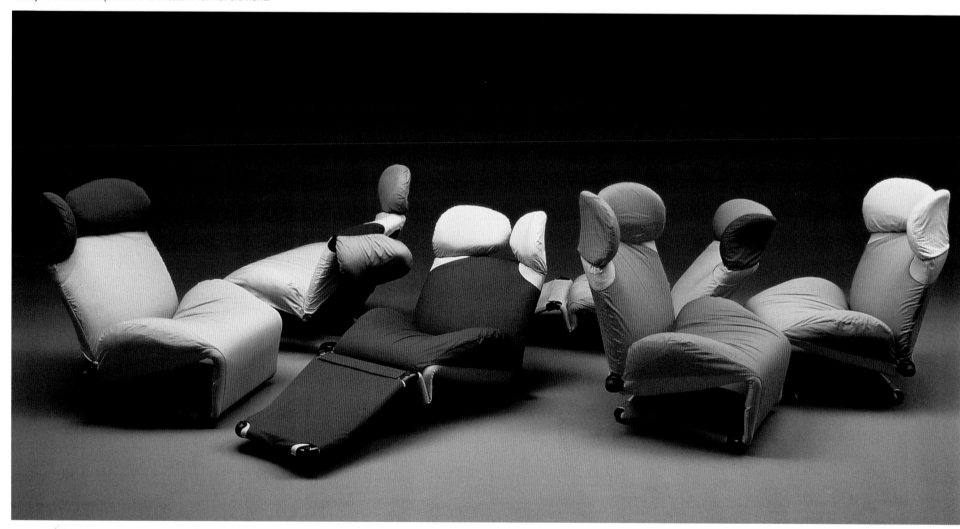

146

in an ingenious, original way. Compared by one critic to a
gigantic but friendly insect, the low chair, enveloped in
cheerful zip-fastened covers, draws on the Japanese
practice of sitting on the floor while offering the Western
comfort and support of a recliner. With but a few
components, which include springs of a type used in the
automotive industry, Wink can be flexed and bent in
several positions: the chair can become a chaise to support
the whole body by extending the base forward; the back
can be adjusted to any angle; and the headrest, which is
divided into two parts, can be bent in a backward or
forward "winking" position. "On Wink," Kita explained,
"one sits sideways, straddles, or sits cross-legged. I have
often imagined a chair which would gently receive our
changing moods at all times."[35]

147 Child's chair, 1980
Designed by Toshimitsu Sasaki
Made by Tendo Mokko, Tendo
Beech, 30 15/16 x 16 3/4 x 19 1/2"
(77 x 42.6 x 49.5 cm) (each)
Tendo Company, Ltd., Tendo

With this child's chair Toshimitsu Sasaki created a sturdy, comfortable, and amusing piece of furniture that appeals to the playful spirit. The chair has a wide U-shape base, which ensures stability and balance, and an extra rung for

147

climbing up into the seat or to use as a footrest. The seat is at adult height, so that children can sit at the dining table without added support, and can continue to use the chair as they grow. The horseshoe-shape back forms a continuous curve with the sides of the chair; when the chair is turned over, these double as rockers, and the chair can then be used for play.

148 Maisema fabric, 1982
Designed by Fujiwo Ishimoto
Made by Marimekko, Helsinki
Screen-printed cotton, width 54" (137.2 cm)
Philadelphia Museum of Art. Gift of Marimekko

Having worked in Finland for Marimekko since 1974, Fujiwo Ishimoto has introduced a Japanese aesthetic to the firm's printed fabrics in his patterns of visible, uneven brushstrokes and subtle, sometimes dissonant, colors. This squared Maisema (Landscape) design is based on the transfer of the artist's wax crayon sketches to the printed cloth; its soft, varied lines visibly communicate the freshness and spontaneity of the act of drawing, which many artists in the 1980s considered the most important part of the design process. Ishimoto conceived his landscape design as a series on the four seasons, abstractly representing seasonal changes in climate, vegetation, and lighting by changes in color ways and density of the lines. Frequently inspired by nature in his choice of patterns, Ishimoto bridges Finnish and Japanese cultures, both of which value the natural and spiritual beauty of the landscape.

148

149 Box, 1981

Designed and made by Shigeru Akizuki
Gouache on paper over cardboard,
28 1/8 x 82 11/16 x 4 3/4" (71.5 x 210 x 12 cm) (extended)
Collection of Shigeru Akizuki, Tokyo

Designed for the "4 Box-ers" exhibition held at the Ginza
Wako Gallery in Tokyo in 1981, this large box, made to
hold the Japanese folk dolls known as *daruma*, was
inspired by the construction of the folding case (*chitsu*)
traditionally used for books. Akizuki was particularly
interested in the "impression of unexpectedness and
strength that develops continuously as the outer cover is
opened to the right and the inner cover to the left."[36] For
him, the power and expression of its dramatic unfolding
were suited to the character of the folk toys inside
(considered good-luck charms) and to the fiercely stylized
images of the Japanese folk masks that decorate the cover
of the box. Akizuki derived his color scheme from the
daruma dolls, with their red bodies and their facial
features and other details in black.

149

150 Screen, 1980

Designed and made by Samiro Yunoki
Resist-dyed silk, 68 7/8 x 29 3/4" (175 x 75.5 cm)
(each panel)
Collection of Hiroko Iwatate, Tokyo

Samiro Yunoki's consistent theme of the unit versus the
whole is well served by the repetitive possibilities of stencil
dyeing. His style is at once bold, comforting, and playful,
as demonstrated by this screen, which, as he explained,
was inspired by a trip that he once took. "I saw a group of
people waiting at a bus stop," he recalled. "I thought to
myself, I could either be one unit among the many in that
group or an individual existing alone and outside of the
group. Then I thought of how all living creatures have
these two aspects to their existence. Although not
original, this observation amused me."[37]

150

151 Kyo lamp, 1983
Designed by Toshiyuki Kita
Made by IDK Design Laboratory, Osaka
Paper and steel, 21 5/8 x 17 11/16 x 7 7/8"
(55 x 45 x 20 cm)
IDK Design Laboratory, Ltd., Osaka

Toshiyuki Kita, like the sculptor Isamu Noguchi before him, reinterpreted the materials and technique of Japanese paper lamps, inspired by the texture and translucency of handmade *mino* paper to which he was introduced by a paper craftsman. "The paper he made was no longer selling because there was no use for it," Kita recalled. "Hand-made paper was nearing the end of its long history. At that time, I promised him that I would find some use for the paper." Beginning the Kyo series in 1970, Kita developed lamps with different shades in simple, geometrical shapes such as this. "In those days," however, "this kind of light did not sell well in Japan. The reason was that the public, interested in fluorescent lamps because they consume less power, showed no great interest in this kind of light through paper. I decided . . . to find a market for the paper . . . [and] the light was put on the market in Italy,"[38] eventually selling internationally not just in Europe but in Japan as well.

152 Bicycle fabric, 1982
Designed by Hiroshi Awatsuji
Made by Fujie Textile Company, Tokyo
Screen-printed cotton, 78 11/16 x 70 7/8" (200 x 180 cm)
Philadelphia Museum of Art. Gift of Fujie Textile
Company, Ltd.

This screen-printed fabric with a large bicycle against a
rice-grain pattern is from the "art screen" series that
Hiroshi Awatsuji designed for Fujie Textile. Each design
for the series was an oversize stylized representation of
everyday things, from this bicycle to a bunch of bananas, a
cup and saucer, and the moon and stars in the night sky.
Intended for use as roller blinds, each fabric came in a
fixed width and length in a single color way, and Fujie
Textile sold them fully assembled with the roller
mechanism attached.

152

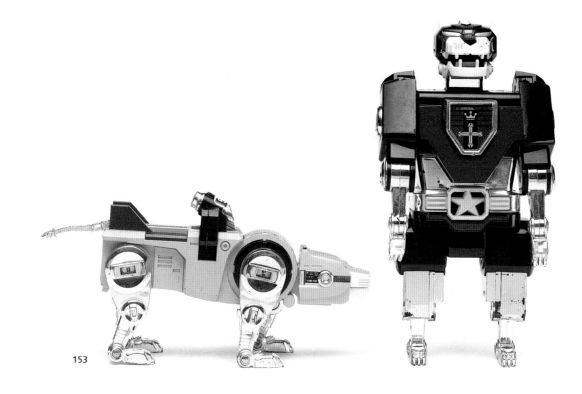

153 King of Beasts Five Lions transformer series, 1981
Designed and made by Popy, Tokyo
Plastic and metal, black lion 7 7/8 x 4 1/8 x 2 3/4"
(20 x 10.5 x 7 cm); yellow lion 2 3/4 x 2 3/8 x 6 11/16"
(7 x 6 x 17 cm)
Collection of William Hiesinger, Philadelphia

Transformers are toys that can be broken down to form
other, smaller toys. These black and yellow figures are
two of the five "lions" (*golion*) that assemble to form
Popy's robot King of Beasts (*hyakuju oh*). The black lion
forms its body and the yellow lion its left leg; red, green,
and blue lions complete the figure. The King of Beasts
Five Lions transformer was developed from a popular
children's television program produced by the Toei motion
picture company, which was aired on Japanese television
in 1981 and 1982. In fifty-two episodes the program
followed the adventures of a group of boys and a princess
who are held captive in a castle and try to regain their
freedom with the help of the five lions. The ability of the
separate figures to combine and transform themselves
into a single robot of great power was a popular feature
of Japanese robot toys during the 1970s and 1980s.

153

154

154 Picnica bicycle, 1981
Designed and made by Bridgestone Cycle Company, Tokyo
Steel, 27 13/16 x 46 7/16 x 16 9/16" (96 x 118 x 42 cm)
Bridgestone Cycle Company, Ltd., Tokyo

With its high-rise handlebars and its seat borrowed from
youth-market specialty bikes, Bridgestone's collapsible
Picnica was advertised as a bicycle built for everyone—a
"second car" for every member of the family to drive and
enjoy. A marvel of compactness, the bicycle collapses by
applying "one-touch" pressure to a mechanism under the
seat. Weighing about twenty-eight pounds, it is light
enough to be carried easily to and from a car, up and
down stairs, and in and out of elevators, making it a
portable, convenient, and economical means of
commuting to work, to the closest train station, or to the
picnics and other leisure activities suggested by its name.

155 Voltes V transformer, 1982
Designed and made by Bandai, Tokyo
Metal and plastic, 12 13/16 x 5 1/2 x 5"
(32.5 x 14 x 12.7 cm)
Zip's Toys To Go, Ardmore, Pennsylvania

Bandai produces many toys known as transformers,
including the series exported under the name Godaikin.
The Voltes V super robot from that series, a samurai
warrior wielding a sword, was based on the television
show "Voltes V" produced by the Toei motion picture
company and broadcast on the Asahi channel in 1977 and
1978. Like the television show, the toy is aimed at older
children who would be captivated by its complexity.
Voltes V can be transformed into a giant space age tank or
broken down into five high-tech, missile-firing transports:
two airplanes, Crewzer and Bomber; the tank Panzer; the
ship Frigate; and the landing vehicle Lander.

155

156

永井一正の世界展　8月1日〈金〉――11月30日〈日〉　池田二十世紀美術館　静岡県伊東市・碧湖畔

156 Bustier, Spring–Summer, 1982
Designed by Issey Miyake
Made by Issey Miyake Inc., Tokyo
Rattan
Miyake Design Studio, Tokyo

For Issey Miyake, "there are no boundaries for what can be fabric, for what clothes can be made from. Anything can be clothing."[39] No designer has ever been bolder or more experimental in his use of materials than Miyake, from high-tech fabrics like polyurethane-coated polyester to traditional materials like *abura-gami*, an oil-soaked handmade paper commonly used for umbrellas and by Miyake, for rainwear. Here, with natural rattan, he created an extraordinary cagelike bustier. Worn with a polyester jacket and skirt, the bustier gives structure and textural contrast to the irregularly pleated fabric. The rattan was woven by Shochikudo Kosuge, a bamboo and rattan craftsman known for his tea-ceremony utensils. "We have clothes made of iron, paper, cane, bamboo, stones," Miyake said in an interview. "There are any number of possibilities once you let your imagination roam. And in that way, the area of my work is expanded. That's enough to achieve, I feel."[40]

157 Exhibition poster, 1980
Designed by Kazumasa Nagai
Silkscreen, 40 1/2 x 28 11/16" (103 x 72.8 cm)
Collection of Kazumasa Nagai, Tokyo

Announcing Kazumasa Nagai's retrospective exhibition at the Ikeda Museum of 20th Century Art, this poster also serves as a summary of his ongoing exploration of abstract, geometric forms and linear patterns that create the impression of spatial depth. The surreal landscapes Nagai designs are based on meticulous hand drawings that parallel the use of lasers and computer graphics to achieve equally precise effects. His highly individual graphic universe of planets and energy forces moving in space represents an alternative, futuristic view of the environment: "We know that the world's resources are limited, that a depleted natural world is crying out in protest. Somewhere in the infinite universe we must find new sources of energy. I have faith in the resourcefulness of the human race and its efforts to accomplish this objective."[41]

158 Spider Web fabric, 1984
Designed by Junichi Arai
Made by Nuno, Tokyo
Silk, width 31" (78.7 cm)
Philadelphia Museum of Art

Often inspired by natural forms, Junichi Arai based his
knitted-silk Spider Web fabric on a drawing by his wife,
the painter Riko Arai, translating the design into this
gossamerlike fabric by using a computerized lace machine.
Instead of making a two-dimensional cloth, he created an
interlacing of silk filaments that evoke the delicate motion
of the insect's web.

158

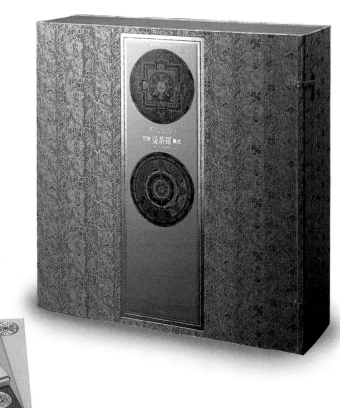

159 *Tibetan Mandalas: The Ngor Collection*, **1983**
Designed by Kohei Sugiura
Published by Kodansha, Tokyo
Printed book, portfolios, and slipcase,
2 3/8 x 21 1/4 x 22 1/16" (6 x 54 x 56 cm)
Freer Gallery of Art and the Arthur M. Sackler Gallery
Library, Washington, D.C.

A mandala (from the Sanskrit meaning "to acquire the essence of things") is a pictorial representation of the Buddhist universe. Mandalas are usually hung in pairs on either sides of the central altar in Buddhist temples during special ceremonies, and they also serve as a focus for meditation. The complexity and symmetry of the geometric disposition of images in a mandala symbolize the rationality and perfection of the cosmos. Kohei Sugiura captured the sense of the infinite and the sublime at the heart of the mandala in his design of *Tibetan Mandalas: The Ngor Collection*, a sumptuous catalogue of a collection of nineteenth-century painted mandalas published in a limited edition. Each of the 139 mandalas is illustrated on a separate leaf, so that it can be contemplated singly or in groups.

160 Poster for the Museum of Modern Art,
Toyama, 1983
Designed by Kazumasa Nagai
Prize: Tokyo Art Directors Club, members prize, 1983
Offset, 40 1/2 x 28 11/16" (103 x 72.8 cm)
Collection of Kazumasa Nagai, Tokyo

Commissioned by the Museum of Modern Art, Toyama, to design a poster advertising its 1983 exhibition of one hundred landscapes of the Toyama area by one hundred different artists, Kazumasa Nagai also created a landscape, incorporating grids of lines to depict the snowcapped mountains of this northwestern prefecture and spiky circular forms to suggest the stars in the sky overhead. The names of the hundred artists in the exhibition rain down between the stars.

161 Jacket, Dress, and Pants, Spring–Summer 1983
Designed by Yohji Yamamoto
Made by Yohji Yamamoto, Inc., Tokyo
Cotton
Kyoto Costume Institute. Gifts of Hiroshi Tanaka
(jacket, pants) and Sumiyo Koyama (dress)

Yohji Yamamoto's "torn" ensemble, consisting of a jacket,
dress, and pants in white cotton, features cut outs in
shapes that look like large flowers or leaves and geometric
motifs. Like REI KAWAKUBO, Yamamoto gave the
spring–summer 1983 Paris show a "black shock." He
explained that he "wanted to make a once in a lifetime
garment that derived a natural, accidental beauty from
being worn and exposed to the sun and rain."[42] The holes
are said to have come from the idea of tearing apart a
garment, unlike the deliberate constructions of Paris
couture. This was the starting point for Yamamoto, who
translated everyday clothes into original creations.

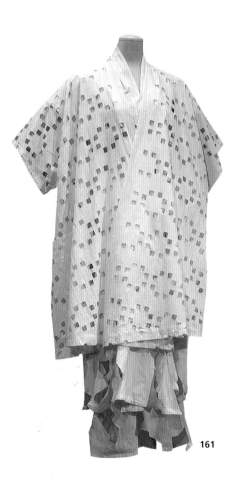

161

162

163

This fabric was developed by Makiko Minagawa as a coat material for the designer ISSEY MIYAKE, with whom she has worked for more than two decades. The linen-and-wool-blend fabric was woven with two different surface textures, one fine and dense, the other loose and larger in scale. The material was then subjected to a felting process of soaping, washing, and tumbling—which affects only scaly animal fibers, in this case the wool—to create a compacted material. The result was that the loose, predominantly wool surface was shrunk and matted, and transformed into a pebbly, cauliflowerlike area, yielding a highly textured, light-color, patterned field offset by a more subtle, dense, dark ground. The large-scale geometric pattern in high relief and the heavily textured surface give the fabric a wild, forceful look.

163 Fire fighter's uniform, 1983
Designed and made by Teikoku Sen-i Company, Tokyo
Aromatic polyamide
Teikoku Sen-i Company, Ltd., Tokyo

In eighteenth- and nineteenth-century Japan, when the most prevalent building material was wood and fire was a constant threat, the fire fighter became a folk symbol of valor. The multilayered cotton or felt uniforms worn by firemen then represented the most advanced technology for protective gear available at the time. Likewise, the latest technology of the 1990s was used in the Teikoku Sen-i company's fire fighter's uniform, with the application of aramide, a new type of nylon fiber that is lightweight, has five times the tensile strength of steel, and is fire-resistant. The company wanted to update the image of the fire fighter by giving the uniform the look and mobility of contemporary sportswear, designing a double-breasted jacket in lime yellow with four off-center snaps and silver trim to match the high boots. But the heroic image of the past is retained in such elements of the uniform as the visored helmet with a neck guard that recalls the hooded felt garments once worn by the feudal lords, or daimyo, who served as the titular heads of the fire-fighting brigades in their domains.

164 Sweater, Fall–Winter 1982–83
Designed by Rei Kawakubo
Made by Comme des Garçons, Tokyo
Wool
Comme des Garçons, Tokyo

Utterly defying convention, Rei Kawakubo attacked the very fabric of fashion design with this apparently shapeless, asymmetrical, oversize sweater full of holes. Hand-knit, it represented to Kawakubo the values of expressiveness, authenticity, and intimacy found in traditional artistic handicraft. "The machines that make fabric are more and more making uniform, flawless textures," she said. "I like it when something is off—not perfect. Handweaving is the best way to achieve this. Since this isn't always possible, we loosen a screw of the machines here and there so they can't do exactly what they're supposed to do."[43] With its holes irregular in their size and placement, this sweater suggested to some the delicate open structure and irregularities of handmade lace; to others, it looked as if it had been eaten by moths.

164

165 Posters for Garo, 1984
Designed by Makoto Saito
Offset, 40 1/2 x 57 5/16" (103 x 145.6 cm) (each)
Collection of Makoto Saito, Tokyo

These two posters for the Garo clothing boutiques were conceived by Makoto Saito as a pair, both featuring the cast shadow of a man wearing a bright-green necktie. Each poster bears a large image of an embroidered letter—representing the syllable *ah* in one, *un* in the other—as well as the Garo logo. "*Ah* and *un* are the first and last sounds of the Japanese syllabary," Saito explained. "The *ah* is the sound of being surprised, and the *un*, the sound of agreement. *Ah*, *un* is the message this company wants to send to those who can understand what they are trying to communicate." For Saito, as well as for his client, image, design, display, and communication are more important in advertising than the specifics of any one product: "Fashion changes as time passes," he said, "which means that even if you produce a great design it signifies nothing if you can't communicate it to the customer."[44]

166 Ariake floor lamp, 1983
Designed by Motomi Kawakami
Made by Bushy Company, Tokyo
Paper, acrylic, and lacquered aluminum,
height 74" (188 cm), diameter 11 13/16" (30 cm)
Bushy Company, Tokyo

Intending to create a contemporary Japanese furnishings style, Motomi Kawakami made this Ariake (Daybreak) floor lamp from a high-tech sandwich of traditional and industrial translucent materials. The shade, which adopts the elegant peaked profile of typical paper lamps, is a laminate of Japanese rice paper (*washi*) and acrylic, a stronger material and one better able than paper to hold its shape. The slender aluminum stand is coated with natural *urushi* lacquer, creating a glossy, rich surface that contrasts with the texture of the shade.

166

167 Blouse and skirt, Spring–Summer 1983
Designed by Rei Kawakubo
Made by Comme des Garçons, Tokyo
Cotton
Comme des Garçons, Tokyo

By sewing folded strips of fraying, woven cotton irregularly onto cotton jersey, Rei Kawakubo created beribboned effects that are both texture and decoration. During the 1980s, materials were often the starting points for Kawakubo's fashion concepts, and she transformed ordinary ones such as cotton by exposing the stitching and by pleating, folding, or wrapping. Here, the light three-dimensional cotton cloth, specifically intended for a summer collection, is shaped into the loosely constructed forms that became associated with Kawakubo's work.

168

168 Dress, Fall–Winter 1984–85
Designed by Rei Kawakubo
Made by Comme des Garçons, Tokyo
Wool
Comme des Garçons, Tokyo

Cinched at the waist by what appears to be a broad *obi*-like band, this dress is actually a patchwork of various knitted pieces, *obi* included, that are twisted and stitched together to form the garment. Rei Kawakubo's fascination with seemingly "found" materials challenged the formality of workmanship and fabric that then defined high fashion, and offers here both a ragpicker's send-up of the classic Western tunic and a new aesthetic. With its loose, complex folds, wide waistband, and somber, monochrome palette, the dress also reflects traditional Japanese ideas of beauty, which prize the rough, irregular, and asymmetrical for their lack of artifice.

167

169 X & I scissors, 1985
Designed by Kazuo Kawasaki
Made by Takefu Knife Village, Takefu
Prize: Japan Design Committee Design Forum,
silver prize, 1983
Laminated stainless steel, length 5 1/2" (14 cm)
Collection of Kazuo Kawasaki, Fukui

Kazuo Kawasaki reinvented the form scissors could take in
this pair that he designed for the Takefu Knife Village.
The scissors, which are named for their shape when open
and closed, open under the pressure of a flat spring
between the handles.

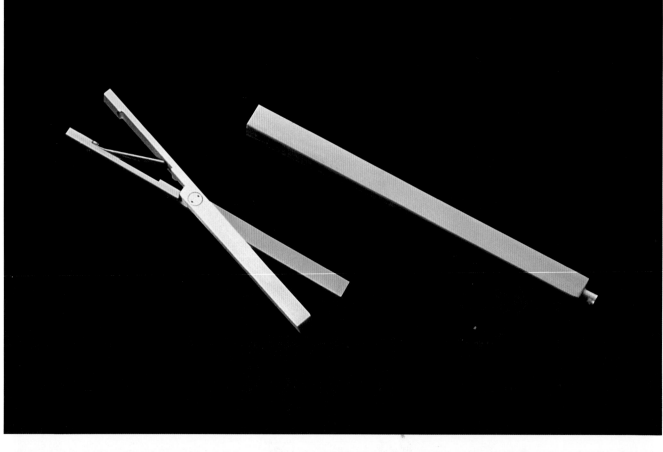

169

170 Artus knives, 1983
Designed by Kazuo Kawasaki
Made by Takefu Knife Village, Takefu
Laminated stainless steel,
length of longer knife 11 13/16" (30 cm)
Collection of Kazuo Kawasaki, Fukui

While industrial designers have designed knives for
manufacturers using modern industrial techniques, Kazuo
Kawasaki created modern designs for a traditional metal-
forging center, the Takefu Knife Village. His Artus (Limb)
knives have cut-out handles that are convenient to grasp;
as the designer suggested in the Latin name he gave to
the series, these were carefully conceived extensions of the
hand for use in the preparation of food.

170

171 Poster for Les Ateliers, Paris, 1984
Designed by Kohei Sugiura
Gravure, 40 1/2 x 28 11/16" (103 x 72.8 cm)
Collection of Kohei Sugiura, Tokyo

Kohei Sugiura created this image of a golden Buddha for "Tradition et Nouvelles Techniques" (Tradition and New Techniques), an exhibition of the work of twelve Japanese graphic artists held in Paris in 1984. Sugiura's exploration of sophisticated technology for this poster reflects the pursuit of new techniques by artists in the exhibition. Printing in gravure on aluminum foil paper with opaque white and laminated black and orange inks, Sugiura was able to give subtle gradations to the gold colors. He represented tradition by his choice of subject, *Evocation of the Mandala on the Summit of Mount Meru*. The mountain at the center of the Buddhist cosmos, Mount Meru, is represented here by the square platform on which a tortoise supports a lotus and *stupa* (reliquary). Superimposed on and encompassing these is the seated Buddha Dainichi (Great Illuminator). The suprahuman character of the idealized figure is manifested in the mathematical perfection of the proportions of the image and in the gold color, which symbolize the Buddha as the radiant source of all knowledge and being. The printing process imbues the complex imagery with a rich texture and a devotional quality. Around the outer border are the Japanese ideographs for the title of the exhibition, alternating with depictions of mandalas, the Buddhist cosmic diagrams of the universe.

172 Box, 1984
Designed by Kozo Okada
Made by Kasukabe Traditional Packaging Industries
Union, Kasukabe
Paulownia wood, 4 1/2 x 35 1/16 x 5 1/16"(11.4 x 89 x 13 cm)
Philadelphia Museum of Art. Gift of Yoshiko Ebihara,
Gallery 91

Returning to traditional kinds of Japanese wood
packaging, Kozo Okada created a series of gift boxes,
including this striped box, that were shown in "The First
Wood Package Exhibition" held in 1984 at Tokyo's Axis
Gallery. By slicing and arranging slender pieces of
paulownia wood on a curved rolled surface, Okada was
able to compose his material in a new, expressive way,
which conveyed his "theme of having one object with two
hidden subjects. One of the subjects is rediscovering the
texture of the material. . . . The other subject is . . .
entertainment. . . . Instead of designing wooden packages
that viewers can 'understand,' the theme of my package
design is for the viewers to feel something."[45]

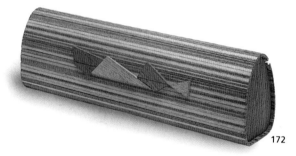

172

173 Yoshiko cabinet, 1984
Designed by Setsuo Kitaoka
Made by Build Company, Tokyo
Plastic laminate, 55 1/2 x 9 1/2 x 9 1/2"
(140 x 24 x 24 cm) (each panel)
Private collection

As he did in his Akira lamp (no. 187), Setsuo Kitaoka
brought a traditional concept, the folding screen, into
contemporary furniture design in his prototype Yoshiko
cabinet. Its five vertical support panels can be arranged in
varying patterns, stretched out to their full width or
folded into a boxlike configuration. The plastic laminate
surface, striped in ten subtle colors, emphasizes the
vertical lines. Kitaoka intended this piece to have multiple
functions: as a screenlike partition, storage space, and
decorative element.

173

In the land of high-tech industrial robots that can assemble complete automobiles, Eiji Hiyama's corrugated-cardboard toy robots are disarmingly unsophisticated. Hiyama's goal in making his Kiddy Scape robots was to use recycled materials and create appealing toys that were easy to assemble yet would withstand the rough-and-tumble world of children's play. Made of six pieces of gray or brown cardboard, his robots are part of an expanding series that includes trains, elephants, and hobby horses.

174

175 Fabric, 1985
Designed and made by Makiko Minagawa
Silk and cotton, with bronze metallic thread,
width 43 5/16" (110 cm)
Miyake Design Studio, Tokyo

The predominant pattern on this fabric, which Makiko Minagawa developed for ISSEY MIYAKE, was created not by the traditional *kasuri*, or *ikat*, dyeing technique but by a chemical process that resembles it. *Kasuri* is a process in which yarns are tied off to resist dyes and the resulting patterned yarns are then woven into cloth. Minagawa created her version by treating the warp yarns chemically, and then weaving them into a fabric. Afterward, the cloth was put through a discharge process with a roller with *kasuri*-like patterns, resulting in a cloth that looks like it was produced with the traditional technique. This method allows for the production of *kasuri*-like fabrics in a considerably greater quantity and at a lower cost than if they were done by the ancient method, making them available for use in ready-to-wear clothing. In developing specific materials for Miyake, Minagawa has sought out the Japanese weaving centers that produced traditional fabrics. This new *kasuri* demonstrates her adaptation of a technique that was originally used in the weaving of country kimonos, as well as Miyake and Minagawa's insistence upon making traditional methods viable for modern manufacturers.

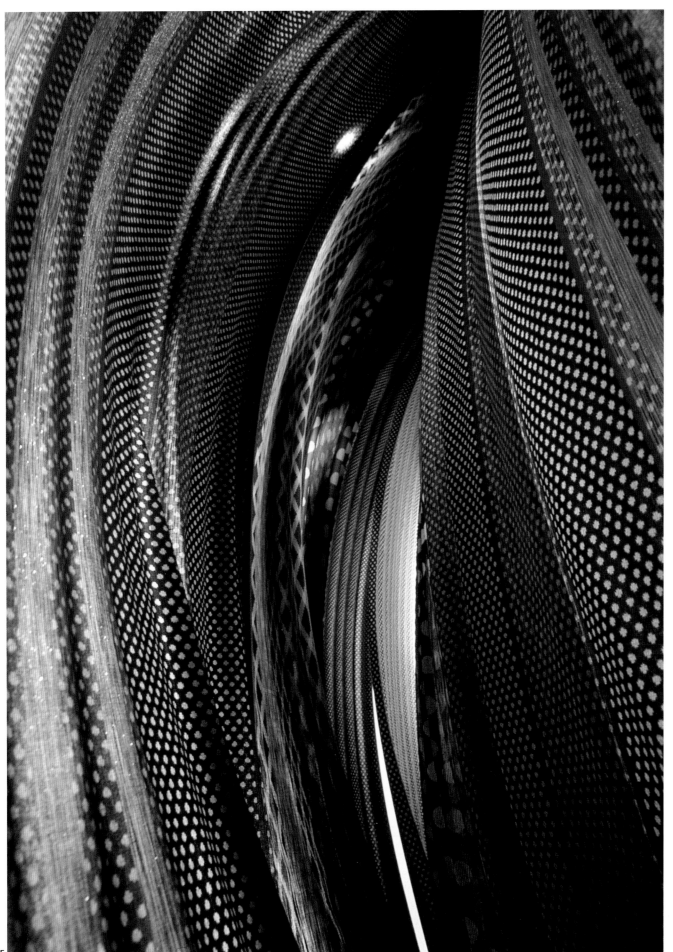

175

176 Pocket television, 1984
Designed and made by Seiko Corporation, Tokyo
ABS plastic, 5 15/16 x 3 1/4 x 1 1/8" (15 x 8.2 x 2.8 cm)
Seiko Corporation, Tokyo

The world's first miniature color television, this pocket
model was based on Seiko's pioneering digital watches of
the 1970s, which, like this set, used liquid-crystal-display
(LCD) technology for their imaging capabilities. By
replacing the conventional television picture tube with
crystals arranged on a tiny integrated circuit, the housing
of the set could be downsized, that is, made thinner and
more compact. Having a two-inch screen and weighing
only about eight pounds, this battery-operated television
could be hand-held for viewing nearly anywhere; because
its tilt-up screen effectively absorbed daylight, it was
targeted by Seiko especially for
use outdoors.

176

177 Team-Demi desk set, 1984–85
Designed and made by Plus Corporation, Tokyo
ABS plastic, 1 1/2 x 4 1/2 x 3 1/4" (3.8 x 11.4 x 8.3 cm)
Private collection

The Plus Corporation design group created this palm-top
portable desk set for use at home, office, school, or on
trips. The compact case incorporates seven items,
including scissors, ruler, and stapler, that fit in openings on
either side of its interior. The case, which has an efficient,
uncluttered look, was manufactured in six colors. The
compactness of the ensemble and the sturdiness of the
miniature utensils made the Team-Demi set an instant hit.

177

178 Metal Wave desk accessories, 1985
Designed by Masayuki Kurokawa
Made by Daichi Company, Tokyo
Stainless steel, tape dispenser 2 1/2 x 8 x 4 1/2"
(6 x 20 x 11 cm)
Collection of Masayuki Kurokawa, Tokyo

"Giving matter form is to breathe life into it," says
Masayuki Kurokawa. "Rubber, which should be soft,
freezes and rigidifies. Metal, which should be hard, twists
and melts."[46] As in his earlier experimentation with rubber
(no. 138), Kurokawa used geometric forms with rounded
edges for his Metal Wave desk accessories, cast in
burnished stainless steel. With the simplest vocabulary of
forms, Kurokawa exploited the reflective quality of the
metal, giving expressive and decorative content to a group
of objects designed for ordinary uses.

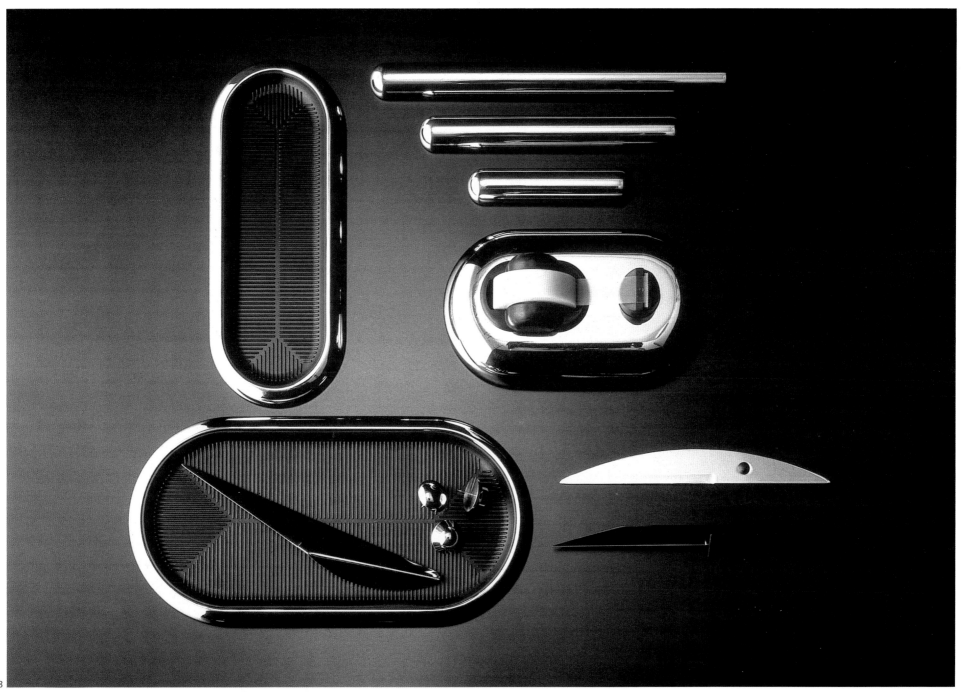

178

179 Clipeus speakers, 1984
Designed by Kazuo Kawasaki
Made by Maruichi Selling Company, Fukui
Lacquered plywood and dyed polyurethane foam,
53 1/8 x 17 11/16 x 16 15/16" (135 x 45 x 43 cm)
Collection of Kazuo Kawasaki, Fukui

Having gained experience by working on Toshiba's Aurex
speaker system during the 1970s, Kazuo Kawasaki
designed his self-polarizing electrostatic Clipeus speaker
system in 1986. The electrostatic technology allows
speakers to be reduced to minimal dimensions; each unit
consists only of a thin panel set in a solid, stabilizing
wedge-shape base. Intended for audiophiles at the high
end of the market, the Clipeus system has a dramatic look
that is as important as its sound. The red polyurethane
circles set off-center on the front of the speakers inspired
its name Clipeus, which comes from the Latin word used
for a shield or solar disc.

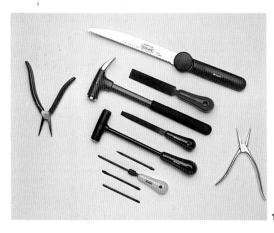

180

180 Tools, 1984
Designed by Michio Hanyu
Made by Takagi Company, Sanjo
Prize: G-Mark, 1984
Plastic and metal, length of saw 13 3/4" (35 cm)
Monopro Designers, Kawasaki

Michio Hanyu designed this set of tools in response to the
growing trend toward do-it-yourself home repairs.
Realizing that almost the only tools available were those
made for professionals, the majority of whom were men,
he broke the stereotype with this series of small tools for
women, adding attractive bright blue, red, and yellow
plastic handles, which were shaped to fit comfortably into
the hand.

179

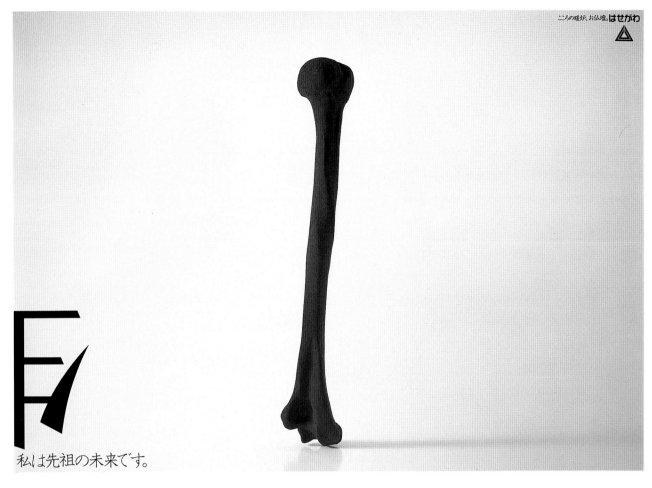

私は先祖の未来です。

私は未来の先祖です。

181 Posters for Hasegawa, 1985
Designed by Makoto Saito
Prizes: Tokyo Art Directors Club, members prize, 1985;
femur: International Poster Triennial, Toyama,
gold prize, 1985
Offset, 40 1/2 x 57 5/16" (103 x 145.6 cm) (each)
Collection of Makoto Saito, Tokyo

With solitary and striking images of a human pelvic bone and femur, Makoto Saito's posters advertise domestic Buddhist altars made by the Hasegawa company. Conceived as a pair, the posters share the *kanji* character *sei* (life), which is only half visible in each, alluding to the cycle of death and rebirth central to Buddhist teaching. The posters are inscribed with different texts: "I am the ancestor of the future" accompanies the pelvic bone, "I am the future of the ancestor," the femur. The directness of Saito's message and image is emphasized by the bright blue color of each bone and the abstract, empty space that surrounds it. Referring openly to death in a commercial advertisement, even one with a Buddhist subject, was risky, and it was for such daring that Saito gained his reputation in the 1980s.

182 Kazenoko stool, 1984

Designed by Sinya Okayama
Made by Interior Object, Tokyo
Painted steel with plastic upholstery, 11 13/16 x 18 x
16 1/2" (30 x 46 x 42 cm)
Philadelphia Museum of Art. Gift of Collab

Giving a specifically Japanese identity to postmodernism,
Sinya Okayama's Kazenoko (Child of the Wind) stool was
the first in his series of original furniture designs that took
their three-dimensional form from *kanji*, the Japanese
written system of pictographs. Like other postmodern
designers, Okayama was interested in creating a dialogue
between the object and the user through the introduction
of familiar symbols, here directed specifically to the
Japanese audience, which can "read" the outline of the
object as a graphic code and understand its literal
meaning, "wind" (*kaze*).

182

183 Table lamp, 1986

Designed by Hiroshi Morishima
Made by Time-Space-Art, Tokyo
Paper and steel, height 30 5/16" (77 cm),
diameter 28 3/4" (73 cm)
Time-Space-Art Inc., Tokyo

Hiroshi Morishima, a papermaker who uses his own
handmade paper (*wagami*) for the shades of his lamps,
joined three layers of mulberry bark (*kozo*) to form the
single sheet of paper that was rolled into a simple cone to
form the shade of this model. The air pockets trapped
between the layers of bark create an irregular pattern on
the textured paper as light diffuses through it.

183

184 Poster for Morisawa & Company, 1985

184 Poster for Morisawa & Company, 1985
Designed by Mitsuo Katsui
Offset, 40 1/2 x 28 11/16" (103 x 72.8 cm)
Collection of Mitsuo Katsui, Tokyo

A leader among the designers who work with computer graphics in Japan, Mitsuo Katsui engaged the broad range of his medium by manipulating and integrating images and typography in this collage poster for Morisawa, a manufacturer of phototypesetting equipment. The ground of the poster suggests a visible bit-map composed of pixels, the smallest possible dot the computer can display and the fundamental building block of computer graphics. Katsui enlarged the normal dot pattern to create a busy ornamental surface into which he positioned an oversize zero. Using an image of reflections on a soap bubble scanned into the computer, he manipulated it to create the flames that leap around and through the zero. With the digital red-green-blue component color system, Katsui subtly controlled not only the levels of color but also his ability to mix them in any combination. They vibrate against the dark ground and make the shapes jump out, in eloquent testimony to the aesthetic and expressive power of electronic technology.

185 Floor lamp, 1985
Designed by Shigeru Uchida
Made by Yamagiwa Corporation, Tokyo
Melamine-coated steel and aluminum,
57 1/16 x 23 5/8 x 20 1/2"(145 x 60 x 52 cm)
Yamagiwa Corporation, Tokyo

Basing the initial sketch for this halogen lamp on "the
image of Japanese men holding up their sun visors to see
who is good looking," Shigeru Uchida adapted a
professional photographer's adjustable light stand with a
perforated aluminum shade to create "dancing grid
patterns of shadow and light"[47] on ceiling or wall. The
lamp uses low-voltage mini-halogen bulbs, preferred by
lighting designers in the 1980s because they required less
electricity, generated less heat, and proved more cost
effective than incandescent bulbs; furthermore they
produced whiter, brighter light from smaller sources than
fluorescent bulbs. Halogen lamps were originally used by
professional photographers and cinematographers, but
this lamp by Uchida converted their technical means and
one of their forms to domestic, and poetic, ends.

186 How High the Moon armchair, 1986
Designed by Shiro Kuramata
Made by Vitra, Basel
Nickel-plated steel, 28 1/4 x 37 3/8 x 32" (71.8 x 94.9 x
81.3 cm)
Philadelphia Museum of Art. Gift of Collab in honor of
Gerard J. Jarosinski, Jr.

Shiro Kuramata used common industrial materials in his
furniture and interior designs, in the process transforming
them into rich and noble components. His How High the
Moon armchair is made of expanded steel mesh, a
structural material that Kuramata plated with nickel
as here (or with copper) to make it sparkle and
which he joined in welded seams of such precision that
the airy mesh is reduced to a series of crossing
points, an apparently weightless, boundless volume.
Dematerialization and the ironies of function were
constant themes for Kuramata, and frequently he used
steel mesh to pursue them, both in furniture and in his
designs of interiors.

185

187 Akira floor lamp, 1984
Designed by Setsuo Kitaoka
Made by Build Company, Tokyo
Enameled stainless steel and plastic laminate,
height 78 3/4" (200 cm)
Private collection

Setsuo Kitaoka's Akira halogen floor lamp is a
sophisticated reinterpretation of the traditional children's
pastime of paper folding (*origami*). When this shade is
closed, it forms a cube made of pyramidal sections
enameled in six bright colors alternating with white. The
shade can be "unfolded" into any number of new
patterns, allowing more or less light to come through.
With this design Kitaoka hoped to "spark the playful
spirit, to have the lamp enjoyed as an 'object,' and enrich
one's living space."[48]

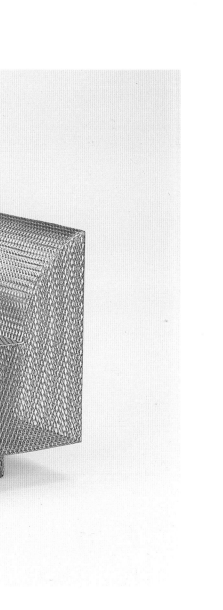

187

186

188 Ojo de Dios hanging, 1985
Designed and made by Junichi Arai
Dyed aluminum- and polyamide-resin-coated polyester,
37 3/8 x 106" (95 x 269.5 cm)
Cooper-Hewitt National Museum of Design, Smithsonian
Institution, New York. Gift of Dianne Benson

The techniques Junichi Arai used in this fabric were
inspired by his interaction with MAKIKO MINAGAWA,
director of textiles for ISSEY MIYAKE, in the early 1980s.
Its sheer, flat, polyester fiber was vacuum metallicized
with aluminum. The metallic yarn was then coated with
ultra-thin polyamide resin, which made it possible to
change the color from silver to black by dyeing it. A
length of this filmlike fabric was accordion-folded into a
small rectangle, folded again into a square, and then
permanently pleated diagonally. When this unique
example was opened, reflected light revealed a
continuous kinetic pattern on the finely pleated surface,
which reminded Arai of the Mexican *ojo de dios* (eye of
God) votive offerings.

189

189 Shopping bag for Yamaya, 1985
Designed by Shozo Kakutani
Printed paper, 11 13/16 x 13 3/4 x 5 9/16"
(30 x 35 x 14.2 cm)
Collection of Hideo Yamamoto, Hakata

When Shozo Kakutani was asked to design a shopping
bag for Yamaya, he sought an appropriate image to
convey the company's role as a purveyor of seafood.
Kakutani used a single, thick brushstroke to represent a
breaking wave in shades of blue to suggest the color and
depth of the sea, superimposing on it a small, silver fish.
Yamaya's logo of three red squares completes the
composition at the lower-right corner like an artist's seal.
Kakutani's design has proved so effective and popular that
after nearly ten years Yamaya still uses this version of its
shopping bag.

188

190 Coquille screen-bench, 1985
Designed by Kazuko Fujie
Made by Fujie Atelier, Tokyo
Plywood with Japanese linden veneer,
71 x 157 x 59" (180 x 400 x 150 cm)
Fujie Atelier, Tokyo

Originally designed for a theater lobby, this screen-bench, like many of Kazuko Fujie's designs, is based on a natural form, here the *coquille*, or shell, for which it is named. Using the edges of plywood panels rather than the flat planes as her surface, Fujie achieved a boldly irregular textured effect unlike that of any other furniture. Created and conceived as unique objects, most of Fujie's work, like this, is site-specific, and the designer lights her pieces in place to dramatize their installation.

191

191 Komako *shochu* bottle, 1985
Designed by Shuya Kaneko and GK
Made by Kikkoman Corporation, Noda
Frosted glass, height 10" (25.4 cm), width 3" (7.6 cm)
Private collection

Shuya Kaneko has called the Coca Cola bottle the quintessential example of American culture in packaging, with its curvaceous forms that remind him of Marilyn Monroe. In the design for his bottle for *shochu*, a vodka-like beverage, Kaneko alluded to other lovely profiles, those of the limpid beauties in eighteenth-century Japanese woodblock prints. The name Komako itself recalls both the sad heroine and the frosty setting of the twentieth-century novel *Snow Country*, by Yasunari Kawabata.

192 Bubble Boy speakers, 1986
Designed by Tomoyuki Sugiyama
Made by Inax Corporation, Tokoname
Ceramic, 11 x 8 5/8 x 7 7/8" (28 x 22 x 20 cm)
Inax Corporation, Tokoname

Although the name and the egg shape of Tomoyuki Sugiyama's Bubble Boy speakers may be whimsical, the system is the result of serious and exacting research and development. The concept occurred to the designer after he had spent time working with Inax, a manufacturer of sanitary fittings, and realized that the same material used for wash basins and toilets could be adapted to speakers. An architectural acoustician, Sugiyama understood the special qualities ceramic would bring to a speaker: it would be strong, dense, nonmagnetic, and nonconductive. The Bubble Boy speaker is not only as powerful as a speaker enclosed in other materials three times its size but it also emits a sound that is uniform throughout the room. Intending the speaker, cast in white, blue, black, or yellow ceramic, to be decorative, Sugiyama stored all of the wiring in the bottom so that the one-piece ceramic shell would remain uninterrupted.

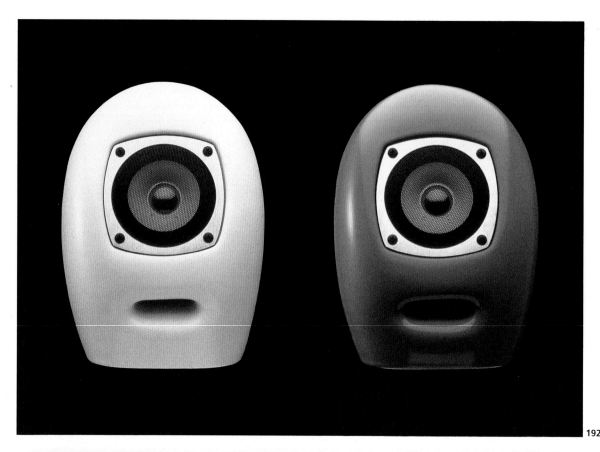

192

193 Rice cooker, 1985
Designed by Masayuki Kurokawa
Made by Paloma Company, Nagoya
Enameled steel and aluminum, height 10" (25.5 cm),
diameter 12 3/4" (32.5 cm)
Collection of Masayuki Kurokawa, Tokyo

Masayuki Kurokawa took on one of the early icons of postwar Japanese industrial design (no. 7) with his elegant update of the rice cooker. Kurokawa's compact version features a lid that is hinged to the cooker, eliminating the need to find a place to put it while serving. The ascetic purity of the clean lines and white enamel finish, relieved by the bright yellow plastic release button on the handle, give it a sleek sophisticated appearance suited to the most contemporary of Japanese interiors.

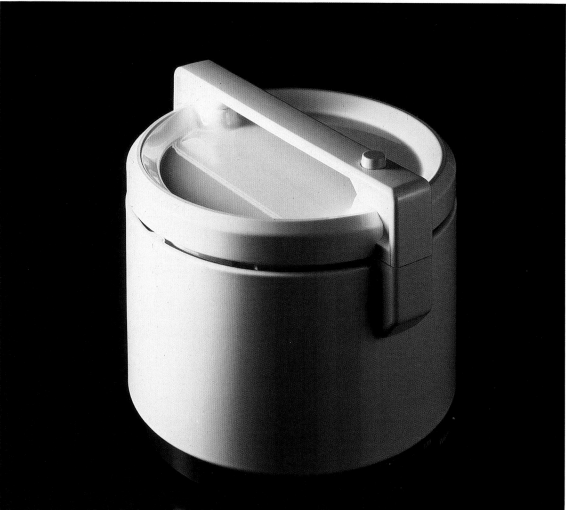

193

194 Thindy speakers, 1984
Designed by Masayuki Kurokawa
Made by Seidenko Company, Tokyo
Steel and plastic, 23 5/8 x 23 5/8 x 5 5/16"
(60 x 60 x 13.5 cm)
Seidenko Company, Ltd., Tokyo

Taking advantage of the new possibilities opened up by
technology through the miniaturization of speaker
components, Masayuki Kurokawa devised his Thindy
speakers, which can be hung on a wall or placed on the
floor in any part of the room, unencumbered by bulky
casing or cords. Kurokawa attempted to create the visual
equivalent of free-floating sound waves with his design
for these gracefully curved speakers, giving them an airy
lightness that liberates them from the solidity of the
traditional earthbound versions.

195 Wind MIDI controller, 1986
Designed and made by Yamaha Corporation,
Hamamatsu
Prize: G-Mark, 1987
ABS plastic, acrylic, and aluminum,
21 1/2 x 2 3/4 x 2" (54.6 x 7 x 5.1 cm)
The Museum of Modern Art, New York. Gift of the
manufacturer

Evolved from an acoustic wind instrument, Yamaha
Corporation's Wind MIDI controller is an electronic
saxophone that uses Musical Instrument Digital Interface
(MIDI), a standard digital language, to convert breath, lip
pressure, and fingering into synthesized musical tones that
cover the entire range of the saxophone family. Governed
by the emission of breath, MIDI translates every aspect of
a note, from pitch and volume to the speed of its vibrato,
into a digital code. Keys, a wind sensor, a plastic reed, and
small switches located on the underside of the instrument
work in combination to control and modulate these notes.
The ability of the Wind MIDI to record and rework musical
phrases further expands and enhances its capabilities.
Although this instrument was in production for only a
short time, critics hailed its successful blending of modern
ergonomic design with the traditional and recognizable
features of the saxophone.

195

196 Copy-Jack copying machine, 1986
Designed and made by Plus Corporation, Tokyo
ABS plastic, 6 3/4 x 2 3/4 x 2" (17.2 x 7.1 x 4.7 cm)
Collection of Reiko Sekiya, Fukushima

Plus Corporation's battery-powered, pocket-size copying
machine was one of its most successful products, selling
50,000 examples in two months in 1986, when they were
first put on the market. The thermal-process copier is
about the same size as the Olympus X-A camera (no. 125),
and has a similar no-nonsense styling; it is equally
portable, and with its thirty-foot paper roll, useful for
making good-quality copies at school, library, or tax office.

196

**197 Coat, sweater, shirt, and pants ensemble,
Fall–Winter 1984–85**
Designed by Kenzo
Made by Kenzo, Paris
Wool
Kenzo, Paris

Among Kenzo's fashion innovations was his way of layering garments to create an outfit, mixing colors and textures, plaids and stripes in unique combinations that had a loose, bulky silhouette mimicking regional dress. The interchangeable layered tops and accessories worn over pants, or sometimes with boots and thick woolen socks or leg warmers, were popular with women who wished to rearrange or recombine the various components for different purposes. "I always begin by conceiving the pieces independently, autonomously in a way," Kenzo explained. "Only afterward do I think about relating them to each other. This manner of composing a garment after the event might seem curious, but it suits my sensibility. It excites my imagination and I take pleasure in thinking that my silhouettes acquire in that way a particular personality."[49]

198 Communicator communication aid, 1985
Designed and made by Canon, Tokyo
ABS plastic, 1 3/16 x 5 1/8 x 3 3/8" (3 x 13.1 x 8.5 cm)
Canon Inc., Tokyo

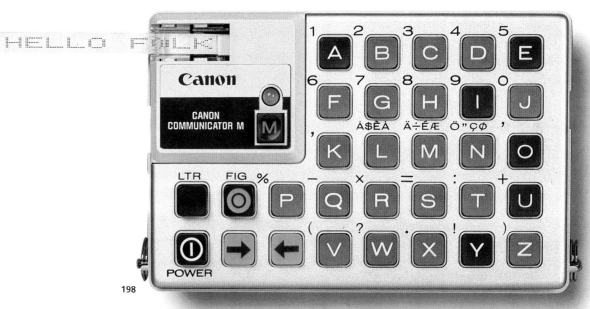

198

Canon's commitment to enhancing all modes of communication and information transferral led in 1977 to the development of the Communicator, a communication aid for people with speech, hearing, or motor impairments, designed with the assistance of two Dutch psychologists. Compact, lightweight, and watertight, this durable product can be strapped to the wrist or to a wheelchair, making it convenient to use in almost any situation. This model, introduced in 1985, is a revised version with enlarged memory capacities and added functions that make it compatible with electronic typewriters and computers.

197

199 Tae writing cabinet, 1986
Designed by Motomi Kawakami
Made by Yamaha Corporation, Hamamatsu
Dyed and polyurethane-lacquered maple,
47 1/4 x 47 1/4 x 17 1/8" (120 x 120 x 43.5 cm)
Collection of Motomi Kawakami, Tokyo

Borrowing the traditional Western furniture form of a
drop-front writing desk, Motomi Kawakami constructed
and finished this sleek Tae (Exquisite) writing cabinet with
the meticulous precision and rich materials of a
preindustrial Japanese or Western craftsman. But in the
creation of this luxury object, he used synthetic materials
along with natural ones and machine production along
with hand labor. Known for his broad knowledge of
materials, Kawakami emphasized the texture and pattern
of the cabinet's spotted bird's-eye maple by dyeing the
wood a deep-gray color and then applying layers of clear
polyurethane coating over it, which contrast with the
natural maple used for the Japanese-style grid door and
for the stringing that decorates the front.

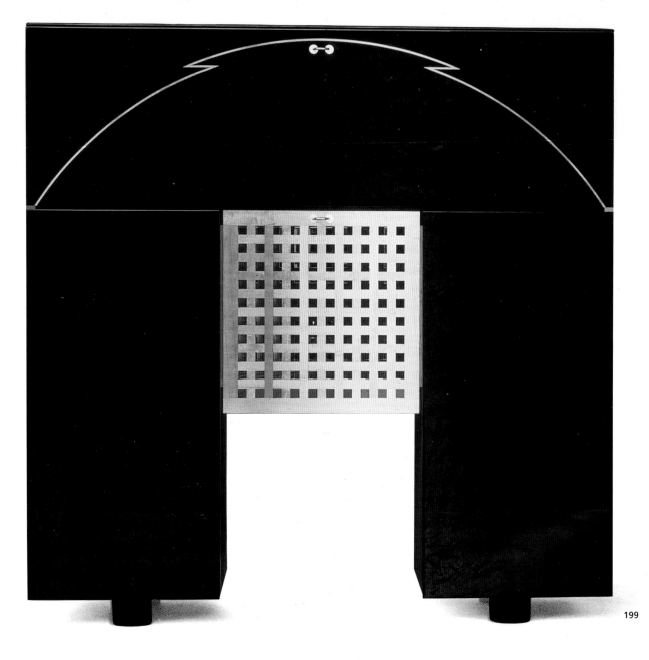

199

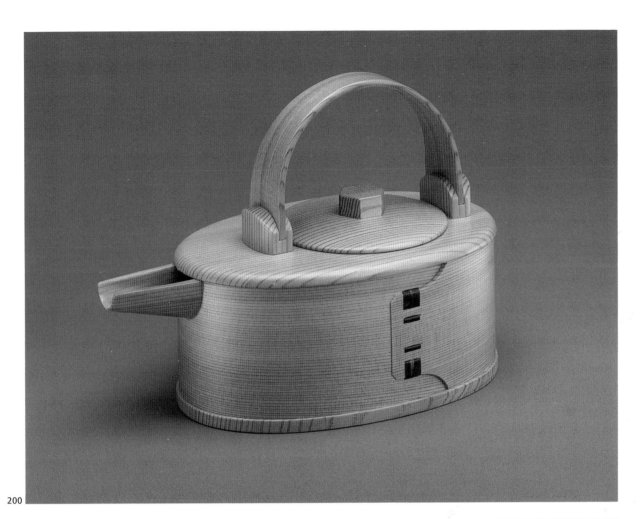

200 Sake pourer, 1987
Designed by Akiraka Takagi
Made by Odate Craft Company, Odate
Lacquered cedar, 5 1/2 x 7 1/2 x 3 15/16"
(14 x 19 x 10 cm)
Odate Craft Company, Odate

The lacquer designer Akiraka Takagi was commissioned by the Odate Craft Company of Akita Prefecture in northern Japan to design a set of sake serving items, using the company's specialty, cedar bentwood (*magewappa*). This elegant oblong sake pourer is part of the series, which also includes cups, trays, and an ice bucket. A clear lacquer coating highlights the natural grain of the cedar, punctuated by darker accents along the seam at the side. The company developed the technology to mass-produce their bentwood containers, formerly handmade, bringing the cost of quality craft items within reach for the average consumer.

201 Urushi tableware, 1986
Designed by Toshiyuki Kita
Made by Omukai Koshudo Company, Wajima
Lacquered wood, height 3 3/16" (8.1 cm), diameter 5 1/8"
(13 cm)
Omukai Koshudo Company, Wajima

In 1985 Toshiyuki Kita was invited by the Omukai Koshudo lacquer company at Wajima to design a line of tableware that would establish a modern identity and market for this traditional industry. The result was the Urushi (Lacquer) series, which includes trays, covered soup bowls, nesting bowls, a large covered dish, condiment dishes, chopsticks, and chopstick rests, distinguished by their sleek, broad silhouettes and fine proportions as well as the high quality of the lacquer technique evident in the depth of the color and attention to detail. "Every Japanese household has at least one piece of lacquer ware," wrote Kita. "They are used to hold traditional New Year's dishes and everyday dishes like miso soup. Thus they are still alive in the everyday lives of the Japanese people. . . . Lacquer ware, simply as tableware, is no match for comparable plastic containers, because the latter are less costly and easier to use. Lacquer ware is easily damaged. For this reason, I thought that if [my] lacquer ware was made only as a functional object, it would not last for the next generation. So I decided to make it as an 'object for the spiritual world.'"[50]

202 Evening dress and cape, Spring–Summer 1986
Designed by Hanae Mori
Made by Hanae Mori Haute Couture, Paris
Printed and embroidered silk with cording, sequins, beads, and rhinestones
Hanae Mori Tokyo

Borrowing her theme for this evening dress from the peacock, the showiest of birds, Hanae Mori used intricate beadwork, sequins, and rhinestones to suggest a long train of feathers, which, spread diagonally around the body almost to the floor, is brilliantly marked with bold eyelike spots. This display of plumage mimics the metallic glint of peacock feathers. Using gold and silver iridescent beads and metallic threads in a technique inspired by Japanese *makie* lacquer decorations of gold and silver powders, Mori built up the decorated surfaces in layers, creating a rich textural effect that she used often in the 1980s. A severely plain black cape in silk georgette provides a dramatic enclosure for the vivid colors and elaborate three-dimensional decorations of the dress.

203 Poster for Matsushita, 1986
Designed by Koichi Sato
Prize: Tokyo Art Directors Club, grand prize, 1986
Offset, 40 1/2 x 86" (103 x 218.4 cm)
Collection of Koichi Sato, Tokyo

Koichi Sato joined three standard-size poster sheets in a triptych to create this unconventional, large-format advertisement for MATSUSHITA ELECTRIC INDUSTRIAL COMPANY's new National "α" tube giant-screen floor-model television. With subtly graded colors and luminous auras surrounding both the large television tube and the moon beyond, Sato translated Matsushita's new technology into a metaphysical, even poetic, event. "I was trying to link traditional ideas of Japanese beauty with industrial high tech in an image associating the light of the picture tube with that of the moon," Sato explained. "I used the three-line format of haiku poetry with its [syllabic] rhythm of five, seven, and five, in the three sheets of paper to represent the vast extent of space."[51] The lines of the haiku written by Sato himself appear beside streaks of light on each sheet: "In the pale color of the moon, a city where no one has gone before—a reflection of the future."

202

α

大気の中へ、環境映像テレビジョン

National

203

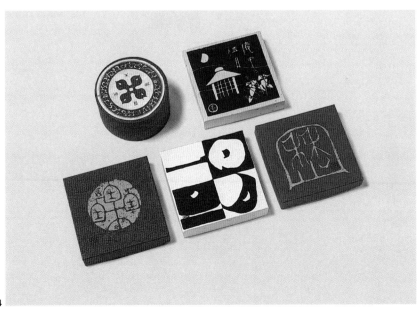

204

204 Packaging for Muho-an-Sohke, 1987
Designed by Hirosuke Watanuki
Printed paper, square package 1 1/8 x 5 3/16 x 5 3/16"
(2.8 x 13.5 x 13.5 cm); height of round package 4 7/16"
(11.3 cm), diameter 4" (10.2 cm)
Muho-an-Sohke, Kobe

Hirosuke Watanuki was inspired by his own practice of the
tea ceremony to design packaging for tea candies made
by a Kobe confectioner. The line of candies, in fact, takes
its name from Watanuki's teahouse, the Muho-an (The
Hut without Rules). Watanuki designs the packages, does
the calligraphy on the labels, and even names the items,
giving them allusive designations such as Passing Time
(Reki) and Deep Chestnut (Mikuri). His designs are also
used for small ceramic serving dishes and the cloth
wrappers (*furoshiki*) used for carrying traditional
foodstuffs and gifts. Watanuki's designs, distinguished by
his highly individual style of calligraphy, have made the
Muho-an products popular and immediately recognizable.

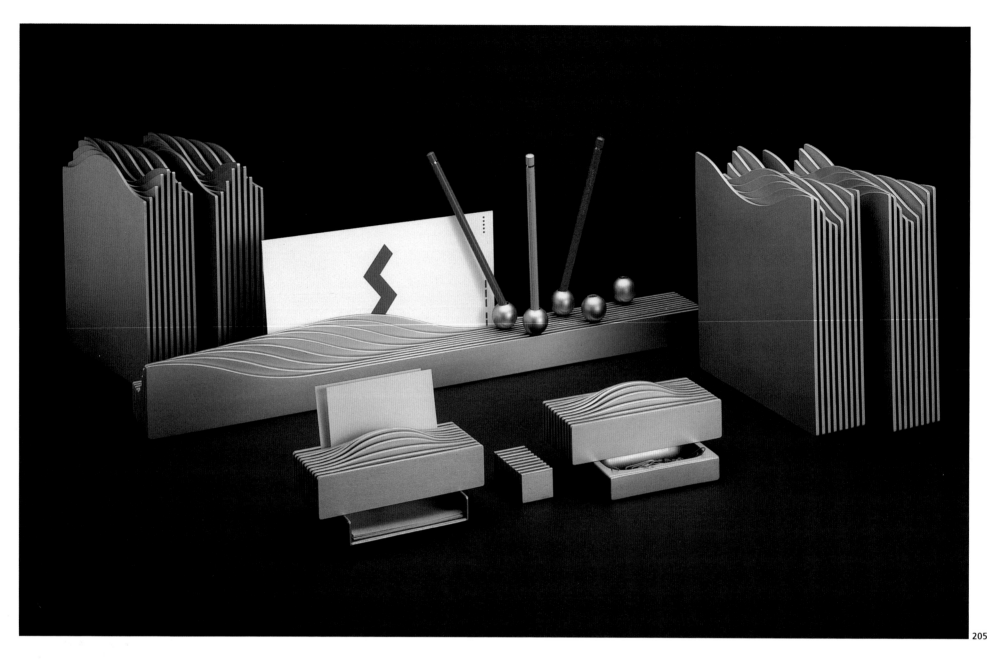

205 Wave desk set, 1987–88
Designed by Tadao Shimizu
Aluminum, envelope and pencil holder,
2 1/2 x 2 3/8 x 22 1/16" (6.3 x 6 x 56 cm);
bookend 7 7/8 x 2 x 7 7/8" (20 x 5.1 x 20 cm)
Cranbrook Academy of Art Museum, Bloomfield Hills,
Michigan. Gift of Tadao Shimizu

Tadao Shimizu used computer-aided machining to create
the wave pattern on his matte silver and gold Wave
(Nami) desk set prototype. The long, low horizontal
components—envelope holder, business-card holder, and
small storage drawers—can be aligned as a single unit or
arranged as separate elements, and are balanced by
bookends that are also speakers. Behind his design,
according to Shimizu, is the dualism of yin and yang, dark
and light, female and male: "They are mutually
complementary, balancing yet interdependent of one
another. . . . The dynamic flow of the waves is balanced by
the calm expanse of water arranged side by side."[52]

LISA-II ::: CARBON

CERAMICS YAMAHA

LISA-II ::: CARBON

CERAMICS YAMAHA

206

206 Graphic program for Lisa II skis, 1987
Designed by Shin Matsunaga
Yamaha Corporation, Hamamatsu

The crisp and clean black, white, and gray lines of Shin Matsunaga's graphic program for Yamaha's Lisa II skis contrasted sharply with the flashy colors and designs of contemporary skis. Here, Matsunaga employed the stepped letters of digital typography, a type of letter form that appeared often in his designs for logotypes, books, and wrapping paper during the 1980s. Yamaha intended the computer-inspired lettering and simple, polished appearance of its Lisa II skis to appeal to the young fashion-conscious consumer.

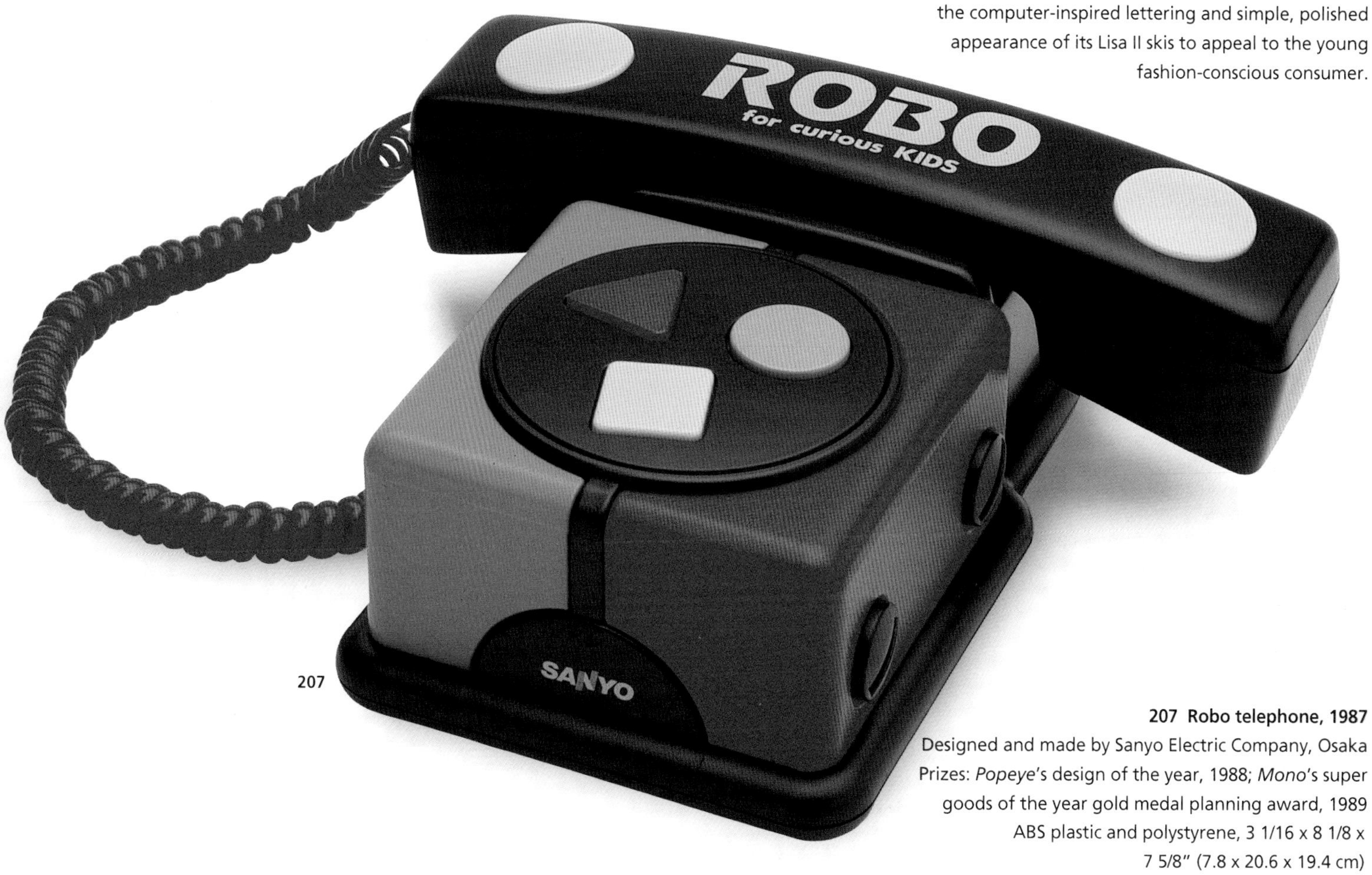

207

207 Robo telephone, 1987
Designed and made by Sanyo Electric Company, Osaka
Prizes: *Popeye*'s design of the year, 1988; *Mono*'s super goods of the year gold medal planning award, 1989
ABS plastic and polystyrene, 3 1/16 x 8 1/8 x 7 5/8" (7.8 x 20.6 x 19.4 cm)
Sanyo Electric Company Ltd., Osaka

Part of the Robo series of electronic products "for curious kids," this telephone, like Sanyo's other Robo designs, is extremely child-friendly, but equipped with memory, a hold function (which plays two melodies: "It's a Small World" and "The Mickey Mouse March"), handset volume control, and automatic redial, it can do as much as a telephone designed for adults. Made of ABS plastic, the body of the telephone is fitted with panels in bright, primary colors held in place with extra-large screws, which may be removed and reinstalled using a special Robo screwdriver. The telephone also has a keypad with three programmable geometric push-buttons in different colors that locks in place over the standard pad and allows even the smallest child to use the telephone. Although Robo products were designed for use by children under five, soon after their release they became popular among young Japanese women, who liked their features and their fresh, colorful style.

208 Spirit of Darkness II, 1988
Designed and made by Natsuki Kurimoto
Lacquer, 13 x 8 1/4 x 13 3/8" (33 x 21 x 34 cm)
Philadelphia Museum of Art. Gift of the artist

Natsuki Kurimoto's lacquerwork, one of a series in the
form of headgear (*kaburimono*), is reminiscent of
medieval Japanese military helmets whose shapes are
often abstractions of natural forms such as snail shells.
Producing a piece of lacquer is an exacting, time-
consuming process, requiring the application of multiple
layers of the substance, which are then polished to a sheen
to achieve rich effects of warmth and depth. "It is
precisely because the work is monotonous and requires
repetition," Kurimoto has said, "that it becomes imbued
with a certain spirituality."[53] His title for the work,
Spirit of Darkness, reflects this sense of mystery, which
is attained through the Zenlike discipline necessary
to the craft.

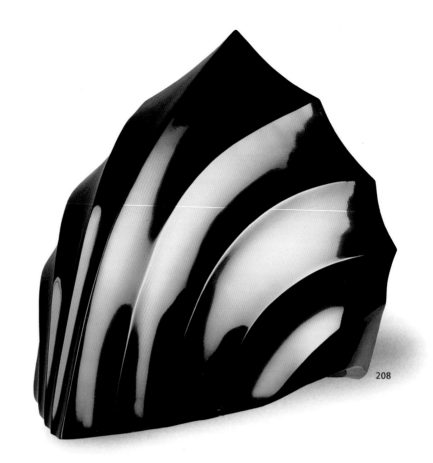

208

209 Organizer, 1987
Designed and made by Sharp Corporation, Osaka
Metal and plastic, 6 9/16 x 5 9/16 x 7/16"
(16.7 x 14.2 x 1.2 cm)
Sharp Corporation, Osaka

The Sharp Corporation's first electronic organizer, with
calendar, scheduler, memo pad, telephone book, and
calculator, went on sale in 1987, replacing the separate
items that were once needed to be carried each day. It
contains a two-thousand-character *kanji* dictionary, and its
telephone book can hold over three hundred entries.
Easily portable, it offers all these features in a product the
size of a wallet. With the addition of integrated-circuit
(IC) cards, its functions can be enlarged: one translates
from Japanese into English; another contains an address
book; and a third is a technical calculator. The organizer
may be connected to a printer, as well as to a cassette tape
recorder, to print or tape information.

209

210 Piedra portable television, 1987
Designed and made by Matsushita Electric Industrial
Company, Osaka
Prize: G-Mark, 1988
Plastic and steel, 7 15/16 x 8 7/8 x 12 9/16"
(20.2 x 22.6 x 31.9 cm)
Design Museum, London

A Japanese interpretation of the "soft tech" industrial design popular in the West during the 1980s, Matsushita's Piedra 8 portable color television has an asymmetrical form, designed to echo the curves of the human body, and finger grooves in the side and top intended to make it appealing to the touch. Intended for use by young singles living in studio apartments, it has an antenna and power cord in one, and the controls on its top are easily accessible. The television comes in three colors, urban black, urban gray, and urban green, and the screen is equipped with a "color looming function," which glows magenta, blue, or yellow when the set is not in use.

211 Audio conference unit, 1988
Designed and made by NEC Corporation, Tokyo
Prize: G-Mark, 1988
Plastic, 2 11/16 x 9 7/16 x 10 7/8"
(6.8 x 24 x 27.6 cm)
NEC USA, Inc., Melville, New York

NEC Corporation's sleek, flat-bodied, undulating audio conference unit, looking as if it should live on the bottom of the ocean rather than on a desktop, offered technology as advanced as its styling. Unlike conventional teleconferencing equipment, the telephone allowed several persons at different locations to talk at the same time without experiencing line noise and echoes (through the use of echo cancelers, a type of digital signal processor used in satellite and overseas telecommunications systems). Weighing only about three pounds and no bigger in size than a notebook, the speaker phone was also more compact and portable than its competitors, most of which required additional microphones and antennas, as well as calibration to match the acoustics of the rooms in which they were installed. Looking to the future of teleconferencing, it was designed to combine with video equipment for the growing market for video conference calls.

212 Hibiki fabric, 1988
Designed by Hiroshi Awatsuji
Made by Design House Awa, Tokyo
Screen-printed cotton, width 51 15/16" (132 cm)
Collection of Hiroshi Awatsuji, Tokyo

Yearning "for a piece that was serene yet unyielding,
pulsing with energy,"[54] Hiroshi Awatsuji designed his
Hibiki (Echo) fabric along with other similar monochrome
textiles for the exhibition "Independent Layer" at the Ma
Gallery in Tokyo in 1988. The fine, multiple, parallel lines
in this fabric form patterns resembling those created by a
resonating sound moving through the air, suggesting both
the energy of movement and serenity at the same time.

212

213 Sou fabric, 1988
Designed by Hiroshi Awatsuji
Made by Design House Awa, Tokyo
Screen-printed cotton, width 51 15/16" (132 cm),
repeat 53 1/8"(135 cm)
Collection of Hiroshi Awatsuji, Tokyo

Sou (Strength), Hiroshi Awatsuji's monochromatic printed
furnishing fabric, which resembles a section of an
enormous leaf, is from a series of such textiles he designed
in the late 1980s that feature oversize natural forms
including leaves, flowers, and bubbles. Awatsuji liked his
naturalistic patterns so much that he also adapted them
for a tableware series he designed and produced through
Design House Awa in 1988, his first work that went
beyond textile production and a characteristically
Japanese application of textile designs to other mediums.

213

214

214 Poster (*Japan*), 1988
Designed by Kazumasa Nagai
Silkscreen, 40 1/2 x 28 11/16" (103 x 72.8 cm)
Collection of Kazumasa Nagai, Tokyo

Japan was the theme for the annual Japan Graphic Designers Association exhibition in 1988. Although Kazumasa Nagai is best known for his abstract designs, in his series of posters for this exhibition he chose to depict animals featured in Japanese folklore: a tortoise, frog, and dragon. "In ancient Japan," he explained, "we had a concept of symbiosis—all living creatures, that is human beings, animals and plants coexisted peacefully. . . . I am creating designs featuring living things because my creative philosophy is now based on my belief that the earth is for all living things, not just for human beings alone."[55] The figure of the tortoise in this poster, as well as the surrounding space, is filled with traditional ornamental patterns such as concentric rings and floral designs, which are like those used in Japanese kimonos and screens.

215 Packaging for Irokuen, 1988
Designed by Rokuhei Inoue
Made by Irokuen, Kyoto
Prize: Kyoto Best Design Conference gold prize, 1989
Printed paper, 6 5/16 x 2 3/16 x 2 3/16"
(16 x 5.5 x 5.5 cm)
Collection of Yoichi Yoshikawa, Tokyo

The custom of drinking tea was introduced to Japan by
Zen Buddhist monks who came from China in the twelfth
century. Rokuhei Inoue returned to the origins of tea
drinking for the inspiration for his Zen Poem green-tea
packaging. Each of the six different fine teas from the Uji
region near Kyoto that are sold in the series is named after
a phrase from a Zen poem or aphorism. The packages are
made from hand-crafted paper (*washi*) in muted colors,
and are tied in the style of cloth wrappers (*furoshiki*). The
labels seem to have been brushed in ink, and recall the
Zen calligraphic scrolls in ink that are hung in the tea
room for the tea ceremony.

215

216 Kimono and *obi* wrappers, 1988
Designed by Keiko Kumagai
Made by Noie Company, Tokyo
Paper, kimono wrapper 34 1/4 x 13 3/4" (87 x 35.5 cm); *obi*
wrapper 24 3/4 x 13 3/4" (63 x 35.5 cm)
Collection of Keiko Kumagai, Tokyo

Keiko Kumagai gave a contemporary interpretation to the
traditional paper wrapper (*tatou*) used for storing folded
kimonos and *obi* sashes. Silk kimonos and brocaded *obi*s
are the most cherished items in a woman's wardrobe, so
care is taken to see that they are individually wrapped.
Kumagai chose handmade mulberry paper, free from
chemical additives that could harm the silk, and layered it
to make the wrapper resistant to insects, fire, and
humidity.   The subtle effect of the dark-red and sand-
color layering at the flap was inspired by the multiple
layers (*kasane*) of colors at the sleeves of robes worn by
noblewomen of the ninth and tenth centuries. Kumagai
added a sense of modernity to the wrapper with the bold
dark-red diamond on the outside, while retaining a
sophisticated elegance in the textured paper and silk ties
that are appropriate to the kimono tradition.

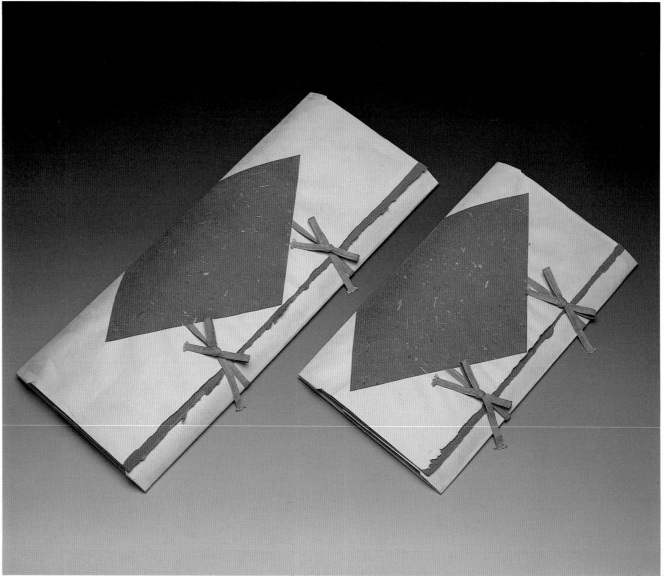

216

217 Alpha whisky bottle, 1988
Designed by Kozo Okada
Made by Nikka Whisky Distilling Company, Tokyo
Prize: Japan Package Design award for encouragement,
1989
Glass and metal, height 10 5/8" (27 cm)
Collection of Kozo Okada, Tokyo

This gift bottle is one of a pair designed by Kozo Okada for the Nikka Whisky Distilling Company and is inscribed with a Greek text that celebrates drink, "encouraging the users," as Okada said, "to enjoy their drinking time with these poems."[56] Okada decorated the spare architecture of his bottle with ancient Greek as a learned or witty quotation from the past, much as a postmodern architect might have applied a classical detail to his building.

218

217

218 Sake bottle, 1988
Designed by Shin Matsunaga
Made by Fukumitsuya Sake Brewery Company, Kanazawa
Glass and printed paper, height 11 7/16" (29 cm)
Fukumitsuya Sake Brewery Company, Kanazawa

Shin Matsunaga's logo for the Fuku-Masamune line of sake alludes to a device known as a *Genji-mon* (Genji crest), which was originally devised as a sort of counting system for an ancient Japanese game. These crests were identified by the titles of the fifty-four chapters in the eleventh-century courtly novel *The Tale of Genji*. While the logo does not duplicate a specific *Genji-mon*, and the graphics can also be interpreted simply as an appealing abstract design, the reference to *The Tale of Genji* conveys a sense of refined connoisseurship. The nuance of well-bred, literary taste implies that this is sake fit for the most discriminating consumer.

219 Carna wheelchair, 1989
Designed by Kazuo Kawasaki
Made by Sig Workshop, Tokyo
Prizes: Mainichi design prize, 1990; ICSID award of
excellence (gold medal), 1992; State of Baden-
Württemberg international design prize, 1993
Titanium, 33 1/16 x 22 1/16 x 35 1/4" (84 x 56 x 89.6 cm)
Collection of Kazuo Kawasaki, Fukui

This wheelchair, with its elegant silhouette, bright colors,
and oversize wheels like those used for racing, addresses
both aesthetics and the social implications of medical
apparatus, which until recently has lagged behind the
technology this type of equipment houses. Made of
strong but lightweight titanium and designed to fold up
for transport, it serves the varied needs of its user, who in
this case, is the designer. "Older people, handicapped and
normal people are separated in today's Japan," Kawasaki
wrote, "so designers need to make designs that are kind
and caring and need to treat more handicapped people
equally in society. As a handicapped person myself, I have
this perspective; to be a visionary designer I want to
design products for myself first."[57] Kawasaki named the
wheelchair after Carna, the Roman goddess of family life.

219

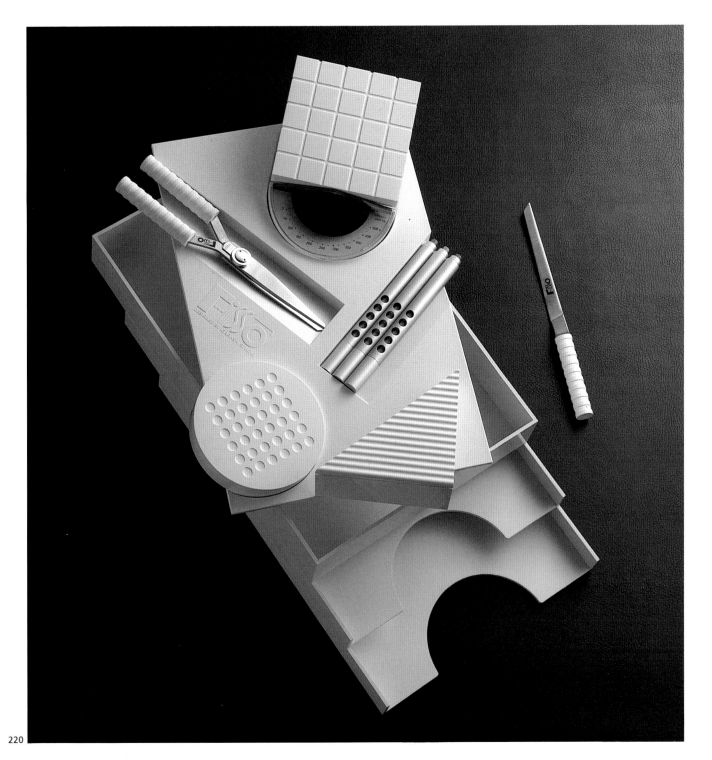

220

220 Fisso desk set, 1988
Designed by Takenobu Igarashi
Made by Raymay Fujii Corporation, Tokyo
ABS plastic, 3 15/16 x 10 1/8 x 14 15/16"
(7.5 x 25.8 x 37.9 cm)
Igarashi Studio, Tokyo

Comprising a pen set, scissors set, postal scale, stamp pad, utility case, and fitted holder, Takenobu Igarashi's Fisso set of desktop accessories was conceived as an exercise in the basic geometric shapes of circle, triangle, and square. The elements are decorated variously with dots, a grid, and lines, which are not only mathematical components of the larger shapes but also reflect the traditional Japanese pattern designs that Igarashi cites as among his influences.

221

221 Shoulder Sattle vacuum cleaner, 1989
Designed and made by Sanyo Electric Company, Osaka
Plastic, 13 7/8 x 10 1/4 x 5" (35.2 x 26 x 12.8 cm)
Sanyo Electric Company Ltd., Osaka

This cordless molded-plastic portable vacuum cleaner belongs to Sanyo's It's range of household products designed to appeal to young Japanese singles living in small apartments. When not in use, the vacuum cleaner and its attachments (including a small brush, a flat head, and hose) stand on a plug-in recharger. The flexible hose, which projects from the top of the machine, fits into a groove in the carrying handle, and the vacuum is held at the hip by a strap slung over the shoulder.

222 Walkman video-cassette recorder and player, 1988
Designed and made by Sony Corporation, Tokyo
Plastic, 8 3/8 x 5 1/16 x 2 5/8"
(21.3 x 12.9 x 6.7 cm)
Sony Corporation, Tokyo

Combining the video recording and miniaturization
technologies that Sony and other Japanese firms
pioneered, Sony Corporation's personal Walkman
combines a video-cassette recorder and player and a
three-inch liquid-crystal-display television set in one pocket
notebook-size, lightweight unit. This Walkman plays and
records television programs and accommodates 8-
millimeter videotapes.

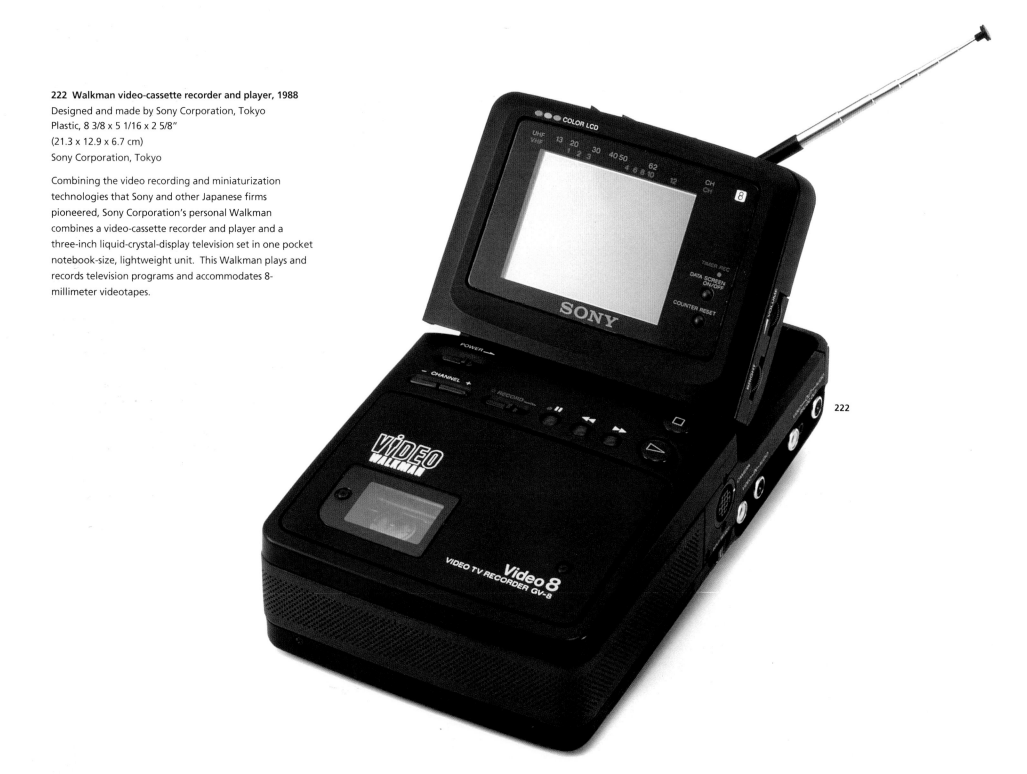

222

223 Game Boy video-game system, 1989
Designed and made by Nintendo Company, Kyoto
Plastic, 5 3/4 x 3 1/2 x 1 1/4" (14.6 x 8.9 x 3.2 cm)
Nintendo of America Inc., Redmond, Washington

Video games, developed by Nintendo and other toy companies during the 1970s for amusement arcades, paved the way for their use on television monitors in the home and as hand-held portable video-game systems with interchangeable software. Game Boy became the most popular hand-held computer toy ever made, outstripping rival products made by Sega, Atari, and NEC CORPORATION, with its playful design (including asymmetrically placed operating buttons and a yellow-green liquid crystal display), lower cost, and popular software, particularly its Tetris game—a puzzle with moving pieces that the player attempts to arrange. Housed in a gray plastic case the size of a transistor radio, Game Boy uses small interchangeable cartridges and combines the attributes of portability, miniaturization, and entertainment, appealing to adults as well as children.

224 Navi personal work station, 1988
Designed and made by Canon, Tokyo
ABS plastic, 12 13/16 x 14 1/2 x 13 7/8"
(32.5 x 37 x 35.2 cm)
Canon Inc., Tokyo

Telephone, facsimile, word processor, and computer in one, Canon's Navi personal work station (distributed as Navigator in the United States) is a sleekly compact, multifunctional machine that is easily positioned on a desktop and offers the convenience and freedom of a wireless keyboard. An extremely complicated software system was simplified and made user-friendly through the use of icons, which appear on Navi's angled touch-sensitive screen. The Canon design team was especially interested in creating amusing and characteristic icons that make performing Navi's numerous functions simple and fun. Wiper Boy, for example, glides across the screen with his squeegee, allowing it to be cleaned without the touch-sensitive functions being activated.

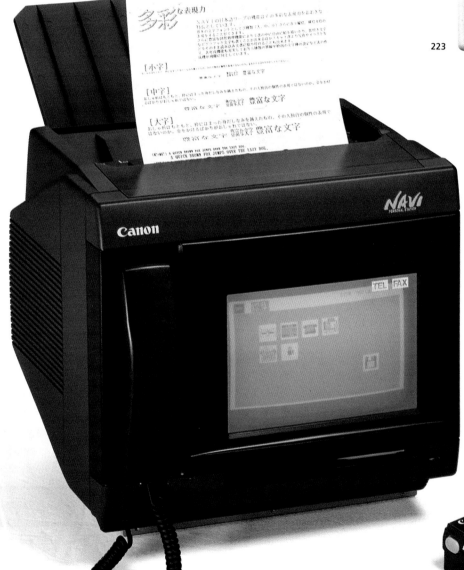

224

225 Boxes, 1989
Designed and made by Shigeru Akizuki
Prize: Japan Package Design award for encouragement,
1989
Lacquered wood with marquetry decoration, 11 x 7 7/8 x
1 3/4" (28 x 20 x 4.5 cm) (each)
Collection of Shigeru Akizuki, Tokyo

Shigeru Akizuki took as the theme of these boxes the
ancient Japanese idea of *fugetsu*, a comforting breeze
with the moon behind thin clouds. Decorating both the
boxes and the dishes they contain with horizontal, circular,
and vertical images to represent wind, moon, and star,
Akizuki reversed the idea that a box is secondary to its
contents. Meticulously crafted, the boxes are fitted with
covers that unexpectedly slide up like theater curtains to
reveal the dishes within.

225

226 Packaging for Pola, 1989
Designed by Hiroyuki Matsuyama
Printed paper, 3 11/16 x 6 5/16 x 6 5/16" (9.4 x 16 x 16 cm)
Private collection

Charged with designing packaging for gift soaps that was
artistic yet practical for mass production, Hiroyuki
Matsuyama turned for inspiration to the traditional
wrapping cloths (*furoshiki*) used to carry gifts in Japan.
On the box, delicate patterns mix with the bold simplicity
of one unpatterned leaf in an abstract representation of a
knot, while each soap within is individually wrapped as if
it were a candy or a sweet.

226

227 Evening dress, Spring–Summer 1989
Designed by Hanae Mori
Made by Hanae Mori Haute Couture, Paris
Printed silk
Hanae Mori Tokyo

Inspired by traditional Japanese *sumie* (black ink painting),
Hanae Mori translated the hand-brushed calligraphy of a
Japanese love note into the screen-printed pattern that
decorates both the transparent chiffon overdress and the
opaque twill underdress of this dramatic evening gown.
Exploiting the properties of black ink on the white silk
ground as a Japanese painter or calligrapher might do,
Mori used the contrast of silk weaves and varied
brushstrokes to create patterns of black and white that
shift and change as the wearer moves. Where the broad
black brushstrokes of over- and underdress happen to
overlap, the effect is very dark and dramatic; where the
white chiffon falls over the black calligraphy, the effect is
subtly grayed.

228 Kotobuki shelves, 1989
Designed by Sinya Okayama
Made by Daichi Company, Tokyo
Lacquered wood, 64 3/16 x 35 7/16 x 15 3/4"
(163 x 90 x 40 cm)
Philadelphia Museum of Art. Gift of Daichi Company

Sinya Okayama's Kotobuki (Celebration) shelves is the
masterwork in his series of furniture designs that
translates Japanese pictographs into three-dimensional
forms, allowing the object to be "read" (see no. 182). It
was created for his one-man exhibition at Tokyo's
Yurakucho Asahi Gallery in 1989, where critics commented
on the expressive value and meaning such furniture had.

228

229 Ikebana speaker, 1989
Designed by Satoru Hibino
Made by Yamaha Corporation, Hamamatsu
Prize: Yamaha audio design competition, 1989
Plastic compound, height 19 11/16" (50 cm)
Collection of Satoru Hibino, Nishikamo

In 1989 Yamaha sponsored a competition for inventive
speakers to replace the conventional black-box designs,
which it advertised with the slogan: "I'll give you the
technology, you give me the imagination." Yamaha
received over 950 entries, including Satoru Hibino's
Ikebana (Flower Arrangement) design, which received one
of the twenty-two prizes. Conjuring the image of a
futuristic flower arrangement, this concept model audio
system has three speakers, one for high tones, one for
medium tones, and the largest for heavy, low tones. The
black matte base and stems, which may be bent to direct
the sound and achieve the desired blending of tones, are
punctuated by the bright colors of the flowers. Witty and
playful, the Ikebana system is an example of the
emergence in the late 1980s of new, humanized and
representational forms for electronic products that conceal
their high-tech functions.

229

230 Big Wave fabric, 1989
Designed by Junichi Arai
Made by Nuno, Tokyo
Aluminum and polyester, width 50" (127 cm)
Philadelphia Museum of Art.
Gift of Nuno Corporation

For his Big Wave fabric, Junichi Arai applied the traditional
tie-resist dyeing (*shibori*) method to a high-tech base
fabric of sheer flat polyester, vacuum-metallicized with
aluminum. Dyeing the metallic fabric was only possible
because the aluminum surface was coated with a minute
amount of polyamide resin. As in the ancient tie-dye
procedure, Arai reserved an area from being dyed by
gathering and tying the material; to the surface patterns
and iridescent variations he created, Arai added the
textural effect of heat-set, permanently pleated wrinkles.
"Paradoxically," one critic has written, "with each stage of
the process, the fabrics lose their impersonal machine-
made quality and gain a singular hand-crafted flavor from
their very advanced technology. Herein lies the heart of
Arai's retro-tech approach to textile creation,"[58] a
postmodern paradigm.

230

231 Skateboard, 1989
Designed by Toru Matsushita
Made by NEC Corporation, Tokyo
Plastic, 6 5/16 x 17 3/4 x 29 1/2"
(16 x 45 x 75 cm)
NEC Corporation, Tokyo

This prototype for a skateboard with a linear motor was shown at the World Design Expo in Nagoya in 1989. Linear motors, used to move low-mass lightweight loads at high speed with rapid acceleration and precise control of position, drive directly without the belts, gears, and other mechanical drive mechanisms of conventional rotary motors. Based on the same electromagnetic principle of operation planned for the linear motorized railway line between Tokyo and Osaka, the motor propels the skateboard forward for a hundred yards after one kick, offering the skateboarder a long, fast, self-propelled ride unavailable on ordinary nonpowered boards.

231

232

232 Morpho II motorcycle, 1989–91
Designed by GK
Made by Yamaha Motor Company, Hamamatsu
Fiber-reinforced plastic and aluminum,
48 13/16 x 79 15/16 x 30 1/8"
(124 x 203 x 76.5 cm)
Yamaha Motor Company, Hamamatsu

Created in prototype in 1989 and further developed as the experimental Morpho II in 1991, this is the quintessential ergonomic motorcycle, designed to be fully adjustable to the rider's needs. With the ability to be changed from a forward-leaning, strong racing position to an upright touring position, the motorcycle adapts to different body types, temperaments, and driving styles. The man-machine interface is supported by high technology, with a liquid-crystal-display panel that provides data in the form of graphs and messages to the rider and an electronic-control suspension system installed with a computer in both back and front wheels. The soft organic forms, low silhouette, and glowing colors that give the Morpho its particular aesthetic character and its name (derived from biological science and the blue South American butterfly) not only are user-friendly and visually appealing but also add to the comfort and safety of the rider: the rounded seat, handlebars, steps, and frame are padded for support, and the finish in reflective sheet, holograph sheet, and fluorescent paint increases the visibility of the motorcycle itself.

233

233 Facsimile machine, 1989
Designed by Keiichi Hirata
Made by We'll Design Something with Emotion, Tokyo
ABS plastic, 7 1/16 x 10 5/8 x 6 11/16"
(18 x 27 x 17 cm)
Collection of Keiichi Hirata, Tokyo

With this conceptual model for a facsimile machine, Keiichi Hirata intended to return meaning as well as psychological and emotional values to design, the soft gray color and curved, scrolling shape of the body a response to the hard lines and finishes still widely associated with electronic products. "These days many machines are black boxes," Hirata said. "But the home fax machine should not be a black box. The fax machine receives and sends important and meaningful messages, and it should be warm and joyful. The machine presents a balance of easiness, understanding, enjoyment, and beauty."[59]

234 Ecru camera, 1991
Designed by Water Studio
Made by Olympus Optical Company, Tokyo
Plastic and metal, 3 9/16 x 4 5/8 x 1 11/16"
(9 x 11.7 x 4.2 cm)
Olympus America Inc., Woodbury, New York

Eschewing the discreet black housing conventionally given
to cameras, this startlingly glamorous off-white Ecru
camera with a shaped profile and shiny silver lens cap
declared itself a fashion statement of the 1990s. A limited
edition product for "collectors," it was meant to look like
a luxury object and be desirable to the cosmopolitan
consumer that its designer, Water Studio's Naoki Sakai,
and its maker, Olympus Optical Company, targeted on the
Japanese market. While the value of the Ecru camera
appeared to depend on its style and image, it also rested
on Olympus's reputation for high-quality technology. It is
an ultracompact point-and-shoot 35-millimeter camera
totally automatic in its functions and manufactured with
the miniaturized components and precision lenses for
which Olympus is known (see nos. 80, 125).

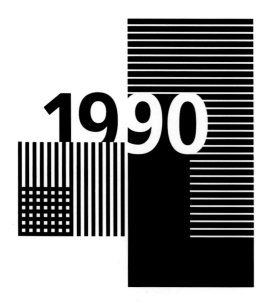

234

235 Rhythm Pleats dress, Spring–Summer 1990
Designed by Issey Miyake
Made by Issey Miyake Inc., Tokyo
Polyester
Miyake Design Studio, Tokyo

Rhythm Pleats belongs to the remarkable series of pleated dresses and tops composed by Issey Miyake from the seemingly unclothing-like elemental geometric shapes of circles, ovals, rectangles, and triangles. Here, two flat ovals sewn together have three slits, two on one side and one on the other, for the head and neck and the arms. Introduced in his spring–summer 1990 collection, Miyake's series reversed the usual method of taking a finished fabric and then cutting and sewing it into final shape. Instead, the fabric was first sewn into shape and then finished with pleats in a heat-set pleating machine. By applying heat and pressure according to the width and thickness of the polyester fabric, Miyake was able to make the pleating permanent. Extra lightweight, hand-washable, fast-drying, and easy to store by simply folding along the pleat lines, these clothes were as functional as they were aesthetically and technically innovative.

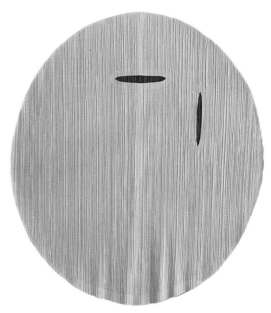

235

236 Palmtop computer, 1990
Designed and made by Sony Corporation, Tokyo
Plastic, 1 3/4 x 8 1/16 x 6 3/16" (4.5 x 20.5 x 15.8 cm)
Sony Corporation, Tokyo

A notebook-size computer with a touch-sensitive screen
instead of a keyboard, Sony Corporation's Palmtop
converts handwriting written directly on the screen with a
special electronic pen to computer type, and then prints it.
Heralded as an entirely new way to communicate with
computers using "fuzzy logic" for its character recognition
and processing system, the Palmtop was designed for
those who were more comfortable writing by hand than
typing on a keyboard, particularly for the Japanese, whose
complex written systems of "alphabets" and some seven
thousand ideographs (*kanji*) proved difficult to reduce to
keyboard characters. The computer also sends messages
by modem as facsimiles. In an article introducing this
notebook computer, the first of several models, the *New
York Times* foresaw that such machines could be expected
to be used like electronic scratch pads, and "as an
architect's sketchpad, a traveling salesman's order and
notebook, and a place for musicians to compose and for
lawyers to update their electronic files."[60]

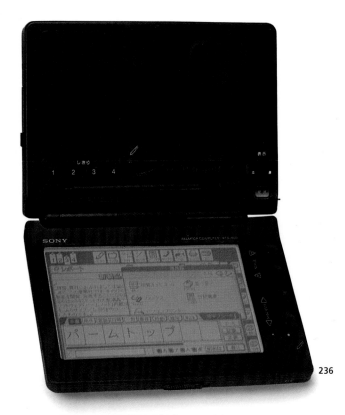

236

237 Compact audio system, 1990
Designed by GK
Made by Yamaha Corporation, Hamamatsu
Prize: G-Mark, 1989
Plastic, system 10 1/2 x 11 13/16 x 11 3/4"
(26.8 x 30 x 29.9 cm); speakers 9 1/8 x 12 3/4 x 9 1/8"
(23.2 x 32.4 x 23.2 cm) (each)
Yamaha Electronic Corporation, USA, Buena Park,
California

While many designers suggested new ways to deal with
speakers in prototypes or limited editions, GK's unusually
descriptive speaker for Yamaha's AST-C30 audio system
was mass-marketed. Moreover, the curved speakers with
their open "ports," which give the system its shaped
profile and submarine look, reflect the audio technology
they house—a resonating chamber so aligned with the
porthole that it can produce strong, powerful sounds from
a very small area.

237

239

238 Symbol for Akebono, 1991
Designed by Tamotsu Yagi

Tamotsu Yagi designed the corporate identity and packaging program for Akebono's sweets and crackers by adopting the Japanese syllables for the company's name written by the daughter of the president of the company. The bright colors and unsophisticated spontaneity of the calligraphy give a fresh and inviting look to the otherwise plain packaging. Akebono's wrapping paper, shopping bags, boxes, and calendars are made from recycled paper, a policy that Yagi convinced his clients to adopt once they saw that ecology-minded design could be sophisticated and attractive.

238

239 Jacket, vest, and skirt, Fall–Winter 1990–91
Designed by Yohji Yamamoto
Made by Yohji Yamamoto Inc., Tokyo
Leather, nylon, and synthetic down
Kyoto Costume Institute, Kyoto

Adapting the light weight and warmth of contemporary down-filled outdoor wear to high fashion, Yohji Yamamoto created this variant of the down-filled jacket, adding his own leather vest (intended to be worn over the jacket) and down-filled A-line skirt. Yamamoto tailored these garments to produce the slimmest possible silhouette; he nipped in and defined the waist of jacket, vest, and skirt, and used the channels for the down both decoratively and to control the placement of the fill so that in the skirt, for example, it falls away from the hips in smooth tiers toward the hemline.

240 Laminaria Seaweed scarf, 1992
Designed and made by Eiji Miyamoto
Silk polyamide and polyurethane, 12 5/8 x 76 3/4"
(32 x 195 cm)
Miyashin Company, Ltd., Tokyo

This triple cloth is composed of two outer layers of sheer
fabric with an inner layer of elastic weft, which when it is
not floating, interacts with the outer layers in two ways: it
catches the outer layers at intervals to create vertical
pleats and it also creates a very fine shirr. The pleats and
shirring alternate vertically, creating a dramatically
textured surface that gives the impression of seaweed
undulating in the waves in the light that filters through
the water. The varied density of the sheer fabric imparts
an ever-changing dimension to the cloth. When it was
woven, the fabric was over 51 inches wide, but once it was
removed from the loom, the elasticity of the weft reduced
the width to 15 3/4 inches.

240

241 Spacy Yarn scarf, 1992
Designed and made by Eiji Miyamoto
Silk, 33 7/8 x 86 5/8" (86 x 220 cm)
Miyashin Company, Ltd., Tokyo

This silk scarf, which appears to be composed of three
separate layers of material stitched together, is actually a
triple cloth—one piece that is woven in three layers. Eiji
Miyamoto designed the fabric with the aid of computer
programs and constructed it on a Jacquard loom. It
exemplifies his creative style in which sophisticated
technological mastery is used to create a fluid, lyrical
image and which gives no hint of the mechanical
complexity of the weaving method.

241

242 Will cutlery set, 1990
Designed by Hiroshi Egawa
Made by Aoyoshi Company, Tsubame
Polymer resin and stainless steel, length of spoon
9 3/16" (23.4 cm); length of fork 8 5/8" (22 cm);
length of spork 8 11/16" (22.1 cm)
Aoyoshi Company, Ltd., Tsubame

Hiroshi Egawa took advantage of the properties of a
polymer resin developed by Mitsubishi Heavy Industries
for the design of his Will cutlery set. This new plastic has
what Mitsubishi calls "shape memory," which allows it to
be formed into and then to retain almost any shape. It is
used in a variety of medical, sports, and toy products.
Egawa adopted the plastic for the cutlery handles, which
come in bright yellow and white; when heated, the
handles can be bent to fit the hands of individuals who
might have difficulty with standard flatware because of
arthritis or physical impairment. "Everyone wants to be
able to enjoy mealtimes in the same way," says Egawa,
who sees his Will cutlery set as "a support tool that
enlarges [each person's] sphere of independence."[61]

242

243 Mold Packo packaging for Tsutsumu, 1992
Designed by Takashi Kanome
Prize: Tokyo Art Directors Club, grand prize, 1993
Paper, height 13" (32 cm), diameter 4 11/16" (12 cm)
Collection of Takashi Kanome, Tokyo

Beginning in 1989 Takashi Kanome explored the use of
recycled paper that he soaked and press-formed into gift
boxes. In 1992 Kanome created his Mold Packo gift
packaging for wine bottles for Tsutsumu, which won him
the Tokyo Art Directors Club grand prize and which like
his earlier works in the medium, were intended to be
enjoyed as objects in their own right after use. Intrigued
by the idea of these cast-paper forms, Kanome exhibited
large sculptural pieces in the same materials in an
exhibition entitled "Box XX" at the Ginza Graphic Gallery
in Tokyo in 1992, where the designer found "beauty" and
"romance" in the empty outer shells, which for him were
"the condensation of a sense of aesthetics."[62]

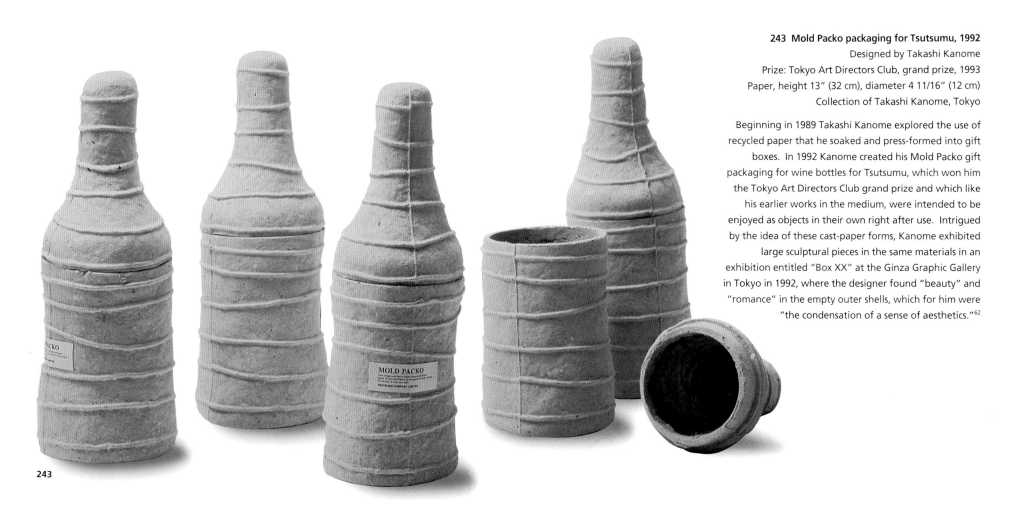

243

244 Viewcam video camera and recorder, 1992
Designed and made by Sharp Corporation, Osaka
Prizes: *What Camcorder*'s best innovation, 1993;
Camcorder User's best innovation, 1993; *Discover*'s award
for technological innovation, 1993; *Popular Science*'s
audio-video grand award, 1993; *Popular Mechanics*'s
design and engineering award, 1994
Plastic, 5 13/16 x 7 13/16 x 3 1/16" (14.8 x 19.8 x 7.8 cm)
Sharp Corporation, Osaka

Applying liquid-crystal-display technology to video
shooting, viewing, and playing, Sharp Corporation's
Viewcam camcorder revolutionized the industry with its
color monitor that doubles as a view finder. More user-
friendly that any other video camera, it allows the
amateur photographer to record the scene as it appears
on the four-inch screen, and by means of a pivot
mechanism, to rotate lens and screen to different viewing
angles. The lens can even be turned toward the user,
allowing the person doing the shooting to get into the
picture. Recognized with many awards for its
technological innovation and compact, portable design
features, the Viewcam took the Japanese industry that
pioneered and dominated the world video market to yet
another stage in its development.

244

245 Liquid Crystal Museum television, 1991
Designed by Water Studio
Made by Sharp Corporation, Osaka
Stainless steel, 18 5/16 x 17 3/4 x 3 1/8" (46.5 x 45 x 8 cm)
Sharp Corporation, Osaka

Replacing bulky cathode-ray tubes with tiny, color, liquid-
crystal displays, the Sharp Corporation was able to develop
the industry's long-sought flat television, a compact and
portable system that produces high-quality, high-
resolution pictures. Using this technology Sharp worked
with Naoki Sakai of Water Studio to produce a new type
of television conceived as an art object, in which the
screen is like an electronic picture hung on the wall in a
frame, which is manufactured in several styles. These
range from sleekly modern models in stainless steel to a
gilt rococo version aimed at more traditional collectors.

245

246 Xshax tissue boxes, 1991
Designed and made by Yosei Kawaji
Prize: Japan Package Design award for fine achievement
in experimental packages, 1991
Plastic-laminated paper, 3 1/8 x 9 7/8 x 4 3/4"
(8 x 25 x 12 cm)
Yosei Kawaji Design Office, Tokyo

Commissioned by the manufacturer Dynic to design
packaging using a plastic-laminated paper that resembles
rubber, Yosei Kawaji came up with these crinkled tissue
boxes, which he made in black, beige, gray, and white.
Their name, Xshax (pronounced kshaks), is a combination
of *kusha-kusha*, a Japanese expression describing
something wrinkled, and *kushami*, the Japanese word for
sneeze. The paper boxes can be wrinkled and crushed
into any desired shape.

246

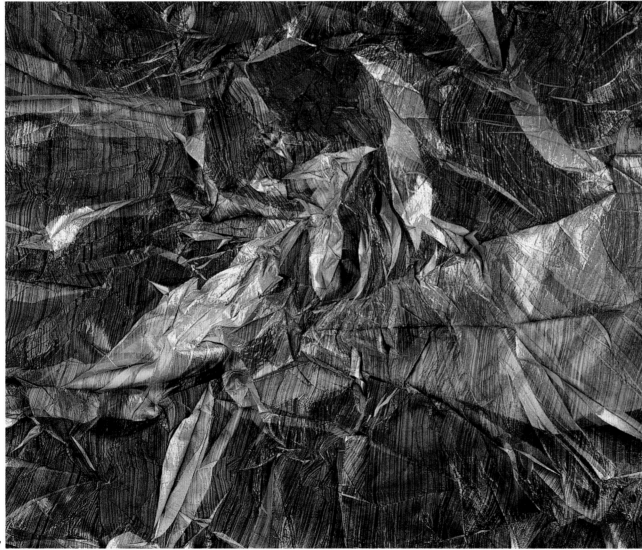

247 Ice Drift fabric, 1993
Designed and made by Junichi Arai
Polyester, width 39" (99 cm), length 174" (442 cm)
Cooper-Hewitt National Museum of Design, Smithsonian
Institution, New York. Gift of the designer

The fabric used for this piece was woven with fine, flat
polyester fibers that Arai calls "slit yarns." These yarns
were made of strips of an ultra-light polyester film that
had been metallicized with aluminum and coated with a
minute amount of polyamide resin. The fabric was
crinkled and then put through a heat-transfer press similar
to a printing press, which simultaneously flattened the
shaped fabric and colored it. By repeating this process
several times, Arai created a pattern similar to that seen in
crystalline minerals or ice.

247

248 Jacket, Fall–Winter 1993–94
Designed by Rei Kawakubo
Made by Comme des Garçons, Tokyo
Acrylic and wool
Comme des Garçons, Tokyo

Developing the principle of what became known in the 1990s as "deconstruction" from her own patchwork clothes and "torn" sweaters of the 1980s (nos. 164, 168), Rei Kawakubo experimented with designs like this jacket that looked as though they were made up of fragments of garments and recycled fabrics. With wool sleeves stitched to a printed acrylic top, this jacket set new rules for constructing clothes and reordered the hierarchy of what can be worn together. The two fabrics combined here neither match nor lay together; the differences in their weight and texture pull the garment out of shape in a tension that Kawakubo exploited for its effects of contrast and contradiction. While other deconstructionist designers have recycled actual thrift-shop remnants, Kawakubo designs the fabrics herself and has them prewashed and decolorized so that they look preworn.

**249 Sweater, jacket,
and pants ensemble,
Fall–Winter 1993–94**
Designed by Kenzo
Made by Kenzo, Paris
Wool, viscose, and velvet
Kenzo, Paris

249

Kenzo's "return to Japan," as the French press hailed his fall–winter collection of 1993–94, reinvented the past by recasting elements of traditional kimono dress in new forms. In this ensemble, the kimono, a long, loose robe tied around the waist with a wide sash, or *obi*, becomes a jacket worn over pants (the collection also included a kimono dinner jacket for men's evening wear). "The kimono frees you from shape," Kenzo explained, "and allows you to focus on details."[63] Mixing sources, patterns, and fabrics with strikingly picturesque results, Kenzo used an allover floral print inspired by French Impressionism for the pants and kimono jacket and small-scale geometric motifs adapted from traditional Japanese textile patterns for the long, loose-sleeved sweater shown with them.

248

250 Coat and dress, Spring–Summer 1993
Designed by Issey Miyake
Made by Issey Miyake Inc., Tokyo
Ramie and cotton
Miyake Design Studio, Tokyo

Just as Issey Miyake in the early 1990s explored the embellishing effects of a heat-set pleating machine (no. 235), he also experimented with the traditional manual technique of tying portions of cloth for resist dyeing. This coat of *kibira*, a fabric made of ramie fiber traditionally used for summer wear in Japan, was like Miyake's pleated clothing, which was first cut oversize and sewn into shape. It was then tied all over, leaving only the top section exposed, and dipped into a shrinking solution. After it had dried and the ties were removed, the untied top had shrunk to a normal size, while the section that was tied remained loose and flowing. Miyake also used the shrinking technique for the sheer cotton dress worn under the coat, the tied-off sections appearing like large, gathered bubbles on the finished surface.

251 Dress, Fall–Winter 1993–94
Designed by Yohji Yamamoto
Made by Yohji Yamamoto Inc., Tokyo
Wool and nylon
Kyoto Costume Institute

251

Taking postmodern delight in complex historical references, Yohji Yamamoto combined Western and Japanese traditions in this elegantly draped dress. Using two black wool rectangles as if he were constructing a kimono, Yamamoto joined them partially up the center and tacked them into folds, creating a draped effect reminiscent of Western dress from about 1910—when the fashion for imitation and for evoking Japanese costume was much in vogue.

250

252 Photura camera, 1990
Designed and made by Canon, Tokyo
Prizes: G-Mark, 1990; *Popeye*'s design of the year,
bronze prize, 1990; Osaka Design House, most valuable
product, 1990
Glass-reinforced polycarbonate resin,
2 7/8 x 3 15/16 x 6 1/8" (7.3 x 9.9 x 15.5 cm)
Canon Inc., Tokyo

Inspired by what Canon perceived as a consumer need for
a fully automatic 35-millimeter camera with greater
creative flexibility, its Photura model is equipped with a
powerful zoom lens extending to 105 millimeters,
autofocus, and a built-in automatic zooming flash, yet it
remains compact, lightweight, and much easier to use
than other cameras with equivalent zoom lens capabilities.
While the high magnification lens accounts in part for its
technological advantage, it also lends the Photura its
characteristic shape and ergonomic appeal. Resembling a
hand-held video camera, Photura (marketed as Autoboy
Jet in Japan and Epoca in Europe) has a cylindrical form
that is a total departure from conventional T-shape
lens/shutter (point and shoot) 35-millimeter cameras,
contoured at the back to fit comfortably and securely in
the right hand. In order to maintain its elongated shape
and compactness, Canon designed an S-shape loading
system to replace the conventional straight film path, as
well as a flash that is incorporated in the lens cover and
swings open on a hinge when the camera is in use. Canon
hoped the Photura would attract the amateur
photographer who would find its technical superiority as
easy to use as it looks.

252

253 Bullet Train model, 1993
Designed and made by Tomix, Tokyo
Plastic, 1 3/16 x 13/16 x 5 7/8" (3 x 2 x 15 cm)
Private collection

Tomix, a division of the toy company Tomy, designs and
manufactures faithful miniature models of Japanese
trains. The Tsubasa type 400 series Shinkansen (Bullet
Train) is based on a train designed by KENMOCHI DESIGN
ASSOCIATES, which began running between Tokyo and
Yamagata in 1992, almost thirty years after the first bullet
train linked Tokyo and Osaka. The Tsubasa 400 is
designed to run on two gauges, as a bullet train and a
conventional train; its wheel base allows the train to
negotiate both flat straightaways and mountainous
curves, and the extensive use of aluminum in its
construction makes it lighter than other bullet trains.
Tomix describes its miniature as a very precise copy of the
full-scale train, from the exteriors of silver and gray with
green trim, to the curved aerodynamic windows and snow
plow, the distinctive pantagraphs on the roofs of the cars
(which connect to overhead electric wires and provide
power), and the working head and tail lights.

253

254 Jacket and skirt, Spring–Summer 1994
Designed by Rei Kawakubo
Made by Comme des Garçons, Tokyo
Rayon, cotton, and polyester
Comme des Garçons, Tokyo

Mixing Western historical sources, Rei Kawakubo's romantic evening ensemble joins a slim, elegant, close-fitting skirt in thin rayon crepe georgette, which suggests an evening dress of the 1930s, with a jacket that has detachable lace sleeves, which refer to the fifteenth- and sixteenth-century fashion for treating sleeves as separate entities, laced (but not joined) to the bodice to permit movement. The stand-up collar and stiff turned-back cuffs are also inspired by medieval dress. Deconstructionist in its patchwork of different fabrics and fragments that look as though they had been detached from finished garments as well as in its contrast of textures (the soft, fluid, jerseylike rayon set against the stiff lace of the sleeves), this was one of the outfits inspired by historic dress that Kawakubo showed at the Kyoto Costume Institute in 1993 alongside period costumes. "Things which are old have their own unique power," she said at the time, "the essence of which is completely transformed when perceived within the context of a different age. I destroy that which already exists. As I destroy, I create from the destruction. I create something entirely new and different. I owe my ability to create the new to the existence of the exquisitely old. I must destroy to transcend, but it is in playing off the old that I am inspired to create the new."[64]

255 Feather Flurries fabric, 1993
Designed by Reiko Sudo
Made by Nuno Corporation, Tokyo
Silk and feathers, width 45 1/4" (115 cm)
Nuno Corporation, Tokyo

Inserting real feathers—goose, peacock, and guinea hen, imported from China and India—into rectangular pockets of transparent silk organdy, Reiko Sudo created a delicately embellished fabric that suggests the airy weightlessness of flight. A traditional ornament of hats and clothes, feathers are captured individually here and their varied shapes, colors, and lusters can be admired like a nonrepeating pattern under the glassy transparency of the silk. The silk itself is a double weave, the two layers producing an iridescent moiré pattern, which plays against the zigzag stripes of the feathers cased inside.

255

Yusuke Aida
Born 1931

Ceramic craftsman and designer, Yusuke Aida was trained as a city planner at Chiba University in Tokyo (1956) and afterward studied ceramics with Ken Miyanohara. From 1961 to 1964 he worked in the United States teaching at the Boston Museum School (1961–62) and as chief designer for Bennington Potters in Vermont (1962–64), creating the Classic series of tableware that is still produced by the company (no. 74), although with an alternate painted decoration. On his return to Japan in 1965, Aida turned from the design of industrially produced tableware to studio and architectural ceramics. His work has included large-scale ceramic walls for the Keio Plaza Hotel, Tokyo (1970–71), Nagoya Kanko Hotel (1971–72), Bank of Japan, Osaka (1980), Funabashi city hall (1981–82), Yamagata city hall and civic center (1983), and Shizuoka city hall (1989). Aida has served as director of the Japan Designer Craftsman Association (1965–76) and of its successor, the Japan Craft Design Association (1976–79). Between 1986 and 1990 Aida was a member of the Export Inspection and Design Encouragement Council of the Ministry of International Trade and Industry, and in 1992 he became professor of art at Tohoku University of Art and Design in Yamagata. His prizes include the Yoshida Isoya memorial prize (1977) and the medal of honor (1993).
References: Tokyo, The National Museum of Modern Art, Crafts Gallery, *Contemporary Vessels: How to Pour* (Feb. 10–Mar. 22, 1982), nos. 119–20, 182–83; Kathryn B. Hiesinger and George H. Marcus, eds., *Design Since 1945* (Philadelphia, 1983), pp. 102, 204; *The Best Selections of Contemporary Ceramics in Japan,* vol. 1, *Yusuke Aida* (Kyoto, 1992).

Yusai Akaji
1906–1984

Yusai Akaji was born in Kanazawa, near the Japan Sea, an area famous for its lacquerwork, particularly the wares made in the town of Wajima. His early training was with local lacquerers, from whom Akaji learned the *kyushitsu* (lacquer varnish) technique for which he would become known. In 1928 he moved to Tokyo to study further and master the *magewa* (bent ring) craft for making wood bowls and trays. The precise application of lacquer varnish to complex bent-ring constructions was to become his chief trademark when he set up his own lacquer studio in 1930, specializing in wares for the tea ceremony, including tea caddies, water-container lids, and serving trays (no. 89). From 1955 Akaji showed his pieces in the annual Traditional Japanese Crafts exhibition of the Japan Craft

Association, winning prizes in 1959, 1960, 1961, and 1967. In 1962 he became a member of the association's board of directors, and served as a juror for the exhibitions. Akaji was also on the board of the Japan Lacquer Craft Association, and taught at the Wajima Lacquer Techniques Research Institute in his native Ishikawa Prefecture. Akaji received numerous honors for his achievements as a lacquer craftsman, including the medal of honor (1972) and the Order of the Rising Sun (1978). In 1974 he was named a Living National Treasure (Holder of Intangible Cultural Property).
References: Tokyo, The National Museum of Modern Art, *Lacquer Art of Modern Japan* (Nov. 1–Dec. 9, 1979), pp. 89–91; Boston, Museum of Fine Arts, *Living National Treasures of Japan* (Nov. 3, 1982–Jan. 2, 1983), p. 272, nos. 192–93.

Shigeru Akizuki
Born 1930

One of Japan's foremost package designers, Shigeru Akizuki has adapted traditional package construction to modern purposes, using, for example, the old type of Japanese folding box *(chitsu)* in his independent work as well as in commercial packaging. Born in Tsingtao, China, Akizuki moved to Niihama, Japan, in 1945. He studied design at the Japan Advertising Art College, Tokyo (1955), after which he became a free-lance package designer. By the 1960s his products ranged from shopping bags for the Mona department store and salad-oil cans for Nisshin Oil Mills, to whiskey and wine bottles for Godo Shusei. His broad vocabulary included Western abstract shapes and Roman letter forms as well as simplified, modern *kanji* characters and traditional patterns (no. 225). Akizuki was a participant in the "4 Box-ers" exhibitions held periodically at the Ginza Wako Gallery, Tokyo, where he received particular notice for his creations using Japanese folk toys as their motifs (no. 149). Akizuki has received many awards from the Japan Package Design Association and is a member of the Japan Folk Toys Society and Japan Graphic Designers Association. He has also lectured at the Kuwasawa Design School, Tokyo.
References: *Boxes by Four: Package Design* (Tokyo, 1982), pp. 17–63, 208; "Shigeru Akizuki," *Graphis,* no. 239 (1985), pp. 52–61.

Alps Shoji Company
Established Tokyo, 1947

The Alps Shoji toy company is best known for its production of battery-operated toys. One of the three largest toy companies in Japan during the 1950s and 1960s, Alps Shoji established the Alps Trading Company in 1956 to export toys to the United States. These toys ranged from the spaceman inspired by NASA's Apollo project (no. 71) to a battery-powered model of a Coney Island ride. Today the firm produces both battery-powered and plush toys and animals, and concentrates its sales in the domestic Japanese market.

Reference: New York, Christie's East, *The Davidson Collection of Robots and Space Toys*, Nov. 11, 1989, no. 134.

Junichi Arai
Born 1932

One of the most technologically innovative textile designers anywhere (see nos. 119, 158, 247), Junichi Arai was born to a family of weavers in the traditional textile center of Kiryu. He started his career in his father's kimono factory and audited textile courses at Gunma University in Kiryu. In the mid-1950s Arai began experiments with metallic yarns and chemically altered fibers, eventually receiving thirty-six patents for his inventions, and in the 1970s he first explored the possibilities of applying computers to fabric design and production. Using highly complex computer-generated punch cards to drive Jacquard looms, Arai brought new possibilities to the textile industry, weaving quadruple-cloths with different designs on each surface, and other fabrics of unparalleled complexity (no. 188). Since the 1980s, he has also experimented with a number of high-tech finishing techniques, using the heat-transfer print machine, for example, to produce extraordinary patterns and textures, such as those created by overlapping layers of permanently pleated wrinkles (no. 230). A number of his fabrics have been used by the fashion designers ISSEY MIYAKE and REI KAWAKUBO. "Arai's conscious rejection of the conventional," the textile designer REIKO SUDO wrote, "leads him to deny total and dispassionate perfection. His subversions of heat-set and chemical-finished processes allow for spontaneity and ever-changing expressions. He sometimes, for example, ties fabrics for tie-dyeing, then purposely omits all dyeing, putting the tied fabrics directly into the tumble-dryer for finishing. This culminates in unpredictably shrunken and buckled textures. He is truly the enfant terrible of Japanese textiles."[1] In 1983 Arai was awarded the Mainichi fashion grand prix special prize and in 1987 named an Honorary Royal Designer for Industry by the British.

References: Maria Tulokas, *Textiles for the Eighties* (Providence, 1985), p. 61; Chloë Colchester, *The New Textiles: Trends and Traditions* (New York, 1991), pp. 20–21, 39–40, 170, figs. 61–63, 117, 119; Tokyo, Yurakucho Asahi Gallery, *Hand and Technology: Textile by Junichi Arai '92* (Mar. 7–25, 1992).

Michio Arai
Born 1938

Chief executive officer of Arai Helmet since 1986, Michio Arai joined the firm as a product designer and engineer in 1963, following graduation from the engineering department of Keio University, Tokyo (1961), and study of chemical engineering at Indiana Institute of Technology, Fort Wayne (1961–62). The helmet company had been founded in 1950 by his father, Hirotake Arai, who was the first to use fiberglass and expanded polystyrene in the production of Japan's safety helmets. Michio Arai, who has been awarded twenty patents for his inventions and designs, developed these through his own work (no. 128), to which his avocation as a semiprofessional racing driver and motorcycle racer gave practical test.

Reference: Andy Saunders, "Helmet Guide 1991," *Motorcyclist*, no. 1135 (Nov. 1991), pp. 24–25.

Masakichi Awashima
1914–1979

Designer of utilitarian and decorative items of glass, Masakichi Awashima graduated from the design department of the Japan School of Fine Arts in Tokyo, where he remained as a lecturer until 1935, when he was hired as a designer by the Kagami Crystal Glassworks (see KOZO KAGAMI). In 1946 he joined the Hoya Glass company as its chief designer, remaining until 1950, when he founded his own workshop, the Awashima Glass Design Institute (which in 1956 became the Awashima Glass Company). Inspired by the desire to create a glass form uniquely his own and uniquely Japanese, Awashima developed a process for *shizuku* (dripping water) glass, which he patented in 1954 (no. 22). Several one-man exhibitions of Awashima's work were held in the 1950s, and in 1956 he was awarded the Inventors Association prize.

References: Dorothy Blair, *A History of Glass in Japan* (New York, 1973), pp. 298–99; Tokyo, The National Museum of Modern Art, Crafts Gallery, *Modern Japanese Glass: Early Meiji to Present* (Sept. 22–Nov. 28, 1982), n.p.; Kathryn B. Hiesinger and George H. Marcus, eds., *Design Since 1945* (Philadelphia, 1983), pp. 103, 205.

Hiroshi Awatsuji
Born 1929

One of Japan's best-known designers of printed textiles, Hiroshi Awatsuji graduated from the design department of the Kyoto University of Fine Arts in 1950, and worked as a designer for the Kanegafuchi spinning company (now Kanebo) in Kyoto. In 1953 he moved to the Kenjiro Oishi Studio in Tokyo, where he worked for four years, and in 1958 opened his own design studio; since 1963 he has designed furnishing fabrics for Fujie Textile and others (nos. 90, 152). During the 1960s his works were shown regularly in the Good Design exhibitions held annually at the Matsuya department store in Tokyo, and in 1968 Awatsuji was made a member of the Japan Design Committee, which organized them. Involved in a number of official projects, such as designing carpets and curtains for two of the pavilions at Expo '70 in Osaka, he has also designed tapestries, carpets, and furnishing fabrics for banks, hotels, and other businesses across Japan. In 1971 he received the Mainichi industrial design prize for his achievements in textile design for interiors and in 1972, the Japan Interior Designers Association prize. He received a silver prize in industrial arts at the Third Textile Triennale in Lodz, Poland, in 1978. In 1988 Awatsuji founded his own manufacturing company, Awa, through which he has produced a series of black-and-white-patterned textiles and tablewares (nos. 212–13). Awatsuji taught at the Otsuka Textile Design Institute (1963–85), and has been a professor at Tama Art University in Tokyo since 1988.

References: *The Textile Design of Hiroshi Awatsuji* (Tokyo, 1990); *The International Design Yearbook* 5 (New York, 1990), pp. 159–61; Yuki Yoshihara, "Awatsuji Hiroshi," *Designers' Workshop*, vol. 9, no. 57 (Oct. 1992), pp. 62–66.

Kiyoshi Awazu
Born 1929

Kiyoshi Awazu's design work spans a broad spectrum, including graphics for posters (no. 76) and books, stage and film sets, and exhibitions. After leaving Hosei University, Tokyo, in 1947, Awazu began his career as an independent designer, deliberately avoiding commercial clients. Recognition first came in 1956 when he won the Japan Advertising Artists Club grand prize for his poster *Give Back the Sea*, which protested American military activity off the coast of Japan. Awazu was one of sixteen designers invited to participate in the important 1965 Tokyo group show "Persona," sharing the 1966 Mainichi industrial design prize awarded to the exhibition. His activities expanded further: to the design of the playground for Expo '70 at Osaka; art direction for films, winning the Japanese equivalent of an Oscar in 1980 and 1987; and the plan and design for the National Museum of Ethnology in Osaka, which opened in 1977. His work in exhibition design and graphics became familiar to Western audiences through his collaboration on the design of "The Great Japan Exhibition" at the Royal Academy of Arts, London, in 1981, and his participation in "Tokyo: Form and Spirit," organized by the Walker Art Center, Minneapolis, in 1986. He also had solo exhibitions in Los Angeles (1992) and Aspen, Colorado (1993). Awazu has taught design at Musashino Art University, Kodaira (1964–70), and Kyoto Junior College of Art (1980–91).

References: *The Works of Kiyoshi Awazu, 1949–1989* (Tokyo, 1989); *Best 100 Japanese Posters, 1945–89* (Tokyo, 1990), pp. 78–79, 180–81, 241; Richard S. Thornton, *The Graphic Spirit of Japan* (New York, 1991), pp. 87–89, 105, 108, 129, 138, 180, 213.

Bandai
Established Tokyo, 1950

During the 1980s the Bandai company led the Japanese market in creating toys based on characters in films and television series (no. 155), from Godzilla and Ultraman to the Mighty Morphin Power Rangers. In 1983 Bandai acquired the toy company Popy, which specialized in manufacturing television character toys and had been a leading contributor to the transformer market (no. 153). Today Bandai is diversifying into other products for children and young adults, from word processors and electronic organizers to clothing, toiletries, and candy.

Reference: D. J. McCann and others, "Asia: A Major Player in the Toy Industry," *Asian Business*, vol. 22 (Aug. 1986), pp. 47–56.

Bridgestone Cycle Company
Established Tokyo, 1949

The Bridgestone Cycle Company was founded as a subsidiary of the Bridgestone tire corporation to produce bicycles for economical transportation in postwar Japan. Offering the consumer high-quality bicycles that it promoted for multipurpose use (no. 154), Bridgestone became Japan's largest bicycle manufacturer by 1970. It remains the largest bicycle manufacturer in Japan and one of the five largest in the world, producing some one and a half million bikes each year at the five factories it operates in Japan. During the 1970s Bridgestone also made bicycles for companies abroad, notably for Schwinn in the United States; since the 1980s, it has distributed its bicycles under its own name in the United States, Europe, and parts of Asia.

Reference: Michael Gamstetter, "Cutbacks Force Bridgestone to Quit U.S. Market," *Bicycle Retailer & Industry News* (Apr. 1994), pp. 1, 22–23.

Canon
Established 1933

Known as a leader in technology, in particular for its high-quality cameras (no. 252), business machines (no. 224), and optical products (including medical and industrial equipment), Canon defines itself as a company concerned with "information in visual form, created, edited, stored, reproduced, transmitted." Canon was founded in 1933 as Precision Optical Research Laboratories, and in the next year developed Japan's first 35-millimeter camera, the Kwanon, named for the Buddhist goddess of mercy. Precision Optical soon ventured into the production of medical diagnostic equipment, designing another first for Japan, the indirect X-ray camera (1940), which has been credited with contributing to the eradication of tuberculosis in Japan. The success of its precision cameras inspired the company to change its name once more, becoming Canon Camera (anglicizing "Kwanon" for export) in 1947; in 1969 the corporate name Canon was adopted. Throughout the 1950s and 1960s, Canon developed international markets for its products, opening offices in New York (1955) and in Geneva (1957). The company continued its policy of diversification, developing 8-millimeter movie cameras (1956, no. 20; 1963, no. 82) and Japan's first zoom lens for television cameras (1958), and, perhaps most importantly, expanding into the fields of micrographics in 1959 and business machines in the 1960s. Canon introduced the now-familiar ten-keypad electronic calculator in 1964 (Canola 130; no. 81), NP (New Process) plain-paper copier technology in 1968, and bubble-jet printing in 1981, using a completely new technology. By 1987 business machinery accounted for over 70 percent of Canon's global sales. Canon has consistently invested substantial amounts in its research and development divisions, stressing the importance of cooperation between designers and engineers. Designers are required to have a significant knowledge and understanding of the technical constraints of the products they design, and at the same time, respond to human-factor issues (no. 198): "Consideration of life-style is extremely important. Pulling together elusive human feelings can't be done with words. That's why we need designers, because they can express things with forms which resist verbal definition."[2]

References: *The Canon Story 1986 / 87* [corporate brochure]; Roxanne Guilhamet, "In-House Rules," *Design*, no. 489 (Sept. 1989), pp. 42–45.

Daisaku Choh
Born 1921

Daisaku Choh considers his work as architect and interior designer two sides of the same coin. "Where in a single building does the interior end and the exterior begin?," he asks. "The fundamental essence of Japanese architecture . . . is that the exterior and interior construction . . . is almost identical."[3] After graduating from the Tokyo School of Fine Arts in architecture, Choh joined the architectural firm of Junzo Sakakura in 1947. In 1960 he was responsible for the Japanese section of the Triennale in Milan, which won a gold medal for his exhibition design on that occasion. He also designed furniture, manufactured by TENDO MOKKO (no. 49), and served as a juror for the company's annual furniture design competition (1960–66). Choh opened his own office in 1972, specializing in residential architecture and furniture design. His furniture has received numerous awards, including the Mainichi industrial design prize in 1971 and the Japan Interior Designers Association prize in 1977.

References: Japan Design Committee, *Design 19* (Tokyo, Matsuya Ginza, Sept. 3–8, 1982), n.p.; Japan Industrial Designers Association, ed., *Seiichi no Kozo: Nihon no Indasutoriaru Dezain* (Structure of Dexterity: Industrial Design Works in Japan) (Tokyo, 1983), p. 130.

Chubu Design Research Center
Established Nagoya, 1956

The Chubu Design Research Center was established by seven designers in Nagoya in 1956. Since its beginnings, its energies have been directed toward design activities in the *chubu*, or central, region of Japan, including the prefectures of Mie, Ishikawa, and Aichi, where Nagoya is located, and during the 1960s members of the center began organizing exhibitions featuring the work of regional designers at the Aichi Prefectural Museum of Art. In 1983–84 the firm directed the Aichi Prefecture industrial design development project, and since then has acted as technical advisor to the governments of the three prefectures. The Chubu Design Research Center has engaged in a wide variety of product, graphic, package, and interior design, ranging from a motor scooter in 1957, its first design, to a wireless microphone (no. 107), office accessories, corporate logos, and interiors. Its products have received many G-Mark citations, as well as a silver prize at the Matsuya department store's Design Forum exhibition in 1981 and most recently, the 1991 Yamagata (Prefecture) Green Design prize. In 1988 the firm became a founding member of the Free-lance Designers Club in Nagoya.

Reference: Japan Industrial Designers Association, ed., *Man and Tool—Discovery of New Interrelations* (Tokyo, 1973), pp. 52, 80.

Hiroshi Egawa
Born 1964

Product designer Hiroshi Egawa graduated from the Tokyo Designers Institute in 1984, where he majored in industrial design, and in 1989 joined Aoyoshi as a designer. He has concentrated on the design of cutlery, including that for use by the elderly and physically impaired (no. 242).

Reference: "Hiroshi Egawa / Product Designer," *Axis*, no. 40 (Summer 1991), p. 3.

Eitaro-Sohonpo
Established Tokyo, 1857

Founded in 1857, the Tokyo confectionery firm Eitaro-Sohonpo gained an early reputation for creating original sweets, from the plum candy that the company first produced to *kintsuba*, a bite-size sweet made from the adzuki bean. The Eitaro store was destroyed in the earthquake of 1923, and the firm was not fully rebuilt until after World War II. Eitaro products are now sold in Eitaro shops as well as boutiques in Japanese department stores and elsewhere throughout Japan (no. 111), an expansion that adheres to the firm's founding philosophy of trying to reach increasing numbers of people.

Reference: "Eitaro-Sohonpo," in Takeo Yao, ed., *Package Design in Tokyo* (Tokyo, 1987), pp. 124–27.

Kazuko Fujie
Born 1947

After completing her studies of industrial design at Musashino Art University (1968), Kazuko Fujie worked for the architect Dan Miyawaki (1969–73) and then for Endo Planning (1973–77). In 1977 she established the F Atelier and in 1987 incorporated her design office as the Kazuko Fujie Atelier Company. Fujie is known for the site-specific furniture she has created for public spaces and buildings (no. 190), including that in the Mita library of Keio University in Tokyo, the National Museum of Modern Art, Kyoto (1986), and Kirin Plaza in Osaka (1987). She has worked on numerous design projects for her native city of Toyama, including the Toyama municipal airport, and has exhibited in one-woman shows in Tokyo (1989) and Toyama (1990). Fujie has won awards from the Japan Product and Environmental Planning Association (1989) and the Japan Interior Designers Association (1990).

References: *The International Design Yearbook* 2 (New York, 1986), pp. 98, 223; Toyama, Shimin Plaza, *Fujie Kazuko Kagu Dezain-ten* (Kazuko Fujie Furniture Design Exhibition) (May 8–29, 1990).

Kenji Fujimori
1919–1993

Designer of Japan's first and best-known chair for *tatami* use (no. 58), Kenji Fujimori reinterpreted the traditional Japanese custom of sitting on the floor. Fujimori graduated from Waseda University, Tokyo (1941), and worked in the general planning division of Toshiba Machinery Company before opening his own design office in 1948. In 1954–55 he was sent by the Japanese government to study product design in Helsinki, initiating a life-long involvement in international—particularly Finnish—design institutions and programs, including the directorship of the Japan-Finland Association between 1955 and 1972. From 1963 until his retirement in 1985, Fujimori was professor at the Kanazawa University of Arts and Crafts; he also lectured variously at Aoyama Gakuin University, Tokyo (1966–68), and Waseda University (1968–70). Fujimori has studied and promoted traditional handcrafts both in Japan and abroad: a founder of Japan Foundation's craft center in Tokyo in 1959, he served as its director from 1959 to 1972. In 1972 he was named Japanese delegate to the World Crafts Council, subsequently serving as vice president and honorary director of that organization (1980, 1984) and organizing a number of its conferences, including one held in Kyoto in 1978. Between 1974 and 1984 Fujimori was advisor to the Japanese government for the promotion of traditional crafts. His many honors include medals from the Finnish government (1962, 1967) and the World Crafts Council (1978).

References: Shukuroo Habara, "Fujimori Kenji no Kinsaku Isu" (Kenji Fujimori's Recent Chair), *Design*,

no. 111 (July 1968), pp. 56–57; Japan Industrial Designers Association, ed., *Seiichi no Kozo: Nihon no Indasutoriaru Dezain* (Structure of Dexterity: Industrial Design Works in Japan) (Tokyo, 1983), p. 130.

Shigeo Fukuda
Born 1932

Recognized internationally as one of the most innovative graphic designers of his generation, Shigeo Fukuda won his first poster contest at age eleven, and his first published work was a comic strip drawn while he was in high school. After graduating from the Tokyo National University of Fine Arts and Music in 1956, he joined the advertising department of the Ajinomoto firm, where he worked with MITSUO KATSUI. In 1958 he helped found the Deska graphic design firm, working with TAKASHI KONO, before starting out on his own as a free-lance designer in 1959. At the Tokyo World Design Conference (WoDeCo) in 1960 Fukuda met Bruno Munari, the Italian designer whose graphic work and toys Fukuda cites as a major influence on his career. It was in 1963, after the birth of his daughter, that Fukuda began designing toys (no. 77) and illustrated books, which were the subject of an exhibition, "Toys and Things Japanese: The Works of Shigeo Fukuda," organized in 1967 by the noted American graphic designer Paul Rand at New York's IBM gallery. The exhibition subsequently traveled to Amsterdam, London, and Milan. Fukuda's graphic work was also shown in Europe at the Biennale of Graphic Design in Brno, Czechoslovakia, where it took prizes regularly from 1966 to 1980. In Japan, Fukuda's extraordinary talents were quickly recognized as well when he was invited to participate in the 1965 "Persona" exhibition of young graphic designers, for which he shared the Mainichi industrial design prize. His second Mainichi prize came in 1970 for his series of three-dimensional toys. His seemingly limitless creative spirit ranged widely over the following decades, and his work included the design of posters (nos. 112, 115), books, and calendars; graphics and signage for Expo '70 and the Sapporo Winter Olympics in 1972, for which he also did the commemorative medal; toys and *origami* (no. 106); sculpture; and installations—such as his "100 Smiles of Mona Lisa" shown in the United States and Europe (1972–74)—as well as exhibitions and displays. Among the many honors he has received are gold and silver medals at the International Poster Biennale, Warsaw; the book design prize sponsored by Kodansha publishing company (1980); Japan Display Design Association prize (1980, 1981); Mainichi art prize (1986); and New York Art Directors Club Hall of Fame award (1987). Shigeo Fukuda's manifold, creative output is memorable for its playfulness and ingenuity, the work of a graphic magician.

References: Shigeo Fukuda, *Fukuda Shigeo no*

Rittai Zokei (Shigeo Fukuda's 3-D Creations) (Tokyo, 1977); Japan Design Committee, *Design 19* (Tokyo, Matsuya Ginza, Sept. 3–8, 1982), n.p.; Shigeo Fukuda, *Fukuda Shigeo Hyohonbako* (Shigeo Fukuda's Specimen Box), rev. ed. (Tokyo, 1987).

Saburo Funakoshi
Born 1931

From 1957 Saburo Funakoshi made his career at Hoya Glass, where he became head of the design department at its Musashi factory in 1982, creating sleek, refined, glass tableware for mass production and decorative pieces for manufacture by hand in limited editions. Funakoshi retired from Hoya in 1993 to pursue his work as a glass artist, but continues to provide the firm with designs on a free-lance basis. His functional glassware (nos. 54, 87) raised industrial tableware to a new level of quality in Japan and won many distinctions there, including recognition from the Osaka Design House in 1970. His work has been seen in solo exhibitions at the Matsuya department store (1969, 1982), Museum of Modern Art, Kamakura (1974), National Museum of Modern Art, Tokyo (1980), and Hokkaido Museum of Modern Art (1982). Funakoshi was trained in the crafts department of Tokyo School of Fine Arts (1954) and worked from 1954 to 1957 at the Shizuoka Prefectural Industrial Testing Institute.
References: Corning, New York, The Corning Museum of Glass, *New Glass: A Worldwide Survey* (Apr. 26–Oct. 1, 1979), p. 255, no. 63; Tokyo, The National Museum of Modern Art, Crafts Gallery, *Modern Japanese Glass: Early Meiji to Present* (Sept. 22–Nov. 28, 1982), no. 142; Tokyo, The National Museum of Modern Art, Crafts Gallery, *Contemporary Vessels: How to Pour* (Feb. 10–Mar. 22, 1982), nos. 189–90; Japan Design Committee, *Design 19* (Tokyo, Matsuya Ginza, Sept. 3–8, 1982), n.p.; "Saburo Funakoshi," in *Design Collection Matsuya Ginza* (Tokyo, 1989), pp. 10–11; Kogei Zaidan, ed., *Nihon no Indasutoriaru Dezain: Showa ga unda Meihin 100* (Japanese Industrial Design: 100 Masterpieces Produced during the Showa Period) (Tokyo, 1989), p. 92.

Tetsuya Furukawa
Born 1920

Tetsuya Furukawa graduated from the Tokyo College of Industrial Arts in electrical engineering in 1943, continuing his studies and receiving a doctorate in engineering in 1959. He joined Fuji Electric Company in 1945, where he worked in various capacities, from 1950 as head of the product development group for home electronics design, and from 1964 until his retirement in 1977 as head of the development department. Furukawa has received numerous patents for his designs, particularly in the field of home electronics (no. 16). The Inventors Association prize was awarded to Furukawa a number of times, and he

received its grand prize in 1968. Since 1971 Furukawa has taught at Kashima University, and since 1985, at Kumamoto Institute of Technology.
References: Tetsuya Furukawa, "Shinseihin no Sempuki ni tsuite" (About the New Production Fan), *Kogei Nyusu* (Industrial Art News), vol. 18 (Sept. 1950), pp. 22–23; Fuji Electric Company Archives, ed., *Fuji Electric Company History II (1957–1973)* (Tokyo, 1974), pp. 46–47; Fuji Electric Editorial Committee, ed., *Fuji Electric Cooling Machines: 25 Year History* (Tokyo, 1990), p. 9.

GK
Established Tokyo, 1953

One of Japan's largest industrial design firms, GK was founded in 1953 (and incorporated in 1957) by four former students of IWATARO KOIKE at the Tokyo National University of Fine Arts and Music. The four, Kenji Ekuan, SHINJI IWASAKI, Kenichi Shibata, and Harutsugu Ito, took the name of their office from the initials for "Gruppe Koike." As their first project in 1953 the group entered a competition for the design of an upright piano organized by Nippon Gakki (later YAMAHA CORPORATION), which became one of the firm's most important long-term clients. Until about 1965 the office concentrated on the design of individual objects using new production technologies, including a series of motorcycles for Yamaha, designed by Iwasaki (no. 35); bicycles for Maruishi, with whom GK also had a long-term contract; and the well-known Kikkoman soy-sauce bottle (no. 53). Between 1965 and 1975 GK engaged in comprehensive design projects, such as the planning and design for Expo '70 in Osaka and for the Kyoto Trust Bank in 1972, a program that included architecture, interior-design graphics, and even employee uniforms. After 1975 GK established several affiliated companies to provide a variety of complex and interconnected design services. These include five companies in specialized fields: GK Inc., for product design and planning; GK Sekkei, for architecture, urban, and environmental design; GK Graphics, for a range of graphic services, from corporate identity and packaging to computer software; GK Dynamics, for motorcycle, sports, and leisure products; and GK Tech, for technology and transportation. In addition, there are four GK companies outside Tokyo: GK Kyoto; Design Soken Hiroshima (a joint venture with Mazda Motor Corporation); GK Design International in Los Angeles; and Global Design in Amsterdam—as well as three companies for research and information: GK Institute of Doguology, Voice of Design, and GK Institute of Industrial Design. Kenji Ekuan, chairman of GK, has impressed his personality most strongly on the company and has been credited with its present diversified organization and global outlook (see nos. 191, 232, 237).
Reference: *A Design Declaration—GK's 30 Year History*, *Space Design*, special issue, no. 19 (Jan. 1988).

Michio Hanyu
Born 1933

A 1956 graduate of the Kuwasawa Design School in Tokyo, Michio Hanyu stayed on as an assistant and in 1959 became a full-time lecturer. During his summer vacations, he traveled to regional production centers, where instead of finding the refined work of highly trained craftsmen he was expecting, he was dismayed to discover, for example, that a town once renowned for its cutlery was now making cheap pocket knives for export. These experiences moved him to open his own design studio in Tokyo in 1966, which in 1972 he expanded into his present firm, Monopro Designers. Monopro initiated cooperative ventures with regional manufacturers to design and produce high-quality cutlery, tableware, glassware, ceramics, kitchen wares, and tool sets (no. 180). The Monopro approach was unusual for its time; the firm participated in every step, including product planning, design, production, packaging, store-display design, and marketing. Hanyu aims to create a total concept rather than single products; the Monopro tableware, for example, includes an entire line of cutlery intended to be used with matched glassware and serving dishes. Hanyu received the Mainichi industrial design prize in 1976 for Monopro's tableware and kitchen ware series, and his kitchen wares also received a special G-Mark citation in 1986. The Kunii Industrial Art Award went to Hanyu in 1988 as well, in recognition of his twenty years of work to encourage regional production of quality goods for the domestic market.
References: "Jimusho Homon: Nihon no mo Hitotsu no Sangyo o Tsukuritai" (Visit to an Office: Monopro Co., Ltd.), *Industrial Design*, no. 118 (July 1982), pp. 24–27; Kogei Zaidan, ed., *Nihon no Indasutoriaru Dezain: Showa ga unda Meihin 100* (Japanese Industrial Design: 100 Masterpieces Produced during the Showa Period) (Tokyo, 1989), p. 100; Michio Hanyu, "Monopro," in Kogei Zaidan, ed., *Nihon no Kindai Dezain Undoshi 1940 Nendai–1980 Nendai* (History of the Japanese Modern Design Movement, 1940s–1980s) (Tokyo, 1990), pp. 156–59; "Purofuiru: Hanyu Michio" (Profile: Michio Hanyu), *Nikkei Design*, no. 53 (Nov. 1991), pp. 138–39.

Hiromu Hara
1903–1986

Revered as a graphic designer (no. 45), art director, teacher, and design activist, Hiromu Hara was one of the pivotal figures in the emerging field of Japanese design. After graduating from the Tokyo College of Industrial Arts in 1921, he began teaching at his alma mater. In 1933 he helped establish the publishing firm Nippon Kobo, becoming art director of its journal *Nippon* in 1938; from 1942 to 1945 he was also art director for the magazine *Front*, which became known for its layout

and dramatic use of photomontage. Hara was a founding member of the Japan Advertising Artists Club in 1951 and, with YUSAKU KAMEKURA and RYUICHI YAMASHIRO, of the Nippon Design Center in 1960, serving as president in 1969. In the 1960s Hiromu Hara's graphic work included art direction for *New Japan* and *Taiyo* magazines, receiving the Tokyo Art Directors Club prize in 1964 for his work on the latter. Hara was one of the team of graphic designers who worked on the program for the 1964 Tokyo Olympics, for which he was coordinator of typography and advertising design. Throughout his career, Hara was particularly respected for his book (no. 102) and paper designs, for which he received many awards, including the Mainichi industrial design prize in 1960 for the series he designed for the Takeo paper company. Hara was professor of graphic design at Musashino Art University in Kodaira from 1952 to 1970. In 1971 he received the medal of honor from the Japanese government for his contributions to the field of graphic design, and in 1988 the Tokyo Art Directors Club prize was awarded to him posthumously for his book designs.
References: *Hara Hiromu* (Tokyo, 1985); *Best 100 Japanese Posters, 1945–89* (Tokyo, 1990), pp. 66–67, 241.

Tokio Hata
Born 1911

Born to a master of traditional landscape gardening, Tokio Hata recalls the strong impression on his youthful mind of the beauty of the trees, plants, and natural surroundings of his native Kanazawa, located near the seacoast of western Japan. As a child Hata loved to draw and paint, developing these talents further when he was apprenticed to a master of the paste-resist-dyeing (*yuzen-zome*) technique in his hometown in 1925. Six years later he went to Kyoto to continue his training, and in 1937 established his own workshop there. During the war, Hata participated in a government program developed to help further traditional crafts as one of the artists designated to preserve the *yuzen* dyeing technique. Hata first exhibited his work in 1955 in the Traditional Japanese Crafts exhibition, and has continued to show there annually. The resist-dyed and hand-painted patterns of Hata's kimonos are painterly renditions of subjects taken from nature, such as pheasants and ducks, plum blossoms and snow-covered pine trees, and the rippled sands on the shorelines of the seacoast of Kanazawa (no. 56). He has also received commissions for the huge, gorgeous textile hangings that decorate the carts in the annual Gion festival in Kyoto, and for a kimono presented by the government of Kyoto to H.R.H. Diana, Princess of Wales, in 1986. Hata has been decorated for his work with the Order of the Sacred Treasure (1982). In 1988 Hata's skill in *yuzen-zome* was further recognized when he became a Living National Treasure, and a retrospective exhibition of his work was held at the Kyoto Municipal Museum of Art in 1992.

Reference: Kyoto Municipal Museum of Art, *Hata Tokio-ten* (Exhibition of Tokio Hata) (May 22–June 20, 1992).

Yoshio Hayakawa
Born 1917

By means of simple, abstracted designs, Yoshio Hayakawa played an important role in shaping the development and character of postwar commercial design in Japan. Following graduation in 1936 from the Osaka Municipal School of Industrial Arts, Hayakawa began to design window displays at the city's Mitsukoshi department store, a career soon interrupted by the war. Afterward he became art director at the Kintetsu department store, also in Osaka (no. 4), and by 1952 had established himself as an independent designer there, where he continued to work primarily for department stores until he moved to Tokyo in 1971. During the 1950s Hayakawa designed packaging and stage sets as well as the posters for which he is best known. His more recent work includes posters for Ina Seito (1984) and Allstate Automobile and Fire Insurance (1985). A lecturer at the Kyoto University of Fine Arts from 1953 to 1970, he has also taught at the University of Fine Arts, Osaka, since 1964. Hayakawa was a founding member of the Japan Advertising Artists Club in 1951, served as a panelist in the World Design Conference (WoDeCo) of 1960, and was instrumental in developing the graphic program for Expo '70 in Osaka. Among his many awards are the Mainichi industrial design prize (1955), medal of honor (1984), and Order of the Sacred Treasure (1988).

References: W. H. Allner, *Posters* (New York, 1952), pp. 50–51; R. Yamashiro, "Yoshio Hayakawa," *Graphis*, no. 55 (1954), pp. 358–67; Hans Wichmann, *Japanische Plakate: Sechziger Jahre bis Heute* (Munich, 1988), pp. 65–67; *Best 100 Japanese Posters, 1945–89* (Tokyo, 1990), pp. 34–39, 90–91, 202–3, 242; Colin Naylor, ed., *Contemporary Designers*, 2nd ed. (Chicago, 1990), pp. 241–42; Richard S. Thornton, *The Graphic Spirit of Japan* (New York, 1991), pp. 72–75, 81–82, 119, 232–33; Masaaki Tanaka, "Nihon o Daihyo suru 'Osaka no' Gurafikku Dezainaa" (Yoshio Hayakawa: The Graphic Designer in "Osaka" Represents Japan), *JSSD*, vol. 1 (1993), pp. 22–24.

Hayakawa Metal Works, see Sharp Corporation

Satoru Hibino
Born 1951

A graduate of the Tokyo Design School in 1973, industrial designer Satoru Hibino has received numerous awards for his designs for cameras, cutlery, and audio products, including the Yamaha international audio design competition prize for his Ikebana speaker in 1989 (no. 229). A member of the International Design Center and Design Net International (DNI), Hibino is presently employed by Maruwa Electronic and Chemical Company, which develops and produces stereo equipment for automobiles.

Reference: "Yamaha no Odio Conpe" (Yamaha's Audio Design Competition), *Axis*, no. 34 (Winter 1990), pp. 76–79.

Takuo Hirano
Born 1930

Takuo Hirano's designs for industrial products received 440 G-Mark prizes between 1957 and 1993, including one for the Opus 8 sewing machine, which he designed in 1979 for Brother Industries (no. 133). After graduating from the Tokyo National University of Fine Arts and Music in 1953, Hirano became an examiner for the government patent bureau, and a few years later was sent by the Ministry of International Trade and Industry to study at the Art Center College of Design in Pasadena. Hirano's studies abroad convinced him of the importance of government involvement in design and design education; he became active in establishing Japan's G-Mark system in 1957, and has since been a juror for the annual G-Mark selection and advised Japanese universities on developing their design departments. Having formed his own design firm in 1970, Hirano has also done interior design, notably for the supreme court building in Tokyo, and environmental graphic design, developing the logo and graphics for the Hotel Okura in Tokyo. Hirano is a professor at Tama Art University in Tokyo.

Reference: Japan Industrial Designers Association, ed., *Seiichi no Kozo: Nihon no Indasutoriaru Dezain* (Structure of Dexterity: Industrial Design Works in Japan) (Tokyo, 1983), p. 104.

Keiichi Hirata
Born 1958

Industrial designer Keiichi Hirata graduated from Tokyo University of Art and Design in 1981. He was employed in the industrial design department of Isuzu Motors until 1986, and then worked in Ricoh's design center, where he collaborated on the design of word processors, personal computers, and telephones. In 1989 he left Ricoh to establish his own design office, We'll Design Something with Emotion, the name signaling Hirata's intent to bring psychological and emotional values to modern technology, as his elegant, scroll-shape fax machine

(no. 233), created the same year, visibly demonstrated.

Reference: "Keiichi Hirata / Industrial Designer," *Axis*, no. 34 (Winter 1990), p. 4.

Keiko Hirohashi
Born 1931

In her designs for packaging and wrapping paper, Keiko Hirohashi seeks to express the Japanese ideals of gift giving through a Japanese aesthetic, which she finds primarily in traditional textile patterns (no. 140). One of the first women to enter the field of package design, Hirohashi graduated from the Tokyo National University of Fine Arts and Music in 1954 and was hired by the Morinaga confectionery company in Tokyo as a graphic and package designer. She received recognition early in her career, when she was awarded a Ministry of International Trade and Industry prize for her posters and the Mainichi prize for advertising design in 1954, and an honorable mention from the Tokyo Art Directors Club prize committee in 1955. Hirohashi established her own firm in 1962, and three years later won the members prize at the Japan Package Design Association annual exhibition. From 1966 to 1976 Hirohashi taught at the Tokyo National University of Fine Arts and Music, and from 1978 to 1992 she worked at the design research center at Tamagawa University in Machida.

References: "Our Cover Artist Hirohashi Keiko," *Kogei Nyusu* (Industrial Art News), vol. 25 (Oct. 1957), p. 56; Japan Package Design Association, ed., *Package Design in Japan: Its History, Its Faces* (Tokyo, 1976), pp. 53, 54, 57, 75, 133, 149, 175; *Boxes by Four: Package Design* (Tokyo, 1982), pp. 162, 172, 211; Keiko Hirohashi, "Pakeeji Dezain no Bigaku" (The Aesthetics of Package Design), *Design News*, no. 215 (1991), p. 51; Richard S. Thornton, *The Graphic Spirit of Japan* (New York, 1991), pp. 156–57.

Eiji Hiyama
Born 1946

Eiji Hiyama, a package and toy designer, sees himself as a paper craftsman. A graduate of the Kuwasawa Design School in Tokyo (1969), Hiyama opened his own design studio, Papyrus, and built his reputation through the creative use of cardboard. Although he has used the material for everything from packaging to clocks, it is particularly in the field of toys that he has been most active. One of his series, Unit Houses, was shown in the Design Forum exhibition in 1979 at the Matsuya department store in Tokyo. Simple cardboard structures—square, hexagonal, and in the conventional shape of a barn—they can be put together easily by children themselves (using rubber bands for the joints), and then be stored flat again afterward. Hiyama's other cardboard toys include hobbyhorses, trains, carousels, and robots (no.

174). His toys have generally been featured yearly in Matsuya's craft and design galleries, and were included as well in the biennial Creative Toy Festival at the Japan Regional Toy Museum in Kurashiki (1991) and at the Azabu Museum of Arts and Crafts in Tokyo (1993). Hiyama also published an instructional book for making cardboard toys in 1988.

Reference: Dorothy Mackenzie, *Green Design: Design for the Environment* (London, 1991), pp. 74–75, 167.

Honda Motor Company
Established Hamamatsu, 1946

Founded by motor and racing enthusiast Soichiro Honda, the Honda Motor Company originated in 1946 as the Honda Technical Research Institute, taking its present name two years later after incorporation. Initially Honda was a manufacturer of motorbikes, its earliest models being no more than bicycles with military-surplus engines attached. By 1949, however, Honda had produced its first motorcycle, the Dream Type D, the first example with engine and frame designed by the same manufacturer. Honda controlled Japan's annual domestic motorcycle production by 1955, and in 1959, following a policy of international expansion, established its first subsidiary outside of Japan, American Honda Motor Company in Los Angeles, followed closely by ten more overseas sales companies throughout Europe. Along with its motorcycles (no. 84), Honda began production of four-wheel vehicles, the T360 light truck and the S600 sports car being marketed in 1963, but it was in 1972, with the introduction of the CVCC (Compound Vortex Controlled Combustion) engine installed in the Civic model (no. 113), that Honda broke most forcibly into the American market. The low-pollution engine, which burned a leaner mix of gasoline without using a catalytic converter, was, along with Toyo Kogyo's rotary engine, the first to meet the United States Environmental Protection Agency's 1975 standards. Initially popular with Americans for the compactness, fuel efficiency, and low emissions of models such as the Civic and Accord (introduced in 1976), Honda has more recently entered the luxury car market. By 1983, with the establishment of a manufacturing plant in Ohio, Honda became the first Japanese automobile manufacturer to extend its production beyond Japan's borders.

References: Tetsuo Sakiya, *Honda Motor: The Men, the Management, the Machines* (New York, 1982); Obituary [Soichiro Honda], *New York Times*, August 6, 1991, pp. A1, A15.

Yoshisada Horiuchi
Born 1937

Born in Nagano, in the Alpine region of Japan, Yoshisada Horiuchi opened the prefecture's first independent industrial design office in 1969. One of his first commissions, to design a pair of roller skates that simulate ice skating, was intended for summer practice of the region's hockey players (no. 93). Like many of his other products, including a snowmaking machine and a six-wheel snowmobile, the skates were geared to Nagano's winter sports. Active in promoting industrial design in his native region, Horiuchi served as chairman of the Nagano regional delegation to the 1973 conference of the International Council of Societies of Industrial Design in Kyoto (ICSID '73), taught design at the Nagano Technical College (1974–76), and was an officer of the Nagano Prefectural Design Association (1969–74). In addition to sporting goods, Horiuchi has designed audio and camera equipment, tractors, medical equipment, and electrical appliances such as small electronic telephone-answering machines and hair dryers. Eager to avoid overspecialization, he believes that a designer can learn from local, traditional products whose forms have been developed with time and usage. "Some of my colleagues in the cities may be thinking that I'm just off living in the mountains, [so] what could I be designing out here? What I do is real, it's concrete. Although I cannot turn back the clock, at least I can try to make something harmonious with that life and with that way of thinking and [of] doing things."[4] In recent years Horiuchi has become increasingly concerned with preserving the natural environment. In 1990 he became the director of the Community Ecology Research Association of Nagano.
References: Kitanishi Misako, "Hito Mono Tojo: Horiuchi Yoshisada" (Personality / Works: Yoshisada Horiuchi), *Industrial Design*, no. 118 (July 1982), pp. 36–38; Yoshisada Horiuchi, *Coffee Break for You* (n.p., 1984).

Gan Hosoya
Born 1935

In 1954 the graphic designer Gan Hosoya joined Light Publicity, an advertising agency that had been established in 1951; he has remained there since then and is now its president. He worked with a young team that included Sho Akiyama, considered one of Japan's foremost copywriters and still with the firm as vice president, and later with IKKO TANAKA. Hosoya's talents were quickly acknowledged in the awards he received from the Japan Advertising Artists Club (1955, 1956). He received the Tokyo Art Directors Club silver prize in 1959 for the first of a series of advertisements for YAMAHA MOTOR COMPANY, which was to become one of his major clients (no. 63), and in 1961 for his book design. A dramatic use of photography became the signature of Hosoya's

work for his corporate clients, and he was awarded the Mainichi industrial design prize in 1963 for his achievements in applying photography to advertising design. In 1965 he was one of the young graphic artists invited to participate in the famed "Persona" exhibition, for which he designed an exhibition poster and shared the Mainichi industrial design prize. Hosoya's graphic work was the recipient of three Tokyo Art Directors Club prizes (1971, 1984, 1988). Hosoya serves on the prize committee for the Mainichi industrial design prize.
References: Kodaira, Musashino Art University Museum, *Imeeji no Tsubasa: Hosoya Gan Sakuhinten* (The Wings of Image: Exhibition of Works by Gan Hosoya) (Nov. 21–Dec. 17, 1988); *Best 100 Japanese Posters, 1945–89* (Tokyo, 1990), pp. 72–73, 243.

Takenobu Igarashi
Born 1944

One of a group of Japanese designers who work both in the West and in Japan, Takenobu Igarashi has studied and taught at Tama Art University in Tokyo and the University of California at Los Angeles. His graphic work is distinguished by the use of axonometric Roman alphabets, which he deploys decoratively according to his own Japanese sensibility. Born in Takikawa, Hokkaido, Igarashi was educated in Tokyo and studied design and design theory privately with Masato Takahashi, a professor at Tokyo University of Education (1961–64). After receiving degrees from Tama Art University (1968) and UCLA (1969), Igarashi opened his own office in Tokyo in 1970. His exhibitions at the Fujie Gallery in Tokyo in 1973 and 1975 introduced his work to the Japanese design community and to future clients, including Citizen Watch (1976; see no. 143), Zen Environmental Design (1976), and Summit Stores (1976), all of whom benefited from his three-dimensional graphic style. In 1980–81 Igarashi's career accelerated, particularly with his involvement in several environmental graphics projects, among them signage for the Kitashiobara community center (1980) and for the Parco Part 3 department store (1981). Igarashi designed poster calendars for the Museum of Modern Art in New York for five years (1983–87)—a project he continued for a second five-year period with the Alphabet Gallery in Tokyo—and he also created shopping bags, posters, products, and graphics for the museum's gift shop. In 1986 he began designing products, a field in which his office (with the partnership of Kazuhiro Hayase) has been increasingly active (no. 220). During the 1980s Igarashi also experimented with sculpture, and in 1987 held his first exhibition at the Yurakucho Asahi Gallery in Tokyo. In 1989 he was commissioned by Nissan Motor to design a series of limited edition Infiniti sculptures for the firm's European and American showrooms. From 1979 to 1983 Igarashi taught at Chiba University,

Tokyo, and since 1989 at Tama Art University. Igarashi has received a number of awards, including those from the Japan Sign Design Association (1980) and the Art Directors Club of Los Angeles (1985, 1989), and the Japan Design Committee Masaru Katsumie design award (1989).
Reference: Takenobu Igarashi, *Rock, Scissors, Paper: Design, Influence, Concept, Image* (Tokyo, 1991).

Rokansai Iizuka
1890–1958

Revered as the man who raised the aesthetic level of the bamboo crafts in twentieth-century Japan, Rokansai Iizuka specialized in making baskets for the tea ceremony (no. 30). He was the son of a master bamboo craftsman from Tochigi Prefecture; after completing training with his father, he moved to Tokyo and began exhibiting his work there in 1922. In 1925 he won a bronze medal at the Exposition Internationale des Arts Décoratifs et Industriels Modernes in Paris, and in 1927 the first of a series of solo exhibitions of his baskets was held at the Mitsukoshi department store in Tokyo. At the 1932 annual Imperial Academy of Fine Arts exhibition, Iizuka's work was awarded a special commendation and was purchased by the Imperial Household Agency, the first time this honor was accorded to a bamboo-craft artist; his prestige increased with the visit of the German architect Bruno Taut to his studio in 1933. He was a juror for many of the annual national craft exhibitions, including the Nitten exhibition, which has been sponsored by the Ministry of Education since 1946. Iizuka became the first director of the newly established Japan Craft Association in 1955.
Reference: Utsunomiya, Tochigi Prefectural Museum of Art, *Iizuka Rokansai: Master of Modern Bamboo Crafts* (Feb. 5–Mar. 26, 1989).

Rokuhei Inoue
Born 1934

Rokuhei Inoue entered his family's business, Irokuen, the venerable purveyor of the best Japanese green teas from the region of Uji near Kyoto, in 1957, after graduating from Doshisha University in Kyoto (and succeeded his father as head of the firm in 1977). The conservative world of Japanese green tea producers was hard-pressed to compete in the postwar revolution in the marketing and distribution of products, as coffee and black tea were gaining popularity. Inoue shocked many of his colleagues when he began to produce green tea in tea bags, and distributed them through department stores and supermarkets. In 1974, as a way of promoting interest in his teas, Inoue revived a seventeenth-century festival during which the first tea harvest of the season was brought from Uji to the shogunal residence in Kyoto, which has since become a major annual event. In 1980 Inoue opened the Salon Roku Roku

in the Royal Hotel in Kyoto, which followed the original intent of the tea ceremony by offering a quiet place of respite from the busy world where visitors could sample the Uji teas free of charge. The salon became the site of informal seminars on tea, and to Inoue's surprise, a source of large orders for the firm's teas. Inoue also revived the idea of packaging his teas in traditional Japanese paper (*washi*) instead of the conventional round tins, and Inoue himself has designed some of the firm's special packaging (no. 215).
Reference: Masahiko Nagahama, "Irokuen Ujicha" (Irokuen Uji Tea), *Nikkei Design*, no. 29 (Nov. 1989), pp. 26–31.

International Industrial Design
Established Tokyo, 1962

International Industrial Design was founded in Tokyo in 1962 at the suggestion of Konosuke Matsushita, president of MATSUSHITA ELECTRIC INDUSTRIAL COMPANY, as a free-lance design firm for his company. Although the Matsushita company had its own in-house design department, led by ZENICHI MANO, Matsushita himself felt that an independent firm with young designers coming up with fresh ideas would give its corporate design an added stimulus. International Industrial Design was headed by Ryuichi Takeoka, who came from Matsushita's advertising department, and two Japanese-American designers, Alan Shimazaki and Joseph Nukazawa (now president of the firm). Matsushita became their first—and largest—client, for whom the firm designed everything from the first vacuum cleaner with a plastic body (no. 83) to refrigerators. For Japan Victor Company (JVC), another early corporate client, it created a line of audio equipment and televisions with white plastic housings to attract both the youth and export markets (no. 114). International Industrial Design has a variety of clients, including Texas Instruments and Miyata Industries (for whom it designs bicycles), and its work has now expanded beyond product design to include interior and graphic design as well.
References: Japan Industrial Designers Association, ed., *Seiichi no Kozo: Nihon no Indasutoriaru Dezain* (Structure of Dexterity: Industrial Design Works in Japan) (Tokyo, 1983), p. 104; Shosuke Matsuo, "Senryaku-teki Dezain e no Michi" (The Road Toward a Strategic Design), *Design News*, no. 220 (Dec. 1992), pp. 84–85.

Fujiwo Ishimoto
Born 1941

A textile designer whose career was made at the Marimekko Company in Finland, Fujiwo Ishimoto introduced Japanese-style brushstroke patterns into the firm's vocabulary of printed fabrics. Ishimoto studied graphic design at the Tokyo National University of Fine Arts and Music (1960–64) and subsequently worked as a commercial artist in the advertising department of the Ichida textile company (1964–70). In 1970 he left Japan for Finland, joining Marimekko's sister company, Décembre, where as a textile designer he collaborated on the design of a popular series of canvas carrying bags. Transferred to Marimekko in 1974, Ishimoto began independently to design dress and furnishing fabrics, with his Sumo and Jama fabrics of 1977 and his Taiga of 1978 revealing the calligraphic nature of the designs for which he became known (no. 148). In 1982 the Japan Design Committee exhibited his work as part of a series on designer-manufacturer collaborations, "Fujiwo Ishimoto + Marimekko." Ishimoto was awarded the American Roscoe prize in 1983 and given honorable mentions at the "Finland Designs" exhibitions of 1983, 1989, and 1993.

References: Pekka Suhonen and Juhani Pallasmaa, eds., *Phenomenon Marimekko* (Helsinki, 1986), pp. 130–31; "Fujiwo Ishimoto," in *Design Collection Matsuya Ginza* (Tokyo, 1989), pp. 14–15.

Eiko Ishioka
Born 1938

A pioneer in the field of art direction in Japan, Eiko Ishioka is best known internationally for her work in advertising and her costume and set designs for the stage and screen. Ishioka studied at the Tokyo National University of Fine Arts and Music (1957–61) and in 1961 joined the advertising division of Shiseido (where she remained until 1968), winning gold (1964) and silver (1966, 1968) prizes from the Tokyo Art Directors Club (Tokyo ADC) for advertising for the firm. In 1966 she photographed Shiseido's summer campaign in Hawaii, the first time a Japanese company traveled abroad to create an advertising image. During this period she also worked independently, winning awards from the Japan Advertising Artists Club for record albums (1963) and posters (1965), and opened her own studio in 1970. Since 1971 she has created boldly original media campaigns for Parco (no. 129), which brought her the Mainichi industrial design prize and the Tokyo ADC members prize in 1975. Ishioka has also worked for ISSEY MIYAKE, designing textiles (1971) and boutiques (in collaboration with SHIRO KURAMATA; 1975) and creating and directing fashion shows (1975–77). During the 1980s and early 1990s Ishioka collaborated with several American theatrical and film directors: she was production designer for Paul Schrader's film *Mishima* (1984), winning the Cannes Film Award for Artistic Contribution in 1985; set and costume designer for David Hwang's play *M. Butterfly* on the New York stage (1988), receiving awards from the American Theater Wing and the Outer Critics Circle (1988); and costume designer for Francis Ford Coppola's film *Dracula*, winning an American Academy Award in 1993.

Reference: *Eiko by Eiko, Eiko Ishioka: Japan's Ultimate Designer* (San Francisco, 1990).

Arata Isozaki
Born 1931

One of the most international of Japanese architects, Arata Isozaki studied architecture at the University of Tokyo (1954) and from 1954 to 1963 worked in the office of Kenzo Tange. Around 1960 Isozaki emerged as a member of the Metabolist group, proposing new multiuse metropolitan megastructures with prefabricated components clipped onto supporting structures. Considered metaphors of biological processes, these Metabolist projects included City in the Air Shinjuku Project (1960–61), and Clusters in the Air (1960–62), a forest of high-rise housing units arranged like "leaves" around vertical core "trunks." In 1963 Isozaki established his own office, and his first public commission, the feudal-looking Oita prefectural library (1962–66), brought him international recognition, with its complex mix of Metabolist ideas and historical reference. Isozaki developed a personal language of architecture and design that was symbolic, associative, and sometimes ironic: his Marilyn chair based on the body curves of the American actress Marilyn Monroe (1972; no. 99); Fujimi country clubhouse (Oita, 1973–74) shaped like a question mark; West Japan General Exhibition Center (Kitakyushu, Fukuoka, 1975–77) designed to resemble a boatyard; and Team Disney Building (Lake Buena Vista, Florida, 1987–90) with a canopy in the shape of Mickey Mouse ears. Isozaki has designed furniture and other consumer products for Memphis (1981), Colorcore (1985), and VorWerk (1988–89).

Reference: *Arata Isozaki: Architecture 1960–1990* (New York, 1991).

Kenji Itoh
Born 1915

After graduating from the Tokyo College of Industrial Arts in 1935, Kenji Itoh began a long and successful career in advertising art and commercial illustration. One of his early clients was Daido Worsted Mills; his poster designs for the company's Milliontex products were among the first in Japan to combine photography with graphics (no. 15). His design of the logo for CANON in 1954, as well as a series of posters for the same company, brought him widespread recognition (no. 14). Itoh also designed window displays for the Mitsukoshi and Wako department stores, and his graphic involvement has extended to neon signs for the MATSUSHITA ELECTRIC INDUSTRIAL COMPANY and NEC CORPORATION buildings in the Ginza section of Tokyo. For these works in neon he received a Mainichi industrial design prize in 1965. In 1952, in what was to become a forty-year-long association, Itoh began designing covers for the medical monthly *Stethoscope*; he was awarded a bronze medal for those designs at the Biennale of Graphic Design in Brno, Czechoslovakia, in 1984.

Since 1958 Itoh has designed over a thousand books, primarily for the Kappa publishing company. His work has been shown in numerous exhibitions, including, in 1951, the first solo exhibition for a graphic designer in postwar Japan. In addition to many other prizes, Itoh received the medal of honor in 1983 for his lifelong contribution to the field of graphics.

References: *Best 100 Japanese Posters, 1945–89* (Tokyo, 1990), pp. 46–47, 245; Richard S. Thornton, *The Graphic Spirit of Japan* (New York, 1991), pp. 69, 75, 78–79, 176; Mitsuo Ishida and Shyoji Koide, "Itoh Kenji Eien no Modanisuto: Karei Tasai" (Kenji Itoh: The Everlasting Modernist— Beauty and Color), *JSSD*, vol. 1 (1993), pp. 10–13.

Shinji Iwasaki
Born 1930

Shinji Iwasaki studied industrial design at the Tokyo National University of Fine Arts and Music from 1950 to 1956. As a student of IWATARO KOIKE, and a member of what came to be known as the "Gruppe Koike" (incorporated as GK Industrial Design Associates in 1957), Iwasaki helped design a hi-fi tuner that took first prize in the 1955 Mainichi industrial design competition and was subsequently manufactured by YAMAHA CORPORATION. In 1956 Iwasaki designed his first motorcycle, the Yamaha YD-1, which became the pacesetter for the development of the sport-motorcycle market in Japan (no. 35). After spending 1958–59 studying industrial design at the Hochschule für Gestaltung in Ulm, West Germany, he returned to Japan to resume his career at GK, where motorcycle design was his forte until his retirement as vice president of the firm in 1991. Iwasaki has also been active in the Japan Industrial Designers Association since 1958, and has served on the board of directors of the Japan Design Committee since 1969. In 1993 he accepted the post of professor and chairman of the industrial arts department at Okayama Prefectural University.

References: Iwataro Koike and Shinji Iwasaki, "Purodakuto Dezain no Ichi" (The Status of Product Design), in Japan Design Committee, *Dezain no Kiseki* (Tracks of Design) (Tokyo, 1977), pp. 198–205; Japan Design Committee, *Design 19* (Tokyo, Matsuya Ginza, Sept. 3–8, 1982), n.p.

Yoshiharu Iwata
Born 1924

Yoshiharu Iwata joined Tokyo Shibaura Electric Company (now Toshiba) upon his graduation from the Kyoto College of Industrial Arts in 1944, and remained with the company until his retirement in 1980. At Toshiba he rose to become head of one of the design sections (1965) and then director of the design department (1973), retiring as a managing director. Under his direction, the Toshiba design group produced a wide range of electronic devices for the home, including audio equipment,

toasters, and vacuum cleaners, and Iwata himself holds over three hundred patents for his products. His best-known and most long-lived design was the first electric rice cooker, designed in 1954 (no. 7). As an early pioneer of design in a corporate setting, Iwata was also active in promoting his field, serving as director of the Japan Industrial Designers Association from 1959 to 1967, member of the Japan Industrial Design Promotion Organization from 1973 to 1985, and juror for numerous design competitions. After his retirement Iwata joined the faculty of Kyoto University of Crafts and Textiles, and since 1987 has taught industrial and product design at Musashino Art University in Kodaira.

References: Japan Industrial Designers Association, ed., *JIDA '59* (Tokyo, 1959), pp. 106–7; Kogei Zaidan, ed., *Nihon no Indasutoriaru Dezain: Showa ga unda Meihin 100* (Japanese Industrial Design: 100 Masterpieces Produced during the Showa Period) (Tokyo, 1989), p. 11.

Kozo Kagami
1896–1985

Glassmaker to the imperial household of Japan, Kozo Kagami was internationally renowned for his tablewares and luxury items of fine crystal (no. 19). A native of Gifu, Kagami graduated from the Tokyo College of Industrial Arts in 1920. In 1927 he went to Germany to study at the Kunstgewerbeschule in Stuttgart under Wilhelm von Eiff, an influential figure in the arts of glass and stone engraving in Europe. There Kagami gained an appreciation for the qualities of transparent crystal glass, and after his return to Japan in 1931 continued to utilize the materials and techniques with which he had become familiar in Germany. His work was first shown in Japan the same year, when he exhibited in the twelfth Imperial Academy of Fine Arts exhibition. In 1934 Kagami established the Kagami Crystal Glassworks, the first factory in Japan to specialize in crystal. Since 1943 Kagami Crystal has been executing designs ranging from tablewares to chandeliers for the imperial household, and in 1952 the firm created a cherry-blossom pattern for use in Japanese embassies throughout the world. His international recognition began with a prize at Chicago's Century of Progress Exposition in 1934, and he received awards at the Paris Exposition Internationale in 1937, the New York World's Fair of 1939, and Expo '58 in Brussels in 1958 (grand prize). In 1960 he won the Japan Art Academy prize.

References: Dorothy Blair, *A History of Glass in Japan* (New York, 1973), pp. 297–98; Tokyo, The National Museum of Modern Art, Crafts Gallery, *Modern Japanese Glass: Early Meiji to Present* (Sept. 22–Nov. 28, 1982), n.p.

KAK Design Group
Established Tokyo, 1953

In 1953 three designers pooled their talents to establish the industrial design firm KAK, the initials taken from their surnames, Itaru Kaneko (born 1920), Yoshio Akioka (born 1920), and Junnosuke Kawa (born 1919). Kaneko had worked at the Industrial Arts Institute, where he was editor of the monthly *Kogei Nyusu* (Industrial Art News); Akioka was a free-lance designer; and Kawa had been associated with Kawanami Shoten publishers. One of their first commissions came from Seiko Electric (see SEIKO CORPORATION), which wanted a logo for a new camera. The group convinced Seiko that they should design not only the logo but the camera and related instruments as well, beginning a long partnership that included the design of KAK's award-winning Sekonic line of camera equipment (no. 13). Their success with this project led to work for Minolta, for which they also conceived a total design program, including corporate logo, packaging, and graphics for posters and displays, as well as the cameras themselves. The KAK firm expanded quickly to produce designs for everything "from chopsticks to automobiles,"[5] in Akioka's paraphrase of Raymond Loewy's famous line "from lipsticks to locomotives." Their designs include trucks (Hino), washing machines (MATSUSHITA ELECTRIC INDUSTRIAL COMPANY), radios (Chrysler), and scientific educational toys. Although the founding members have retired, KAK is still in operation, and has been the training ground for many younger industrial designers, such as MICHIO HANYU (who went on to found his own firm, Monopro).
References: Itaru Kaneko, "KAK," in Kogei Zaidan, ed., *Nihon no Kindai Dezain Undoshi 1940 Nendai–1980 Nendai* (History of the Japanese Modern Design Movement, 1940s–1980s) (Tokyo, 1990), pp. 61–62; Kiyoshi Miyazaki, "Akioka Yoshio: Jiko no Tetsugaku to Seikatsu no Hanei to shite no Dezain" (Yoshio Akioka: Design as a Reflection of Philosophy and Life), *JSSD*, vol. 1 (1993), pp. 6–9.

Shozo Kakutani
Born 1928

After working in the advertising division of Daimaru department store in Osaka from 1947 to 1957, Shozo Kakutani became head of graphic design for the Bijutsu Kobo firm in Takamatsu on the island of Shikoku, southwest of Osaka. Kakutani also began doing free-lance work in graphic and package design, and established his own firm in 1963. One of his first commissions, in 1964, was packaging for Japanese rice crackers for a local railway company, and for the past thirty years he has continued to design packaging for traditional foods such as Japanese green tea, candies, noodles, and seaweed for the venerable firms that are the purveyors of these products in western Japan (no. 108). Since

1985 Kakutani has been design consultant for the Yamaya seafood company (no. 189).
Reference: Japan Package Design Association, ed., *Package Design in Japan: Its History, Its Faces* (Tokyo, 1976), nos. 125, 267, 307, 438, 731, 733, 750.

Yusaku Kamekura
Born 1915

No one has influenced the course of Japanese postwar graphic design more than Yusaku Kamekura. Through his work and his participation as a founding member in Tokyo's Japan Advertising Artists Club (1951) and the Japan Graphic Designers Association (1978), as co-founder of the Nippon Design Center (1960), and most recently, as founder and editor of the graphic design magazine *Creation* (1989), Kamekura has set the high standards of design and printed production for which Japanese graphics have become known. Born in Niigata, Kamekura studied at the New Academy of Architecture and Industrial Arts in Tokyo (1935–37), which had been established by Renshichiro Kawakita in 1931 following the pattern of the German Bauhaus. There Kamekura became familiar with the geometric forms and systematic organization characteristic of the European avant-garde graphics on which he would base his later work. From 1938 to 1960 Kamekura worked in the art department of the publishing firm Nippon Kobo, becoming art director in 1940; he provided direction for several of the firm's magazines, including *Nippon*, *Shanghai*, and *Commerce Japan*. By the 1950s Kamekura was already considered one of Japan's leading graphic artists and was the best known internationally. He was given his first one-man show in 1953; designed a cover for the Swiss magazine *Graphis* in 1953; wrote articles for various publications, including *Idea* (1953, 1955); and organized the acclaimed "Graphic '55" exhibition at the Takashimaya department store in 1955. In 1954 Kamekura began his long association with Nippon Kogaku (later called NIKON CORPORATION), designing posters (no. 33), corporate collateral material, packaging, and other commercial graphics, even extending his reach to neon signage (1971). In 1960 he co-founded the Nippon Design Center and was its managing director until he left in 1962 to practice independently, opening his own design office in Tokyo. Kamekura was for many years the designer of choice for government and trade organizations, designing symbol marks for the Japan Industrial Designers Association (1953; no. 3), Ministry of International Trade and Industry's Good Design prize (1957; no. 24), and Japan Architects Association (1960); he designed posters for the Japan External Trade Organization Agency (1960), Tokyo Olympics Organizing Committee (1961, no. 65; 1962–63, nos. 67–68), Japan Design Committee (1965, 1982), Japan Association for the 1970 World Exposition (1967, 1969), Sapporo

Olympics Organizing Committee (1969–70), and Japan Industrial Design Promotion Organization (1973). At the same time Kamekura worked for a variety of corporate clients, most notably, in addition to Nikon, Yamagiwa, a lighting fixtures manufacturer, for which he designed a symbol mark (1966), posters (from 1968), and other advertising materials; and the audio-equipment manufacturer Onkyo (from 1980). While geometry, symmetry, and abstraction have characterized Kamekura's work throughout his career (no. 23), since the 1980s he has returned to Japanese motifs and forms in order to express a sense of tradition.
References: *The Graphic Design of Yusaku Kamekura* (New York, 1973); Masataka Ogawa, Ikko Tanaka, and Kazumasa Nagai, eds., *The Works of Yusaku Kamekura* (Tokyo, 1983); Akiko Moriyama, "Kamekura Yusaku," *Nikkei Design*, no. 36 (June 1990), pp. 42–48.

Shuya Kaneko
Born 1937

Shuya Kaneko joined GK in 1960 after his graduation from the Tokyo National University of Fine Arts and Music and today is vice president and managing director of GK Graphics and managing director of the GK Institute of Industrial Design. Kaneko has worked in a variety of areas during his career at GK, including graphics and packaging (no. 191), and product and environmental design. On numerous occasions Kaneko has been sent abroad as the spokesman for design by the Ministry of International Trade and Industry and the foreign ministry. He currently also serves on the board of directors of the Japan Package Design Association. Kaneko has written several books on design, including *Industrial Design* (1965) and *Package Design* (1989).
References: Takeo Yao, ed., *Package Design in Tokyo* (Tokyo, 1987), p. 147; *A Design Declaration—GK's 30 Year History, Space Design*, special issue, no. 19 (Jan. 1988); *East Meets West in Design: Archaeology of the Present* (New York, 1989), pp. 42–43, 84.

Takashi Kanome
Born 1927

One of Japan's foremost package designers and a winner of the Tokyo Art Directors Club grand prize (1993) for packaging made of recycled paper (no. 243), Takashi Kanome was born in Hokkaido and graduated from the Tokyo National University of Fine Arts and Music in oil painting (1950). In 1951 he moved to Osaka, where in 1966 he opened his own studio for graphic and package design, drawing many of his clients from the Osaka area (no. 131). During the 1970s Kanome gained recognition for packages that combined traditional decorative patterns with modern calligraphy (no. 122) and he received awards from, among others, the Japan Package Design Association (1976). In

the early 1980s Kanome created a series of art experiments in box forms (*Would-but-Would-Not Wood Box*, 1980, Asahikawa Art Museum) and in 1984 moved his office to Tokyo, where he has a broad commercial practice designing for such clients as Shiseido, Marukan, and Nestle.
References: *Boxes by Four: Package Design* (Tokyo, 1982), pp. 65–111, 209; Takashi Kanome, "Aideia ni Jikan o Kakete" (Spending Time on Design Ideas), *Designers' Workshop*, vol. 2, no. 11 (Dec. 1985), pp. 40–46; Takashi Kanome, "Kakutogi to shite no Pakkeeji Dezain" (Package Design as Martial Art), *Nikkei Design*, no. 62 (Aug. 1992), pp. 64–65.

Akemi Kashima
Born 1946

Akemi Kashima graduated from the Women's College of Fine Arts in Tokyo in 1969, and since then has worked for the Honshu paper company, where she has mainly produced graphic and package design for food products. Among the clients she works with are Morinaga candies, Meiji dairy, and Hyobando foods. Kashima seeks to find a Japanese identity for her package designs, particularly for traditional foods and snacks (no. 100). In 1971 she was a prize winner at the seventh annual exhibition sponsored by the Japan Package Design Association.
References: Japan Package Design Association, ed., *Package Design in Japan: Its History, Its Faces* (Tokyo, 1976), no. 259; Japan Package Design Association, *Member's Work Today 1986* (Tokyo, 1986), p. 48.

Mitsuo Katsui
Born 1931

Known for his creative, digital designs (no. 184), Mitsuo Katsui used computer and photograph technology to challenge traditional Japanese graphic concepts as early as 1971, when he was awarded a prize by the Japan Typography Association for his computer-designed posters. Katsui graduated from the Tokyo University of Education (1955), where he studied design and photography. He joined Ajinomoto, a food products company, in 1956 and received national recognition when he was awarded a gold medal by the Japan Advertising Artists Club in 1958. In 1961 he established his own design office in Tokyo, and at the same time began to teach at Tokyo (now Tsukuba) University and at Musashino Art University in Kodaira. Katsui was made art director of the Japanese government pavilion Orgorama at Expo '70 in Osaka, the Okinawa International Ocean Exposition in 1975, and Kodansha's pavilion at Expo '85 in Tsukuba, for which he also designed the official exhibition map. His many awards range from the Mainichi industrial design prize for his participation in the 1965 "Persona" exhibition of graphic design in Tokyo to the gold prize at the International Poster Biennale in Warsaw in 1994.

References: *Best 100 Japanese Posters, 1945–89* (Tokyo, 1990), pp. 94–95, 160–61, 198–99, 228–29, 246; Tokyo, The National Museum of Modern Art, *Graphic Design Today* (Sept. 26–Nov. 11, 1990), pp. 5–53, 112; Richard S. Thornton, *The Graphic Spirit of Japan* (New York, 1991), pp. 109, 224–25; Hiroshi Kashiwagi, "Mitsuo Katsui," *Creation*, no. 9 (1991), pp. 142–65, 167–68; Yusaku Kamekura, "Mitsuo Katsui," *Creation*, no. 20 (1994), pp. 140–45.

Yosei Kawaji
Born 1940

Graduating from the Tokyo University of Art and Design with a degree in graphics in 1962, Yosei Kawaji went to work at the Pola cosmetics design research center designing packaging for a full line of make-up products. In 1970 he became art director of the Ito design research institute, leaving in 1985 to open his own design office. A lecturer at the Women's College of Fine Arts, Tokyo, and a member of the Japan Package Design Association (JPDA) and Tokyo Designer's Space, he has won a number of awards for packaging including the American Clio award for his packaging for sake and tea, the Ministry of International Trade and Industry prize at the 1989 Japanese Packaging Exhibition, and a JPDA award for his Xshax tissue boxes (no. 246) in 1991.
Reference: Japan Package Design Association, ed., *Package Design: JPDA Member's Work Today '90* (Tokyo, 1990), p. 48.

Kyoichiro Kawakami
Born 1933

Grandson of Denjiro Kawakami, a pioneer of modern glass design in Japan, Kyoichiro Kawakami studied at the Tokyo National University of Fine Arts and Music (1953–56) and after his graduation, worked in the advertising department of MATSUSHITA ELECTRIC INDUSTRIAL COMPANY (1956–63). In 1963 Kawakami took up glassmaking, joining the design department of Hoya Glass (where he worked until 1986) and studying at glass factories in Sweden and Finland in 1965. As in other Japanese craft-based factories, Kawakami was able to design both production wares (no. 88) and unique pieces, exhibiting the latter in exhibitions at the Corning Museum of Glass, New York (1979), National Museum of Modern Art, Kyoto (1980), National Museum of Modern Art, Tokyo (1982), Glass Art Gallery, Toronto (1989), Hakone Open-Air Museum (1990), and Notojima Open-Air Museum (1991).
References: Corning, New York, Corning Museum of Glass, *New Glass: A Worldwide Survey* (Apr. 26–Oct. 1, 1979), pp. 120, 260; Tokyo, The National Museum of Modern Art, Crafts Gallery, *Modern Japanese Glass: Early Meiji to Present* (Sept. 22–Nov. 28, 1982), no. 143.

Motomi Kawakami
Born 1940

An industrial and package designer, Motomi Kawakami is best known for his exquisitely decorative furniture and lighting in which traditional craft techniques such as lacquer (*urushi*), or their modern equivalents, in this case, synthetic coating materials, are applied to boldly simplified, contemporary forms for a luxury market (no. 199). Kawakami's craft bias, which appears in his detailed treatment of surfaces and textures (no. 166), extends also to his lines of less expensive office furniture. Kawakami graduated from the Tokyo National University of Fine Arts and Music in 1964, and received his master's degree there in 1966. Like a number of other Japanese designers, Kawakami went to Italy, where between 1966 and 1969, he worked in the office of the architect Angelo Mangiarotti, and he also practiced independently, using the synthetic materials and industrial technology for which Italian design was then so well known (his Fiorenza chair in ABS plastic was manufactured by Bazzani in 1968). In 1971 he opened his own office in Tokyo, gradually developing a practice that included by the 1980s both Italian clients (Skipper, Arflex, Cassina) and Japanese (YAMAHA CORPORATION, Yamagiwa, Bushy). Kawakami has won numerous awards, including those from the Japan Interior Designers Association (1976; 1984, no. 130) and the Japan Package Design Association (1985, 1986), as well as the Mainichi design prize (1991) and the Kunii Industrial Art Award (1992). Since 1970 Kawakami has also lectured at the Tokyo National University of Fine Arts and Music, Aichi Prefectural University of Fine Arts, and Kyushu University, Fukuoka.
Reference: *Gachi: Motomi Kawakami's Furniture* (Tokyo, 1988).

Yoshimasa Kawakami
Born 1945

Package designer Yoshimasa Kawakami graduated from Musashino Art University, Kodaira, in 1969, when he joined the package design center at Honshu paper company, where he worked for clients such as Hyobando (no. 100). He left in 1973 to study at the Art Center School in Pasadena, California, remaining after completing his studies there to work with Southern California Packaging in 1978. Two years later he was hired by his present employer, the marketing department of the University of California at Los Angeles.
Reference: Japan Package Design Association, ed., *Package Design in Japan: Its History, Its Faces* (Tokyo, 1976), no. 259.

Rei Kawakubo
Born 1942

Asymmetric and oversize in monochrome black or white, often folded or wrapped and riddled with holes (no. 164), Rei Kawakubo's clothes defied the traditional conventions of fashion design when they were first shown in Paris and New York in the early 1980s (nos. 167–68). Kawakubo has remained among the most revolutionary of fashion's avant-garde and yet she has marketed her clothing so successfully that global sales under her Comme des Garçons (Like the Boys) label reached one hundred million dollars by 1992. Kawakubo studied fine arts at Keio University in Tokyo and on graduating in 1964, joined the public relations department of Asahi Chemical Industry, a textiles and chemicals firm, where her collaboration with photographers and art directors drew her to the fashion industry. In 1967 she became a free-lance fashion stylist, and when she found that the clothes she needed were not available, she started to design and make them herself. By 1969 she was using the Comme des Garçons name although the company was not formally established until 1973. In 1975 Kawakubo showed her clothes for the first time in Tokyo, opening a shop in the Minami-Aoyama district the following year. Working with the architect Takao Kawasaki, she created a sparse, minimalist, largely white interior that has since characterized the firm's some 260 showrooms that followed. By 1981, when Kawakubo first showed her collection in Paris, she had found her own radical voice as a fashion designer. Her clothes had unique silhouettes that fell in surprising folds or seemingly lacked a component, sometimes requiring diagrams and instructions so that customers could get into them. She continues to create the unexpected (nos. 248, 254). For spring–summer 1992 she sent out annotated models of garments in paper and clothes with unfinished seams, while in 1993 her fabrics were decolorized, and the seams and darts left unpressed to lend her clothes the rough, handmade quality that has been a constant theme of her work. In 1988 Kawakubo launched her own large format magazine, *Six*, which is published to coincide with the Comme des Garçons biannual collections. Kawakubo has also designed furniture, initially for use in her stores and later also for production (Chair No. 1 of 1983). Kawakubo was awarded the Mainichi prize in 1985, and in 1991 she became the first Japanese recipient of France's Veuve Cliquot award as the businesswoman of the year.
References: Leonard Koren, *New Fashion Japan* (Tokyo, 1984), pp. 108–25; New York, Fashion Institute of Technology, *Three Women: Madeleine Vionnet, Claire McCardell, and Rei Kawakubo* (Feb. 24–Apr. 18, 1987), n.p.; Deyan Sudjic, *Rei Kawakubo and Comme des Garçons* (New York, 1990); "Rei Kawakubo: Designer Keeps Eye on Bottom Line," *Asahi Shimbun Japan Access*, Mar. 23, 1992, p. 7.

Kazuo Kawasaki
Born 1949

Kazuo Kawasaki, an industrial designer who works in a wide product range (no. 179), is also known for his interest in traditional crafts techniques. Born in Fukui Prefecture, where he still lives and works today, Kawasaki graduated from Kanazawa University of Arts and Crafts with a degree in industrial design in 1972. He worked at Toshiba from 1972 to 1979, first on the planning and development of its Aurex sound equipment, and then as the creative director for product planning. After leaving Toshiba he did free-lance work, then opened his own design studio in 1981, founding a new design office, Ex-Design Inc., in 1985. Kawasaki has been closely involved with the Takefu Knife Village in Fukui Prefecture, the oldest craft community in Japan, and is director of the Hokuriku Academy of Traditional Industrial Arts. Despite his keen interest in preserving the cultural identity of Japan, in his own work for the Takefu Knife Village Kawasaki moved beyond traditional designs to produce knives and scissors in unusual modern forms (nos. 169–70). In recognition of his work with Takefu, Kawasaki was awarded the Japan Design Committee Design Forum silver prize in 1983, and the Kunii Industrial Art Award in 1993. Disabled in an accident as an adult, Kawasaki now focuses much of his energy on designing for special needs, explaining, "I want to design products for myself first." [6] In 1989 he produced the prototype of a folding wheelchair made of tubular titanium weighing only forty-five pounds (no. 219), and was involved in the exhibition "Design toward Normalization" at Inax in Tokyo. A member of the Japan Industrial Designers Association, he teaches at several schools and universities including Fukui University and Kanazawa University of Arts and Crafts, is a judge for the Good Design selection of the Ministry of International Trade and Industry, and a technical advisor for Fukui Prefecture.
References: "Product Design of the Month," *Japan Interior Design*, no. 294 (Sept. 1983), pp. 80–81; *International Design Yearbook* 2 (1986), pp. 31, 183, 224; *International Design Yearbook* 4 (New York, 1988), p. 214; Masaya Yamamoto, ed., *Product Design in Japan* (Tokyo, 1990), pp. 192–95, 241; "Kazuo Kawasaki Design Message," in Tokyo, Axis Gallery, *Design a Dream* [brochure] (1991).

Isamu Kenmochi
1912–1971

A pioneer of interior and industrial design in early postwar Japan, Isamu Kenmochi was one of the most international in outlook among his designer countrymen; he made a study tour of the United States in 1952, on which he reported in a series of articles for *Kogei Nyusu* (Industrial Art News), and attended the International Design Conference in Aspen in 1953, the first Japanese to do so.

Concerned with creating a modern Japanese design identity, Kenmochi studied and adopted foreign production techniques while retaining a serious interest in native folk craft. Kenmochi was trained at the Tokyo College of Industrial Arts. On graduating in 1932 he joined the Woodwork Technical Section of the Industrial Arts Institute (IAI) at Sendai, where, removed from urban centers, he was inspired by the useful objects made by local craftsmen (no. 12). Bruno Taut's visit to the IAI in 1934 confirmed Kenmochi in his ambition to modernize Japanese interior furnishings and to use technology to lower costs. During the war years Kenmochi and the woodwork section at Sendai studied the application of wood products—particularly plywood—to airplane manufacture, and like Charles Eames's work for the United States Department of the Navy, it led to new uses of the material in the postwar years. Kenmochi was a founding member of several of Japan's principal design organizations, including the Japan Design Committee (1953) and the Japan Interior Designers Association (1956), and his efforts were influential in shaping the Japanese Good Design movement. In 1955, when he left the IAI and established KENMOCHI DESIGN ASSOCIATES, he gave shape to his views by designing the interior of the Japanese pavilion at Expo '58 in Brussels, which won a grand prize. As head of his own design office, he incorporated such international industrial design practices as the use of full-scale drawings and building of prototypes, which had first been introduced to Japan by Taut. Kenmochi's best-known designs came from this period, when he combined the natural materials of Japanese craft with modern forms and manufacturing technology (nos. 38, 52). In 1959 he became a professor at Tama Art University in Tokyo.
References: Kenmochi Isamu no Sekai Henshu Iinkai, ed., *Kenmochi Isamu no Sekai 4: Sono Shiteki Haikei Nenpu / Kiroku* (The World of Isamu Kenmochi 4: Historical Background—Chronological History / Record) (Tokyo, 1961); Masao Hata, "IAI to Kenmochi Isamu" (IAI and Mr. Isamu Kenmochi), *Kogei Nyusu* (Industrial Art News), vol. 39 (Nov. 1971), pp. 70–73; Tetsuo Matsumoto, "Zenin ga Ichinen Kakete Haizara no Dezain o shite mo ii to Omoimasu" (I Think It's Fine if the Whole Firm Spends a Whole Year on the Design of an Ashtray), *Industrial Design*, no. 113 (Sept. 1981), pp. 33–35.

Kenmochi Design Associates
Established Tokyo, 1955

Kenmochi Design Associates was founded by ISAMU KENMOCHI after his retirement from the Industrial Arts Institute in 1955. In the early years of the firm, Kenmochi and his colleagues worked primarily on furniture and interior design projects; they collaborated with the architect Kunio Maekawa on the Japanese pavilion for Expo '58 in Brussels, which won a grand prize, and the same year, designed a series of rattan chairs that made

their debut in the Hotel New Japan in 1960. In 1963 the firm won a Mainichi industrial design prize for their production furniture for clients such as TENDO MOKKO, and received a second Mainichi prize in 1970 for street furniture for Expo '70 in Osaka. The last project on which Kenmochi himself worked before his death in 1971 was the interior of the Keio Plaza Hotel in Tokyo. The firm continued under Tetsuo Matsumoto, who is still its president, and whose interest and architectural training led to commissions for the interiors in the transportation field, including the Japan Air Lines DC-10 and B-747, winning a Japan Interior Designers Association prize for the latter in 1971. Kenmochi Design Associates created a broad range of designs, including interior, street, and sign design, as well as graphic and package design for clients such as Yakult yogurt, and product design (no. 78). "Each person should expand his range as much as possible," says Matsumoto, "so the design of an ashtray, for example, means not merely that one product. The designer should also think of the table on which the ashtray must rest, or how the ashtray is to be cleaned. . . . The ashtray in its context, so to speak. The more experience we have in various areas of design, the better."[7]
Reference: Tetsuo Matsumoto, "Zenin ga Ichinen Kakete Haizara no Dezain o shite mo ii to Omoimasu" (I Think It's Fine if the Whole Firm Spends a Whole Year on the Design of an Ashtray), *Industrial Design*, no. 113 (Sept. 1981), pp. 33–35.

Kenzo
Born 1939

The first Japanese fashion designer to show his work in Paris, Kenzo opened a ready-to-wear boutique there in 1970, when his innovative use of oversize garments, layers, and striking mixes of patterns, bright colors, fabrics, and Western and non-Western styles attracted immediate attention. From his early knit sweaters with kimono sleeves to layered outfits that suggested regional dress (no. 197), Kenzo defined a radical eclecticism distinguished by its richness, variety, and complexity (no. 249). Born Kenzo Takada, he was one of the first male students to be admitted to the Culture Institute, a women's fashion school in Tokyo, and, after graduation, worked for the Sanai department store specializing in junior clothing and created designs for several fashion magazines, including *So-en*. On his arrival in Paris in 1965, Kenzo supported himself by selling designs to Louis Féraud, the fashion magazines *Elle* and *Jardin des Modes*, the department stores Les Printemps and Galeries Lafayette, the clothing manufacturer Pisanti, and the textile company Relations. In 1970 Kenzo opened his Jungle Jap boutique, and (moving several times before establishing his large boutique and headquarters in the Place des Victoires in 1976), presented a show using Japanese cotton fabrics and small-flower-printed remnants he had found at Paris's Marché Saint-Pierre; his daring

combinations of color and pattern aimed at a new youthful market found acceptance in fashion's ready-to-wear mainstream, and were published first in *Elle* and later the same year in British *Vogue*. In 1975 Kenzo showed his work in Tokyo, in 1977 in New York, and in 1979 in Zurich. With a seemingly endless gift for invention Kenzo introduced a succession of new styles, including his "schoolgirl" look (1971–72; no. 97) with mini dresses and coats and berets; "peasant" look (1973–74), inspired by Romanian folk costumes, with short jackets or dresses over long pleated skirts, which he credits for his interest in layering; and "military" look (1978–79), with trousers and long capes. During the 1980s Kenzo expanded his range of design to include menswear (1983), jeans (1986), junior women's clothing (1986), bed and bath linens (1987), and children's clothing (1987). As successful a merchandiser as he is a designer, Kenzo created a market for reasonably priced, individual clothes that combine fun and fantasy. He marketed them aggressively internationally, opening shops in New York in 1983, in five French cities outside Paris and in Copenhagen, London, Milan, and Tokyo in 1984 and 1985, and then throughout Europe and in Asia. His awards include the French *chevalier des arts et des lettres* (1984) and the Mainichi fashion prize (1985).
Reference: Ginette Sainderichin, *Kenzo* (Paris, 1989).

Katsu Kimura
Born 1934

One of Japan's best-known package designers, Katsu Kimura excels at both surface design and paper engineering, creating fresh and often humorous packages for a range of products from cosmetics and foodstuffs to linens and liquor (nos. 137, 141). Born in Tochigi, Ibaraki Prefecture, in 1934, Kimura graduated from the Japan Advertising Art College in Tokyo in 1956. In 1960 he founded Packaging Direction Company, which he continues to run today. He had his first one-man show in Tokyo in 1968, which was followed by others in Tokyo, Osaka, and New York. Kimura was one of four prominent package designers whose work was featured in the "4-Boxer" exhibitions at the Ginza Wako Gallery, Tokyo, in 1981 and 1988. Kimura has received numerous awards, including the Tokyo Art Directors Club prize in 1981 and 1984, and the Japan Package Designers Association members prize (1964, 1974, 1985).
References: Victor and Takako Hauge, *Folk Traditions in Japanese Art* (Tokyo, 1978), pp. 187, 257; Katsu Kimura, *Katsu Kimura: Package Direction* (Tokyo, 1980); *Boxes by Four: Package Design* (Tokyo, 1982); Shigeo Fukuda, "Katsu Kimura: A Japanese Packaging Magician," *Graphis*, vol. 38 (Nov.–Dec. 1982), pp. 78–89; Hiromu Hara, ed., *Contemporary Book Design in Japan, 1975–1984* (Tokyo, 1986); Katsu Kimura, *Box-er:*

Katsu Kimura's Packaging (Tokyo, 1988); Richard S. Thornton, *The Graphic Spirit of Japan* (New York, 1991), pp. 156, 158–59.

Toshiyuki Kita
Born 1942

One of the most international of Japanese furniture and product designers, with offices in both Milan and Osaka, Toshiyuki Kita has nonetheless continued to assert his cultural identity, contributing to the revival of traditional Japanese crafts in a series of lamps with *mino* paper shades (no. 151) and lacquer tableware handmade at Wajima (1986; no. 201). Kita was trained in the industrial design department of Naniwa College, Osaka (1964), and established his first office in Osaka in 1967. In 1969, Kita went to Italy, where he readily found manufacturers for his designs, among them Bernini and Bilumen, although he continued to work in Osaka as well. Kita first gained international notice with his remarkable Wink chair for Cassina (1980; no. 146), a legless lounge with adjustable back, head, and footrest that has remained in continuous production and that prompted further collaboration with Cassina, including his Kick table (1983) and Luck sofa (1986). During the 1980s Kita designed products and objects for numerous firms in Europe and Japan, including Casas in Spain; Tecnolumen and Interprofil in Germany; Alessi, Luci, and Cesa in Italy; and TENDO MOKKO and Seidenko in Japan. In 1989 Kita designed a multipurpose hall for cultural events in the Ginza Sony Building, Tokyo. He received the Kunii Industrial Art Award in 1981 and the Mainichi industrial design prize in 1985.
Reference: *Toshiyuki Kita: Movement as Concept* (Tokyo, 1990).

Yoshiko Kitagawa
Born c. 1940

Yoshiko Kitagawa first became interested in the design possibilities of paper while a student at the Kuwasawa Design School in Tokyo. After graduating in 1962, she experimented with package design, but it was her work in glass, designing a perfume bottle for Kanebo, that confirmed paper as her medium of choice. "Paper does not break, it can be folded, and afterward returned to its original flat state. . . . [Working with glass] made me realize the wonderful qualities of paper, paper as a living material."[8] Soon Kitagawa was exhibiting her paper creations (her first solo show was at Seibu department store in Tokyo in 1967) and winning awards at the annual Japan Package Design Association exhibitions (1967, 1969, 1976). She has designed packaging for posters and wedding bonnets, and for lamps, which she characterizes as packages for light (see no. 109). Her work has been shown in numerous exhibitions at the Seibu, Matsuya, and Wako department stores, and Tokyo Designer's Space, as well as the Hokkaido Museum of Modern Art. In

1974 she opened a boutique, Jack and Betty, dedicated to the sale of packaging as a product in itself. Since 1992 Kitagawa has taught at the Women's College of Fine Arts in Tokyo.

References: Japan Package Design Association, ed., *Package Design in Japan: Its History, Its Faces* (Tokyo, 1976), no. 657; Yoshiko Kitagawa, "Jointo ni Yoru Pakkeeji" (Packages Based on Joints), *Designers' Workshop*, vol. 2, no. 11 (Dec. 1985), pp. 12–17.

Setsuo Kitaoka
Born 1946

Setsuo Kitaoka, a graduate of the Kuwasawa Design School in Tokyo in 1974, established his own Tokyo office in 1977, where he has designed shop displays, furniture (no. 173), lighting (no. 187), and interiors. Among his well-known interiors is the shop he created for the Pour Homme boutique of YOHJI YAMAMOTO in Tokyo. In 1982 Kitaoka received the prize for excellence from the Japan Display Association, as well as the top prize at the first annual shop lighting competition (NASHOP) sponsored by MATSUSHITA ELECTRIC INDUSTRIAL COMPANY (1983), the prize for excellence from the Japan Commercial Building Designers Association for his shop design, and three Japan Sheet Glass Association prizes for shop and window display design (1985, 1987, 1988). Kitaoka participated in the "Models by Ten Interior Designers" exhibition at the Matsuya Ginza department store in Tokyo in 1989. He has exhibited abroad in Belgium, Italy, and the United States (Gallery 91, New York).

References: *International Design Yearbook* 2 (New York, 1986), pp. 50, 136–37, 477; Setsuo Kitaoka, *Kitaoka Setsuo* (Tokyo, 1989); "Setsuo Kitaoka," *Designers' Workshop*, extra issue, vol. 7 (Sept. 1992), pp. 39, 60.

Masakazu Kobayashi
Born 1944

A textile artist and designer, Masakazu Kobayashi studied at Kyoto University of Fine Arts (1963–66), afterward working for the Kawashima textile mills (1966–75). From the 1970s Kobayashi designed both production fabrics and large-scale studio fiberworks (sometimes working in collaboration with his wife Naomi), exhibiting the latter regularly in Lausanne's international tapestry biennials, at the textile triennials in Lodz, Poland, and in such exhibitions as "Moderne Textilkunst aus Japan," at the Museum Bellerive in Zurich in 1984. Kobayashi designed printed fabrics for such firms as Sangetsu (no. 145), and in 1983 opened his own studio. In recent years he has given up designing for industry in order to concentrate on his fiber art, which has been widely exhibited and for which he has won various awards, including prizes at the international textile competition in Kyoto in 1987 and 1994.

Reference: Zurich, Museum Bellerive, *Textilkunst 1950–1990: Europa, Amerika, Japan* (1991), pp. 104–5.

Iwataro Koike
1913–1992

Iwataro Koike holds a special place in the history of Japanese design as an influential and beloved teacher of several generations of designers. A graduate of the Tokyo School of Fine Arts in 1935, Koike was at the Industrial Arts Institute (IAI) in Sendai with KATSUHEI TOYOGUCHI and MOSUKE YOSHITAKE in 1939 and 1940, but left because he found the institute too narrowly focused on the export market. In 1942, after a two-year stint at the Benbo lacquer works in Okinawa, Koike found his niche teaching design at his alma mater (now Tokyo National University of Fine Arts and Music), where he taught for the next thirty-seven years. In his early years at the university Koike gathered an informal salon of his former students, among them SHINJI IWASAKI and Kenji Ekuan, who went on to form the GK industrial design firm in 1953. The firm took its name from the initials for "Gruppe Koike," and Koike himself played an active part in the firm as it started, designing a record player (1954) and stacking ashtrays (1955), and choosing the colors for the YD-1 motorcycle designed by Iwasaki (no. 35). As GK became firmly established, Koike turned his attention to promoting design through the Japan Industrial Designers Association, serving as its director in 1957 and 1965–66. He participated as a speaker in the 1960 World Design Conference (WoDeCo) in Tokyo, and in the 1963 ICSID conference in Paris. Koike also wrote one of the early texts for the new field of industrial design, *Basic Design: Composition and Formation* (1956), and continued to work as a free-lance designer (no. 127). Koike established the College of Urban Design in Tokyo to help promote better design for the city dweller, which became his primary concern after retirement in 1980. In 1981 he led a campaign opposing the new yellow and red buses of the Tokyo metropolitan transit authority, which in his view were a source of visual pollution, and the color was changed to a soothing green. This was in keeping with Koike's basic philosophy, "design is love," by which he meant that designers should show a consideration akin to love for the consumer, creating something honest, useful, and beautiful. Koike's contribution to the design field was doubly acknowledged in 1985, when he received the Kunii Industrial Art Award and the Order of the Sacred Treasure from the government of Japan.

References: Iwataro Koike and Shinji Iwasaki, "Purodakuto Dezain no Ichi" (The Status of Product Design), in Japan Design Committee, *Dezain no Kiseki* (Tracks of Design) (Tokyo, 1977), pp. 198–205; Iwataro Koike, *Dezain no Hanashi* (Talks on Design), rev. ed. (Tokyo, 1985); Nagahama Masahiko, "Koike Iwataro," *Nikkei Design*, no. 22 (Apr. 1989), pp. 86–92; Colin Naylor, ed., *Contemporary Designers*, 2nd ed. (Chicago, 1990), pp. 305–6.

Takashi Kono
Born 1906

Takashi Kono graduated in 1930 from the Tokyo School of Fine Arts, where he studied illustration and traditional crafts, and began to work at the Shochiku Motion Picture Company designing posters and newspaper advertising. He remained in film promotion until World War II, at the same time serving as graphics advisor to the magazine *Nippon*. His first one-man show of graphics took place in 1941, but for the duration of the war he worked in the military propaganda department in northern China and Indonesia. After the war Kono joined the film company Shintoho Productions as an art director. He was one of the founders of the Japan Advertising Artists Club in 1951 (no. 5; see also no. 8), and in 1957 opened his design office Deska in Tokyo. Kono designed the logo for the 1960 World Design Conference in Tokyo (nos. 50–51), and the Japanese displays at international trade exhibitions in Peru (1963) and in Paris and New York (1964); and was a member of the design committees of the 1964 Summer Olympics in Tokyo and the 1972 Winter Olympics in Sapporo, as well as of Expo '70 in Osaka. A lecturer at the Tokyo National University of Fine Arts and Music in 1962, he became a professor at Aichi Prefectural University of Fine Arts in 1966, the university's president in 1985, and professor emeritus in 1990.

References: Itsuhei Itoh, ed., *Kono Takashi no Dezain* (Takashi Kono, Designer) (Tokyo, 1956), n.p.; "Takashi Kono's Graphic Works," *Gurafuikku Dezain*, vol. 87 (Autumn 1982), pp. 61–68; *Mai Dezain: Kono Takashi* (My Design: Takashi Kono) (Tokyo, 1984); Akiko Moriyama, "Takashi Kono," *Nikkei Design*, no. 4 (Oct. 1987), pp. 46–52; *Best 100 Japanese Posters, 1945–89* (Tokyo, 1990), pp. 42–43, 247.

Jiro Kosugi
1915–1981

Jiro Kosugi was the first industrial designer in postwar Japan to have a solo exhibition of his work, which was held at the Bridgestone building in Tokyo in 1954. He was also among the first to establish himself as an independent designer. Born in Tokyo, he initially studied painting with his artist father before entering the industrial arts department of the Tokyo School of Fine Arts. Like many of his contemporaries, Kosugi cut his teeth in the profession during a stint working at the Industrial Arts Institute (in 1944), where during his brief tenure his colleagues included ISAMU KENMOCHI and MOSUKE YOSHITAKE. After the war he began to work as a free-lance designer, and in 1952 was a founder of the Japan Industrial Designers Association. While at first supporting the G-Mark Good Design system of the Ministry of International Trade and Industry, Kosugi soon opposed the policy; he felt that "the system whereby a government-appointed committee

becomes the sole arbiter of 'good design' by slapping a G-Mark sticker onto a product smacks too much of Japan's military past, dictating to the consumer."[9] Beginning in 1948 Kosugi designed a series of trucks for Toyo Kogyo (now Mazda) (no. 42), and also had a long-term association with the Janome Sewing Machine Company (no. 6).

References: "Dai 2-kai Shin-Nihon Kogyo Dezain Nyusho Sakuhin" (Winning Designs of the 2nd Annual New Japan Industrial Design Competition), *Kogei Nyusu* (Industrial Art News), vol. 22 (Feb. 1954), p. 29; Kogei Zaidan, ed., *Waga Indasutoriaru Dezain: Kosugi Jiro no Hito to Sakuhin* (My Industrial Design: Jiro Kosugi, the Man and His Work) (Tokyo, 1983); Kogei Zaidan, ed., *Nihon no Indasutoriaru Dezain: Showa ga unda Meihin 100* (Japanese Industrial Design: 100 Masterpieces Produced during the Showa Period) (Tokyo, 1989), pp. 32–33.

Yoshiaki Kubota
Born 1925

A package designer, Yoshiaki Kubota worked for two years at the Graphic Design Company before opening his own studio in Tokyo in 1955. His graphics and packages, designed primarily for the food industry, have been created for such products as Japanese green tea, seaweed, and sake rice wine. Kubota won prizes at the Japan Package Design Association's annual exhibitions in 1966, 1973, 1974, and 1976, the last two awards for his packaging of sake. His designs are characterized by simple graphics and bold color (no. 135).

References: Japan Package Design Association, *Member's Work Today 1986* (Tokyo, 1986), p. 66; Japan Package Design Association, ed., *Package Design: JPDA Member's Work Today '90* (Tokyo, 1990), p. 54.

Keiko Kumagai
Born 1955

A graphics and package designer, Keiko Kumagai graduated from Musashino Art University in Kodaira in 1976, and began to work for Tokyo Advertising that same year. In 1978 she joined the Mori Design Room before moving to the Noie company in 1979, where she has worked on advertising campaigns for clients such as Toshiba and Marui department stores. Her package designs for Marui won Kumagai a Clio (U.S.A.) award in 1986. Kumagai's work ranges from contemporary designs for Marui's annual spring and summer sales campaigns to traditional paper storage wrappers for purveyors of women's kimonos (no. 216). Since 1992 Kumagai has been with Smith Studio in Tokyo.

Reference: Japan Package Design Association, *Package Design in Japan 1989* (Tokyo, 1989), p. 127.

Shiro Kuramata
1934–1991

A daring inventor of forms both in furniture and interior design, Shiro Kuramata followed a course that was international rather than archetypically Japanese, using a vocabulary of industrial materials—steel mesh, corrugated aluminum, steel cables, and glass—for conceptual, poetic effects. Kuramata was given traditional training in the woodcraft department of Tokyo's polytechnic high school and then worked in the furniture factory of the Teikoku Kizai company (1953). He studied interior design at the Kuwasawa Design School, Tokyo (1956), and worked first for San-Ai (from 1957) and later in the interior design department of Matsuya department store (1964) before opening his own design office in Tokyo in 1965. Kuramata's earliest manufactured furniture designs, Furniture with Drawers (1967), unexpectedly fitted drawers in and around otherwise conventional forms and stacked them in tall pyramids on wheels (1968). By 1970 he was designing furniture in boldly irregular forms (no. 95), and from the mid-1970s, he introduced materials in contexts that subverted their meanings, which resulted in a glass table with bullet-shape rubber legs (1976); illuminated acrylic shelves that appeared to float (1978); his North Latitude table (1985), apparently precariously supported on cone-shape steel-mesh legs; and his How High the Moon chair, constructed entirely of airy steel mesh (1986; no. 186). From his Club Judd, Tokyo (1969), with its walls of overlapping stainless-steel pipes, to the series of boutiques he created for the fashion designer ISSEY MIYAKE in Tokyo, Paris, and New York—most notably Issey Miyake Men with steel-mesh columns in Shibuya Seibu department store in Tokyo (1987)—Kuramata created interior spaces that were as experimental and novel as his furniture. After his death, an exhibition of his last, largely prototypical work, "Sealing of Dreams," was held at the Gallery Ma in Tokyo in 1991.

References: The Work of Shiro Kuramata 1967–1974 (Tokyo, 1976); The Works of Shiro Kuramata 1967–1981 (Tokyo, 1981); Shiro Kuramata, Itsuko Hasegawa, and Minoru Ueda, "A Talk on Materials," Axis, no. 23 (Spring 1987), pp. 6–8; Chee Pearlman, "Shiro Kuramata," Industrial Design, vol. 35 (Sept.–Oct. 1988), pp. 31–33; Shiro Kuramata 1967–1987 (Tokyo, 1988).

Natsuki Kurimoto
Born 1961

Natsuki Kurimoto, a lacquer artist (see no. 208), graduated in 1985 from Kyoto University of Fine Arts, where he completed his master's degree in 1987. His first solo exhibition of lacquer was held at the Suzuki Gallery in Kyoto in 1984, and his work was shown at the gallery of the Tokyo National University of Fine Arts and Music in 1985, that same year winning the Kyoto mayor's prize, which

included the purchase of one of his pieces for the municipal art collection. In 1987 Kurimoto received a prize from the Japan Lacquer Craft Association, and in 1988, the art and culture award from Kyoto, which included a travel scholarship. Kurimoto's study trip to Europe resulted in an exhibition of his lacquers at the Galeries Lafayette department store in Paris. Kurimoto has also shown at the annual craft exhibition sponsored by the Asahi newspaper company since 1985, and his first American exhibition was held in Washington, D.C., at the Sasakawa Peace Foundation gallery in 1993. Kurimoto has taught at Kyoto University of Fine Arts since 1993.

References: "Kurimoto Natsuki Urushi Kaburimono" (Lacquer Headgear by Natsuki Kurimoto), Tokyo Shimbun, Aug. 10, 1991, p. 3; Osaka Contemporary Art Center, Osaka Contemporary Art Fair '91 (Nov. 25–Dec. 7, 1991), pp. 18–19; Washington, D.C., Sasakawa Peace Foundation USA, Natsuki Kurimoto: Urushi Works (Sept. 9–Nov. 16, 1993).

Tatsuaki Kuroda
1904–1982

Tatsuaki Kuroda is known for his wood- and lacquerwork (no. 29), which are admired for their exploitation of the properties of their materials and for their sculptural qualities. Born to a lacquer craftsman in Kyoto, he was apprenticed to the lacquer trade at the age of fourteen. Within a few years he began to study woodworking and ceramics on his own in an effort to go beyond what he considered the confining world of specialized lacquer arts. In the early 1920s Kuroda became involved in the folk-art movement (centering on Japan and Korea) led by Soetsu Yanagi and the potter Kawai Kanjiro. Influenced by their ideas, Kuroda was co-founder of the Kamigamo Folk Art Association in Kyoto and began making furniture and lacquerwork with mother-of-pearl inlay inspired by Korean folk art. In 1935 he held his first one-man show at the Nakamuraya Gallery in Osaka. In the 1940s and 1950s he remained active in promoting crafts in western Japan and helped found the regional section of the Japan Craft Association in 1954. In 1966 he received a commission from the Imperial Household Agency for lacquerwork, in 1970 he was designated a Living National Treasure for his work in wood, and in 1971 was awarded the Japanese medal of honor.

References: Tokyo, The National Museum of Modern Art, ed., Japanese Lacquer Art: Modern Masterpieces (New York, 1982), p. 273; Boston, Museum of Fine Arts, Living National Treasures of Japan (Nov. 3, 1982–Jan. 2, 1983), pp. 28–29, 273, pl. 199; Tokyo, The National Museum of Modern Art, Crafts Gallery, Crafts (1988), pp. 85, 100.

Masayuki Kurokawa
Born 1937

Working as an architect after graduating from Nagoya Institute of Technology in 1961, Masayuki Kurokawa experimented with wood, concrete, and steel to create new types of prefabricated units for residential and commercial use. While in graduate school at Waseda University in Tokyo, Kurokawa also worked for the GK industrial design firm from 1962, before opening his own design office in 1967. His exploration of the possibilities of industrial materials is also characteristic of his work in product design, and he established his reputation for innovative uses of materials (no. 178), as with the Gom series of rubber desk accessories, clocks, and furniture, which he designed from 1973 (no. 138). He has likened himself to a cat with a mouse once he starts toying with ideas, and there is a playful quality to the objects he designs, such as his Cobra lamp (no. 105). Both Kurokawa's architectural projects and his industrial designs were featured in a 1984 exhibition that traveled to New York and San Francisco. In 1985 Kurokawa was one of five designers, including TOSHIYUKI KITA and REI KAWAKUBO, who participated in the exhibition "Hi-Pop Design" (Tokyo) and shared a Mainichi design prize. Kurokawa has also been awarded a prize by the Japan Interior Designers Association (1976), and his furniture was exhibited in Tokyo at the Axis Gallery (1986). Kurokawa continues to expand his repertoire of product design, with appliances such as rice cookers and audio equipment (nos. 193–94).

References: "Masayuki Kurokawa," The Japan Architect, vol. 55 (Aug. 1980), pp. 40–53; Japan Design Committee, Design 19 (Tokyo, Matsuya Ginza, Sept. 3–8, 1982), n.p.; Tadao Ando and others, Design Sympathy (Tokyo, 1985), pp. 284–333, 342; The International Design Yearbook 2 (New York, 1986), nos. 67, 427, 478; Penny Sparke, Japanese Design (London, 1987), p. 136; Fred A. Bernstein, "Design Diplomats," Metropolitan Life (Mar. 1990), p. 86; Masaya Yamamoto, ed., Product Design in Japan (Tokyo, 1990), pp. 162–77, 240.

Yoshihisa Maitani
Born 1933

Yoshihisa Maitani, general manager of the camera development department at OLYMPUS OPTICAL COMPANY, holds nearly two hundred patents for camera-related designs, having obtained the first four while still an undergraduate student in mechanical engineering at Waseda University in Tokyo. After graduating in 1956, he joined Olympus in its camera design department, and his work there has resulted in many innovations, such as the world's first half-frame 35-millimeter single-lens reflex camera in 1963, and the first full-frame compact camera, the Olympus XA of 1978 (no. 125), which changed the face of amateur

photography. Maitani's camera designs have won numerous G-Mark citations, as well as the Governor of Tokyo prize in 1977 and the Japan Photography Association prize in 1980.

References: Yoshihisa Maitani, "Kamera Kapuseru no Shiten" (Capsule Camera Viewpoint), Design News, no. 107 (1980), pp. 14–19; Kogei Zaidan, ed., Nihon no Indasutoriaru Dezain: Showa ga unda Mehin 100 (Japanese Industrial Design: 100 Masterpieces Produced during the Showa Period) (Tokyo, 1989), p. 109.

Zenichi Mano
Born 1916

Zenichi Mano became the first full-time product designer at a Japanese corporation in 1951, when Konosuke Matsushita hired him away from his teaching post at his alma mater, the Tokyo College of Industrial Arts, to design electrical household appliances for the MATSUSHITA ELECTRIC INDUSTRIAL COMPANY. Mano soon proved Matsushita's instincts right, winning the Mainichi industrial design competition in 1953 for his National radio (no. 10). Mano and his expanded design team then won the Mainichi industrial design prize in 1957 for their appliances. By the time Matsushita established its corporate design center in Osaka in 1973, over two hundred designers worked under Mano's direction. The designs produced by his department gained the reputation of being practical and easy to use, in line with the Matsushita policy of providing good design for the ordinary consumer. Mano was active in the Japan Industrial Designers Association, serving as director from 1979 to 1981, and as a juror for design competitions and lecturer at conferences and colleges. He is also the creator of whimsical wood carvings. After retiring from Matsushita in 1977, Mano taught at Musashino Art University, Kodaira, where he is now professor emeritus.

References: "Dai 2-kai Shin-Nihon Kogyo Dezain Nyusho Sakuhin" (Winning Designs of the 2nd Annual New Japan Industrial Design Competition), Kogei Nyusu (Industrial Art News), vol. 22 (Feb. 1954), p. 28; Kodaira, Musashino Art University, Industrial Design Department, ed., Mano Zenichi: Kyoju no Ayumi (Zenichi Mano: A Teacher's Progress) (1986); Akiko Moriyama, "Yoshikazu Mano," Nikkei Design, no. 13 (Apr. 1988), pp. 92–98; Yoko Uga, "Mano Zenichi: Jibun no Dezain shita Mono ga Hiroku Yo no naka ni Yukiwataru Koto o Nozonde" (Zenichi Mano: Hoping My Design Is Accepted in the World), JSSD, vol. 1 (1993), pp. 26–29.

Masudaya Corporation
Established Edo, 1724

Founded in 1724 as a manufacturer and wholesaler of wooden and paper toys, the Masudaya toy company began making wind-up tin toys in the twentieth century. Although it survived the Tokyo earthquake of 1925, the company closed during World War II, reopening in 1946. Incorporated in Tokyo in 1949, Masudaya designed and manufactured some of the most original and popular tin toys of the 1950s, including the first to be radio remote-controlled (no. 40). Known both domestically and internationally for the robots and rockets it produced in the 1950s, Masudaya added space stations to its line in the 1960s. It led the bath toy market in the 1970s, when it marketed a floating frog that proved very popular both in Japan and abroad. Masudaya has won a number of awards over the years for its toys, from the Ministry of International Trade and Industry award in 1951 for a walking penguin toy, to an award for the best toy export of 1962, and awards during the 1970s and 1980s for its international gift designs. Almost three hundred years after its founding, Masudaya is still a family-run business, with factories in Japan and Hong Kong, and retail offices in Japan and Hawaii.

References: *Popular Electronics* (Dec. 1958), cover; Teruhisa Kitahara, *Robots: Tin Toy Dreams* (San Francisco, 1985), pp. 23, 100.

Mashiki Masumura
Born 1910

Named a Living National Treasure in 1978 for his skill in lacquer techniques and decorations, Mashiki Masumura studied traditional lacquer craft at the Kumamoto School of Industry and Commerce (1927) and with a succession of teachers, including YUSAI AKAJI and Shozan Takano in Tokyo. Masumura began his independent career there as a lacquer artist in 1938, first displaying his work at an exhibition commemorating the 2600th anniversary of the founding of Japan, in 1940. After the war, Masumura participated in the Traditional Japanese Crafts exhibitions (from 1956) sponsored by the Japan Craft Association, of which he was a member. Masumura was recognized in these exhibitions particularly for his skill in the *kanshitsu* (dry lacquer) technique, winning the director's award in 1957 and an encouragement prize in 1958. Unusual for a craftsman trained in classical techniques was Masumura's development of original modern vessel forms and abstract decorations (no. 73) for which he is today considered a pioneer.

Reference: Tokyo, The National Museum of Modern Art, ed., *Japanese Lacquer Art: Modern Masterpieces* (New York, 1982), pp. 257–58, 274–75, nos. 180–84.

Katsuo Matsumura
1923–1991

Furniture by the interior designer Katsuo Matsumura is among the best known in Japan. A graduate with a certificate in industrial technology from the Tokyo School of Fine Arts (1944), Matsumura joined the architectural firm of Junzo Yoshimura in 1946. In 1955 he went to work for the Matsuya department store in the Ginza section of Tokyo, taking charge of the store's newly established Good Design Corner. Under Matsumura's direction until 1973, this space became one of the pioneering galleries for the exhibition and sale of interior- and industrial-design items in Japan, and it is still in operation today. In 1956 Matsumura joined RIKI WATANABE at Q Designers, where he helped design a popular line of rattan furniture manufactured by Yamakawa Rattan company. Matsumura established his own firm in 1958, his early commissions including furniture for hotels, restaurants, and exhibition spaces in Japan and abroad, such as the Japanese section of the 1968 "Domus" exhibition in Turin. In the late 1960s Matsumura turned to low-cost, mass-produced furniture for residential use, collaborating with manufacturers such as Johoku, for whom he designed a line of furniture in the until then little-used larch wood (no. 101). This work was awarded the Mainichi industrial design prize in 1971, as well as the Japan Interior Designers prize in 1972. To promote public awareness of interior design, Matsumura helped found the Tokyo Designer's Space in 1973, which remains one of the important showcases for young Japanese designers. Matsumura was also active as a teacher; from his courses in interior design at the Kuwasawa Design School in Tokyo (1956–60 and 1962–65) to his lectures at the Tokyo College of Engineering and Science and Aichi Prefectural University of Fine Arts (1989–91), he remained a committed educator throughout his working life.

References: Katsuo Matsumura and Tadaomi Mizunoe, "Kagu Dezain no Jitsujo" (The State of Furniture Design), in Japan Design Committee, *Dezain no Kiseki* (Tracks of Design) (Tokyo, 1977), pp. 212–18; Japan Design Committee, *Design 19* (Tokyo, Matsuya Ginza, Sept. 3–8, 1982), n.p.; Daisaku Choh, ed., *Matsumura Katsuo no Kagu* (The Furniture of Katsuo Matsumura) (Tokyo, 1992).

Shin Matsunaga
Born 1940

Every Japanese consumer who buys Kibun brand canned vegetables (no. 120), Scottie tissues, or Fuku Masamune sake benefits from the creative talents of the prolific Shin Matsunaga. His work is refreshingly varied (nos. 206, 218), the result of a philosophy that tries not to impose a personal style on his clients' products. A graduate of the Tokyo National University of Fine Arts and Music (1964), Matsunaga worked for the advertising department of Shiseido, where his posters for the company's Bronze Summer suntan oil campaigns won him Tokyo Art Directors Club (ADC) prizes in three successive years (1969–71). In 1971 Matsunaga established his own design office in Tokyo, and while continuing to free-lance for Shiseido, expanded his work to include corporate-identity programs (Mazda, Hankyu), editorial design (*Nonon* and *More* magazines), books and calendars, and package design, winning prizes in every category: Tokyo ADC prizes for calendar (1973) and editorial design (1978), top honors for package design from the Japan Package Design Association (1984, 1985, 1987), and the Mainichi industrial design prize for his graphics and package design (1986). In 1982 Matsunaga attained near legendary status for having designed a book on the Japanese constitution with such an appealing cover and graphics that it became a best-seller. In 1987 Matsunaga succeeded IKKO TANAKA as art director and designer for the Seibu department store's art museum. Matsunaga has also exhibited his work abroad, receiving a gold medal for his posters at the United States–Japan Graphic Design exhibition in New York (1978, 1981), a gold medal at the International Poster Biennale, Warsaw (1988), and an American Clio award for his package designs (1989). He also lectures at the Tokyo National University of Fine Arts and Music, Tama Art University, Tokyo, and the Aichi Prefectural University of Fine Arts.

References: New York, Parsons School of Design, *The Works of Shin Matsunaga* (Oct. 18–Nov. 17, 1989); *Super Market by Shin Matsunaga* (Tokyo, 1992); *Shin Matsunaga* (Tokyo, 1993).

Toru Matsushita
Born 1957

Chief designer in NEC CORPORATION's advanced design division, Toru Matsushita has been responsible for sophisticated products in the fields of computers and communications, including automatic translating equipment (1990), a wearable computer terminal (1990), and a home multi-media station (1991), as well as an experimental skateboard with a linear motor shown at the Nagoya World Design Expo in 1989 (no. 231). Matsushita studied product design at Tama Art University, Tokyo (1977–81), and from 1981 he designed home electronics equipment and personal computers for NEC, including the M-keyboard and the PC-88 series, until he was transferred to the advanced design division in 1988.

Reference: London, Design Museum, *Review* 2: *New Product Design 1990–91* (1990), p. 31.

Matsushita Electric Industrial Company
Established Osaka, 1918

One of the world's largest producers of consumer electronic goods (nos. 10, 83, 210), Matsushita Electric Industrial Company was founded by Konosuke Matsushita in 1918 after eight years' experience at the Osaka Electric Light Company. Established to produce electric light sockets of his own design, the firm by 1923 was manufacturing battery-powered bicycle lamps. The firm was incorporated in 1935 as Matsushita Electric Industrial Company, with nine subsidiary companies, and in 1939 opened its first overseas plant in Shanghai. At the onset of World War II, Matsushita's plants were drafted into the war effort, and after the war, Matsushita was targeted for dissolution by the United States occupying forces, but intense lobbying by Matsushita's employees in support of Konosuke Matsushita and their insistence that he had not been responsible in any way for Japan's entry into the war prevented the company from being dissolved. During the 1950s Matsushita developed electric appliances for the home, and by the end of the decade was supplying Japanese households with many of the "three treasures" (black and white televisions, washing machines, and refrigerators) then in high demand, and selling Japanese-made goods in the United States as well, through the Matsushita Electric Company of America. Listed on the New York Stock Exchange in 1971, the firm expanded abroad throughout the 1970s; today it has manufacturing plants in more than thirty countries around the world. Matsushita's products are sold in North America under the brand names Panasonic, Technics, and Quasar, and worldwide under the National and Technics trademarks (no. 210). Konosuke Matsushita, personally responsible for much of the firm's phenomenal growth, served as the firm's president until 1961, chairman of the board until 1973, and executive advisor until his death at ninety-four in 1988. Throughout his career, he continued to follow traditional Japanese management practices while other Japanese companies adopted American-style management techniques after the war. Matsushita lectured and wrote extensively on his management philosophy, publishing his last book on the subject, *Not for Bread Alone: A Business Ethic, A Management Ethic,* in 1984. In 1987 Matsushita was awarded the Order of the Rising Sun. The company he founded has continued to prosper by following his philosophies and managing practices. Interested in expanding into related fields and increasing its visibility in international affairs, in 1990 Matsushita bought the American movie company MCA, owner of Universal Studios.

References: Japan Industrial Design Promotion Organization (JIDPO), *Design News*, vols. 79–80 (1977), p. 11; Matsushita Electric Industrial Company, *Shape of Matsushita* (Osaka, 1980), p. 64; Obituary [Konosuke Matsushita], *The New*

York Times, Apr. 28, 1988; JIDPO, "Matsushita," *Design Quarterly Japan*, vol. 4 (1988), p. 6; JIDPO, "Piedra 8," *Design News*, vol. 202 (1989), p. 18; London, Design Museum, *Review 1: New Product Design 1989–90* (1989), p. 30; *The International Design Yearbook* 5 (New York, 1990), p. 199; "Matsushita's President Resigns in Scandals," *The New York Times*, Feb. 24, 1993, p. D5.

Hiroyuki Matsuyama
Born 1959

Package designer Hiroyuki Matsuyama graduated in 1984 from Tama Art University in Tokyo, studying in its departments of sculpture and design, and spent a summer at the Art Center College of Design in Pasadena. Matsuyama joined Pola cosmetics in 1984, remaining there until 1990, when he began work for Estee Lauder. Matsuyama received an award of excellence in 1989 from the Package Design Council International for his cosmetic packaging for Pola (no. 226).
Reference: *Package Design in Japan 1989* (Tokyo, 1989), p. 145.

Makiko Minagawa
Born c. 1947

Shortly after graduating from art school in Kyoto in 1970, Makiko Minagawa became a textile designer in the Miyake Design Studio, which had been formed that very year by the then unknown fashion designer ISSEY MIYAKE. Minagawa, now director of textiles for Miyake, became an integral part of the innovative and dynamic fashions that Miyake has created (nos. 162, 175). Despite her influential role in his success, she chooses to stay out of the limelight, firmly convinced that her efforts are but a contribution to the final artistic statement created by the designer. Minagawa has participated in numerous exhibitions of Miyake's work in Europe, the United States, and Japan, including "Les Tissus Imprimés d'Issey Miyake" (Issey Miyake and His Printed Textiles) at the Musée de l'Impression sur Etoffes in Mulhouse, France (1979), which highlighted her fabrics. Miyake has always had specific fashion ideas, and relies on Minagawa to produce the materials he envisions. "Working with Miyake is like a Zen dialogue," says Minagawa. "One day he says, 'White in winter.' That's all he says. I wonder, the white of ice? The white of a salt field? The white celadon of the Korean Li Dynasty? Or simply the white of snow? In ordinary Zen dialogue there is an immediate response. In this case, my responses take at least a year, beginning with the selection of materials for yarn."[10] The entire process, from creating the initial white sample fabrics to the final bolts of cloth for a collection, might take two to three years, with Miyake and Minagawa refining their ideas along the way. Her remarkable way of pushing the boundary of fashion-textile design has influenced the textile industry in Japan. JUNICHI ARAI

developed fabrics for Miyake under Minagawa's direction in the early 1980s. A limited edition book, *Texture*, published in 1987, celebrates Minagawa's career as one of the world's most perceptive and accomplished textile designers. In 1990 she had a solo exhibition at Gallery Ma, Tokyo, and received the Mainichi fashion prize and the Amiko Kujiraoka prize.
Reference: Miyake Design Studio, *Issey Miyake and Miyake Design Studio, 1970–1985* (Tokyo, 1985).

Issey Miyake
Born 1938

Like the architect Le Corbusier and the painter Pablo Picasso, the fashion designer Issey Miyake has redefined the boundaries of his field, changing the shape and form and even the medium of clothing. Exploring new relationships between clothing and the body—even the space between the body and clothes—Miyake broadened the range of what clothes can do and be made from, using traditional Japanese craft materials like bamboo and rattan, for example, to make an open cagelike bustier (no. 156) or oil-soaked handmade paper for rainwear that billows like a cocoon, as well as industrial technologies to produce fabrics like polyurethane-coated polyester jersey that falls away from the body in deeply sculpted leatherlike pleats. Born in Hiroshima, Miyake attended Tama Art University in Tokyo and then worked as a free-lance graphic designer until 1965, when he went to Paris to study fashion design, first at the school of the Chambre Syndicale de la Couture Parisienne and then with Guy Laroche (1966–67) and Hubert Givenchy (1968). In 1969 Miyake went to New York, where he worked with Geoffrey Beene, and then returned to Tokyo in 1970 to found his own design studio. Educated in Western couture, Miyake as an independent clothing designer liberated himself from dressmaking traditions of any kind. "My challenge as a clothing designer has been to create something different," Miyake said later, "not traditionally Japanese nor purely Western fashion. I had to start from the initial concept of clothing as the body covering."[11] In 1973 Miyake presented his first collection in Paris and in 1975 the first of what would become a succession of museum exhibitions of his work ("Inventive Clothes," held at the National Museum of Modern Art, Kyoto). In 1978 he published the book *East Meets West*, a photographic retrospective of his work, and in 1979 his screen prints based on designs by such outside collaborating artists as TADANORI YOKOO (no. 136) and EIKO ISHIOKA were exhibited at the Musée de l'Impression sur Etoffes in Mulhouse, France. During the early 1980s Miyake took his study of the relationship between the human body and the garment directly back to the body, even molding the human form in a plastic bustier designed for outerwear (no. 139), and his garments of this period were presented in "Bodyworks," an exhibition shown in Tokyo, Los Angeles, San

Francisco, and London between 1983 and 1985. While Miyake's fabrics of the 1970s were largely natural, the result of his work with the textile designer MAKIKO MINAGAWA, and often depended on traditional textile techniques such as indigo dyeing, *ikat*, and *sashiko* quiltings, from the mid-1980s he also pursued synthetic fabrics like polyester, which by the late 1980s were finished in a heat-set pleating machine (nos. 235, 250). His machine-pleated clothing was exhibited at the Touko Museum of Contemporary Art, Tokyo, and the Stedelijk Museum, Amsterdam, in 1990 as "Issey Miyake Pleats Please." A commemorative exhibition celebrating two decades of his work, when Miyake was awarded the first Hiroshima Art Prize, was held at the Hiroshima City Museum of Contemporary Art in 1990. Miyake has received numerous other awards for his work, including the Mainichi fashion grand prize in 1984, 1989, and 1993, and the Asahi prize in 1991.
References: *Issey Miyake: East Meets West* (Tokyo, 1978); *Issey Miyake Bodyworks* (Tokyo, 1983); Miyake Design Studio, *Issey Miyake and Miyake Design Studio, 1970–1985* (Tokyo, 1985); Nicholas Callaway, ed., *Issey Miyake: Photographs by Irving Penn* (Boston, 1988); Tokyo, Touko Museum of Contemporary Art, *Issey Miyake Pleats Please* (Sept. 1–30, 1990); Hiroshima, Hiroshima City Museum of Contemporary Art, *Issey Miyake: Ten Sen Men* (1990).

Eiji Miyamoto
Born 1948

In 1975 Eiji Miyamoto joined his family's textile business, the Miyashin Company, located in Hachioji, a centuries-old weaving center near Tokyo. At first he worked in the company's marketing and sales department for traditional kimono fabrics, but then moved on to develop fabrics for Western fashion apparel. Today he heads the company and is its main designer. Miyamoto, along with other forward-looking designers, adapted to the new market demands for Western fashion by boldly applying high technology to the design of fabrics, using computers to provide complex weave structures, employing chemistry to achieve new effects in color, and manipulating and embellishing cloth in its final stages (nos. 240–41). Miyamoto kept a low profile as a behind-the-scenes creator of fashion clothing until his work was included in the "Japan Creative" exhibition at the Seibu department store in Tokyo in 1985 and then introduced his own collection of textiles in 1986. In the last decade Miyamoto has been producing fabrics for fashion designer ISSEY MIYAKE under the direction of MAKIKO MINAGAWA, and he credits her with sparking his imagination and encouraging him to construct unconventional fabrics. In 1988 Miyamoto helped to found the Hachioji Fashion Team, a group of textile producers interested in addressing the need to create innovative materials for the Western fashion

industry. He currently is an executive member of the Hachioji Woven Textiles Industry Trade Association and a committee member for the Hachioji Fashion Center, and works with Tokyo Prefecture's regional industry association to promote and develop the use of high technology in the textile field. His work has been included in numerous group exhibitions, and in two solo shows in Tokyo, "Cloth from the Heart" (1989), and "Eiji Miyamoto" (1991).
Reference: Minoru Ueda, "Interview with Eiji Miyamoto," *EX: Crafts, Architecture & Arts* (Oct. 1991), pp. 1–4.

Hanae Mori
Born 1926

Hanae Mori was the first Japanese fashion designer to show her work in the West (1965), and to be accepted into the organization of French couturiers, the Chambre Syndicale de la Couture Parisienne (1977). Known internationally for her use of gorgeous fabrics, including prints based on traditional Japanese patterns (nos. 92, 227) or her signature butterfly (see no. 134), often embellished with beadwork and sequins for evening wear (no. 202), Mori has designed clothing in largely Western styles for a distinguished clientele ranging from members of the Japanese royal family (including, in 1993, the wedding dress of Crown Princess Masako) to the diplomatic corps of many nations, as well as costumes for productions of *Madame Butterfly* at Milan's La Scala opera (1985) and *Cinderella* for the Paris Opéra ballet (1986). "I have tried to be a link," Mori said, "by introducing to the West the beauty of the East, as well as showing the beauty of the West to the East. Through this link, I may perhaps have been able to keep pursuing my own identity and the identity of Japanese beauty on an international basis."[12] Mori was born in Shimane Prefecture, moving in 1937 to Tokyo, where she studied literature at the Tokyo Women's Christian University (1947). After her marriage to Ken Mori, who was in the textile business, she studied fashion design, opening an independent studio in the Shinjuku section of Tokyo in 1951. By 1954 her work had attracted the attention of the Japanese film industry, and over the next seven years, she designed costumes for some five hundred films, receiving an award for them from the Japan Editors Club in 1960. Mori made her first trips to Paris and New York in 1961, and in 1965 showed her collection abroad for the first time (in New York), ushering in a period of expansion for her firm. She traveled to India in 1966 at the invitation of the Indian government (as she did also in 1987) and in cooperation with its Handicrafts and Handlooms Export Corporation in 1969 used Indian textiles for an international collection that was shown at Expo '70 in Osaka. In 1967 Mori designed uniforms for the flight attendants of Japan Air Lines (and again in 1970 and 1977). She established branches in New York in 1970 and in

Paris in 1976, showing her first collection and opening a haute couture showroom there in 1977, followed by a boutique in 1985. Her work has won her many awards, from Japan's medal of honor (1988) and the Asahi prize (1988) to the French *chevalier des arts et des lettres* (1984) and *chevalier de la Légion d'Honneur* (1989). A pioneer and symbol of the Japanese working woman, Hanae Mori has served on many government committees and was the first female member of the Japan Association of Corporate Executives (1986).
Reference: *Hanae Mori 1960–1989* (Tokyo, 1989).

Masahiro Mori
Born 1927

Masahiro Mori, born in Saga Prefecture near the town of Arita, a center of Japanese ceramic production since the seventeenth century, shares the strong spirit of craftsmanship and regional pride of the potters among whom he was raised and with whom he studied. After two years' apprenticeship with a master potter (1946–48), he entered Tama Art University, Tokyo, receiving his degree in 1952. While still a student Mori served an internship with the Industrial Arts Institute, which gave him his first professional experience; afterward he worked at the research center for the ceramic industry in Nagasaki Prefecture. In 1956 he was hired by the Hakusan porcelain company in Hasami-machi, in his native region, and continued his association with the firm until 1978, when he opened his own design firm. His best-known work for Hakusan is the much-premiated soy-sauce container of 1958 (no. 37), which still remains in production and serves as a testament to Mori's success in helping to modernize regional ceramic production (no. 117). Among his many awards are the Mainichi industrial design prize and the Kunii Industrial Art Award (both in 1973).
References: Tokyo, The National Museum of Modern Art, Crafts Gallery, *Contemporary Vessels: How to Pour* (Feb. 10–Mar. 22, 1982), nos. 172–80; Kogei Zaidan, ed., *Kunii Kitaro Sangyo Kogeisho no Hitobito: 35-nin no Purofuiru* (The Winners of the Kitaro Kunii Industrial Art Award: Profiles of 35 Winners) (Tokyo, 1984), pp. 23–26; Kogei Zaidan, ed., *Nihon no Indasutoriaru Dezain: Showa ga unda Meihin 100* (Japanese Industrial Design: 100 Masterpieces Produced during the Showa Period) (Tokyo, 1989), p. 29.

Kako Moriguchi
Born 1909

At the age of fifteen Kako Moriguchi left his home in Shiga Prefecture for Kyoto, where he became apprenticed to a master of the resist-dye (*yuzen*) techniques, and at the same time studied traditional Japanese painting. When in 1939 Moriguchi established his own textile workshop, his training in both these fields became important elements in his work. Moriguchi's reputation in the field of *yuzen*

dyeing was firmly established when the kimono he submitted to the second annual Traditional Japanese Crafts exhibition took the top prize. He won the highest prize again the following year, and has been invited to show annually since; he now serves as a juror for the exhibition as well. Moriguchi's kimonos are known for their bold, contemporary designs, featuring themes from nature such as plum blossoms, bamboo, and flowing streams. He treats the kimono almost as a flat canvas or screen, with his painterly designs sweeping across its entire surface (no. 57). In 1967 Moriguchi was named Holder of Intangible Cultural Property (Living National Treasure) for his skill in *yuzen* dyeing.
References: Tokyo, The National Museum of Modern Art, Crafts Gallery, *Masterpieces of Contemporary Japanese Crafts* (Nov. 15, 1977–Mar. 19, 1978), nos. 143–46; Boston, Museum of Fine Arts, *Living National Treasures of Japan* (Nov. 3, 1982–Jan. 2, 1983), nos. 63–73, p. 266; Tokyo, The National Museum of Modern Art, Crafts Gallery, *Crafts* (Tokyo, 1988), nos. 226–39.

Hiroshi Morishima
Born 1944

Hiroshi Morishima is particularly passionate about handmade paper, which he uses in both graphic and interior design. A graduate of the Tama Art University, Tokyo (1965), Morishima continued his studies at the Art Center College of Design in Pasadena in 1968. From 1970 to 1973 he worked in New York for CBS and other corporations before returning to Japan, where he joined the Nippon Design Center as a graphic designer. Morishima won a Tokyo Art Directors Club prize in 1979 (and again in 1985), and the same year, a bronze prize at the annual Design Forum exhibition sponsored by the Matsuya department store in Tokyo. Morishima started his own firm, Time-Space-Art, in Tokyo in 1980, where he concentrated on working with paper craftsmen in northern Japan whose specialty is translucent papers made from mulberry bark (*kozo*), which Morishima incorporates in window shades, screens, and lamps (no. 183). In 1983 he was given the Kunii Industrial Art Award in recognition of his creative applications of traditional handmade paper for contemporary interiors, and was invited to participate in the 1987 contemporary craft exhibition in Tokyo sponsored by the Asahi newspaper company.
References: Kogei Zaidan, ed., *Kunii Kitaro Sangyo Kogeisho no Hitobito: 35-nin no Purofuiru* (The Winners of the Kitaro Kunii Industrial Art Award: Profiles of 35 Winners) (Tokyo, 1984), pp. 139–41; *The International Design Yearbook* 2 (New York, 1986), pp. 94–95, 102–3, 169; Time-Space-Art, *Wagami* (Tokyo, 1989); Hiroshi Morishima, "Zanshin na Gendai 'Nippon' Kankaku" (A Fresh, Contemporary Concept of "Nippon"), *Nikkei Design*, no. 52 (Oct. 1991), pp. 114–15.

Shutaro Mukai
Born 1932

One of Japan's leading design educators, Shutaro Mukai is also a designer and a concrete poet. Mukai studied economics at Waseda University, Tokyo, graduating in 1955, after which he studied in Germany on fellowships from the Japan External Trade Organization at the Hochschule für Gestaltung in Ulm (1956–58). He was a research fellow at the Industrial Arts Institute in Tokyo (1958–59), and in 1959 began to work as a designer for KATSUHEI TOYOGUCHI, where he remained until 1963, working on such projects as the Japanese pavilion at the New York World's Fair. Mukai returned to Germany in 1963, when he became a fellow at the school in Ulm (1963–64) and subsequently at the Technische Universität in Hanover (1964–65). On his return to Japan in 1965, he was a founding member of the department of the science of design at Musashino Art University, Kodaira, where he has remained, becoming professor in 1970 and chairman in 1972; he was responsible for its graduate school restructuring plan in 1992. In 1971 Mukai opened his own design office, where he has continued his research into the relationships between linguistics, semiotics, and aesthetics that he first explored in his studies abroad (no. 116). Mukai's many publications, which demonstrate the wide range of his interests, include *Dezain no Genten* (Origin of Design) (Tokyo, 1978), *Kigo to shite no Geijutsu* (Art as Sign) (Tokyo, 1982), *Katachi no Semioshisu* (Semiosis of Form) (Tokyo, 1986), and *Gendai Dezain Jiten* (Dictionary of Today's Design) (Tokyo, 1986–93).
Reference: Osaka, The National Museum of Art, *Design and Art of Modern Chairs* (Isu no Katachi: Dezain kara Aato e) (Aug. 19–Oct. 15, 1978), p. 141.

Reiko Murai, see Reiko (Murai) Tanabe

Kazumasa Nagai
Born 1929

One of Japan's leading graphic designers and art directors, Kazumasa Nagai joined the Nippon Design Center when it was established in 1960 and has served as president (1975–86) and director (since 1986) of what has become the largest and most influential design production company in Japan. Nagai studied sculpture at the Tokyo National University of Fine Arts and Music (1951) but left to work as a graphic designer for the Daiwa Spinning Company in Osaka (1951–60). When he joined the Nippon Design Center, he had already developed his personal style of cool, abstract, geometric forms and linear patterns; it appeared in his early posters for the Nippon Kogaku company (later NIKON CORPORATION; no. 75) and in a poster series created for the life science library of Time-Life books (1966), where it was perfectly

suited to the high-quality optical products and scientific subjects the posters advertised. During the late 1960s, Nagai began to combine photographs, largely of the sea, sky, and horizon, with his geometrical compositions (posters for *GQ* magazine), setting his forms in a vast space. In addition to posters (no. 70), trademarks, and advertisements (no. 79), Nagai independently created prints and drawings that have been widely exhibited in Japan and elsewhere, displaying the technical and production quality for which he is so justly famous (no. 160). Nagai's work with geometric forms in space continued throughout the 1970s and early 1980s (no. 157), but by the late 1980s realistic elements had entered his vocabulary, notably, in the fantastic animal posters he created for the annual exhibition of the Japan Graphic Designers Association in 1988, with each animal composed of traditional, small-scale decorative textile patterns (no. 214), and later in his one-man exhibition, "The World of Kazumasa Nagai," in Toyama in 1990. Nagai has won awards ranging from medals at the International Poster Biennale, Warsaw (1966, 1968), Biennale of Graphic Design, Brno (1984, 1988), and Moscow poster triennial (1992) to the Mainichi industrial design prize (1966, 1983) and the Japanese medal of honor (1989).
References: Tadanori Yokoo, "Nagai Kazumasa no Hito to Sakuhin" (Kazumasa Nagai: His Works and Personality), *Design*, no. 95 (Apr. 1967), pp. 18–33; Tokyo Art Directors Club, *Art Direction Today* (Tokyo, 1984), pp. 162–65, 238–41; *The Works of Kazumasa Nagai* (Tokyo, 1985); *Best 100 Japanese Posters, 1945–89* (Tokyo, 1990), pp. 84–85, 134–35, 138–39, 166–67, 210–11, 220–21, 250; Toyama, Museum of Modern Art, *The World of Kazumasa Nagai* (Sept. 1–Oct. 7, 1990); Tokyo, The National Museum of Modern Art, Crafts Gallery, *Graphic Design Today* (Sept. 26–Nov. 11, 1990), pp. 66–69, 114.

Makoto Nakamura
Born 1926

Makoto Nakamura has provided art direction and graphic design at Shiseido since his graduation from the Tokyo School of Fine Arts in 1948, and in the process has guided the cosmetic company's advertising in its postwar transition from illustration to photography. Serving as Shiseido's advertising director (1959), creative director, and since 1987 as consultant, Nakamura has created highly sophisticated posters and television commercials, some alluding to traditional *ukiyo-e* graphics (no. 123) that have won international recognition and prizes, including those from the Japan Advertising Artists Club, Tokyo Art Directors Club, and the International Poster Biennale, Warsaw, as well as the Mainichi advertising prize, Dentsu advertising prize, Ministry of International Trade and Industry award at the national calendar exhibition, and the Yamaha prize. Nakamura was awarded the Japanese medal of honor.

References: Mill Roseman and Ko Noda, "Japan: The Design Scene," *Communication Arts*, vol. 24 (Jan.–Feb. 1983), pp. 29, 30, 48–55; Richard S. Thornton, *The Graphic Spirit of Japan* (New York, 1991), pp. 97–98.

Tadashi Nakanishi
Born 1917

Tadashi Nakanishi was born in Hokkaido, in the northern part of Japan, where he learned the technique of lacquerwork from his father. In Tokyo in 1954 he established the Musashino Lacquer Polytechnic Institute to continue his research in the field of lacquer and as a workshop for his own productions. In his lacquer pieces, he seeks to use traditional materials and techniques to improve the design of common household utensils, such as soup bowls and soy-sauce containers (no. 96). In 1960 Nakanishi's work was selected for the new artists section of the Traditional Japanese Crafts exhibition, and his lacquers have been seen in these exhibitions almost every year since then. His work has also been shown in numerous solo exhibitions at the Mitsukoshi and Seibu department stores. Nakanishi has also enthusiastically supported efforts to promote the work of lacquer artists, in 1960 becoming managing director of the Japan Lacquer Craft Association (since 1979 serving on its board of directors), and he is also a member of the committee of specialists for lacquer for the cultural properties commission. Nakanishi was a founding member of the Japan Craft Association, and has served on its board of overseers since 1974.

References: Tokyo, The National Museum of Modern Art, Crafts Gallery, *Contemporary Vessels: How to Pour* (Feb. 10–Mar. 22, 1982), nos. 139–40; Tokyo, Mitsukoshi Department Store, *Exhibition of the Lacquer Works of Tadashi Nakanishi* (Nakanishi Tadashi Shitsugei-ten) (Nov. 14–19, 1988).

NEC Corporation
Established Tokyo, 1899

The Nippon Electric Company (renamed NEC Corporation in 1983) was founded in 1899 as a joint venture with the Western Electric Company of Illinois to manufacture telephone sets and switching equipment, the first Japanese joint venture with foreign participation. It remains today Japan's leading manufacturer of communications systems as well as computers, and ranks among the world's largest in this field. From its earliest years Nippon Electric sought to expand into overseas markets and to develop and diversify its product lines. The company established a sales office in Seoul in 1908 and a joint venture with the China Electric Company in Beijing in 1917; and entered the field of radio communications in 1924. Nippon Electric began to research and develop transistors in 1950, computers in 1954, integrated circuits in 1960, and satellite communications in 1963. A subsidiary,

New Nippon Electric Company (presently NEC Home Electronics), was established in 1953 to manufacture and market consumer products, starting with the manufacture of radio tubes. By the late 1950s the company had completed its first computers and in 1964 built a plant in Fuchu to produce them, developing in 1979 its first personal computers. During the 1960s Nippon Electric established subsidiaries overseas, its first marketing affiliate, Nippon Electric New York (now NEC America), in 1963, and its first manufacturing affiliate, NEC de Mexico, in 1968. Today NEC and its affiliates operate ninety-three plants in sixteen countries, including five manufacturing sites in the United States. Since establishing its first research laboratories in 1939, NEC has maintained a consistent policy of reinvesting a considerable portion of its sales into research and development and engineering activities, which since 1977 have been directed to integrating computers with communications systems. NEC's design center creates advanced products based on these technological innovations (nos. 211, 231), developing, for example, Japan's first microcomputer in 1974 and the world's fastest mainframe computer in 1991.

Reference: NEC Corporation, *This Is NEC 1993* (Tokyo, 1993).

Nido Industrial Design Office
Established Tokyo, 1963

Founded jointly by Yoshiyasu Ishii, a 1956 graduate of the Tokyo School of Fine Arts, and Yoshitoshi Ishii, the Nido Industrial Design Office has taken an aggressive approach to effecting change in the technology and design of children's playground equipment and consumer products (no. 124). By developing a computer database of information derived from their comprehensive studies of children and their behavior, Nido has made significant advances in the field, for which it has received many awards, including the Kunii Industrial Art Award in 1986. Yoshiyasu Ishii, head of the Nido Industrial Design Office, has also worked as a special designer for the Children's Science Museum in Yokohama (1983) and on the Tsukuba Expo '85, and served on the selection jury for the G-Mark award (1977, 1978, 1983, and 1984).

Reference: "Kosei Gangu Kapusera no Kaihatsu" (The Development of Capsela—Formation Toy), *Industrial Design*, no. 87 (May 1, 1977), pp. 28–32.

Takeshi Nii
Born 1920

Born into a family that made equipment for the Japanese martial sport of *kendo* on the island of Shikoku, Takeshi Nii entered the family business after graduating high school in 1938, enrolling at the same time in courses in furniture and interior design. By 1952 Nii started making wood furniture, and in 1956 produced his first folding chair in

wood, inspired by the bamboo staves used in *kendo*. Nii was encouraged in his work by KATSUHEI TOYOGUCHI, at the time design chief at the Industrial Arts Institute, and in 1960 Nii produced the first of his folding chairs in steel tubing. That same year three of his chairs were chosen for the first Good Design exhibition at Matsuya department store in Tokyo, and in 1961 Matsuya was the site of a one-man exhibition of Nii's folding chairs. Nii opened his own manufacturing firm, Ny Furniture, in 1969, and the following year began producing his Ny X series of folding chairs (no. 103), one of which won a Japan Interior Designers prize in 1970. Nii calls himself a "craftsman designer" and prides himself on the fact that he personally oversees the production of his designs. His aim has been to create furniture that is as reasonable, comfortable, and ubiquitous in the household as a bowl of curry rice. Since 1982 Nii has held a United States patent for his chairs, which are now sold worldwide. In 1988 Takeshi Nii received the Kunii Industrial Art Award for forty years of service in promoting local industry in his native Shikoku, where his own company is still located.

References: Katsuo Matsumura and Tadaomi Mizunoe, "Kagu Dezain no Jitsujo" (The State of Furniture Design), in Japan Design Committee, *Dezain no Kiseki* (Tracks of Design) (Tokyo, 1977), pp. 212–18; Osaka, The National Museum of Art, *Design and Art of Modern Chairs* (Isu no Katachi: Dezain kara Aato e) (Aug. 19–Oct. 15, 1978), p. 90; Japan Industrial Designers Association, ed., *Seiichi no Kozo: Nihon no Indasutoriaru Dezain* (Structure of Dexterity: Industrial Design Works in Japan) (Tokyo, 1983), p. 129; Kogei Zaidan, ed., *Nihon no Indasutoriaru Dezain: Showa ga unda Meihin 100* (Japanese Industrial Design: 100 Masterpieces Produced during the Showa Period) (Tokyo, 1989), pp. 51–52; Shimazaki Noboru, "Nii Chea Korekushion" (Nii Chair Collections), *Kirkos*, no. 9 (July 1992), pp. 14–18.

Nikon Corporation
Established 1917

Created in 1917 by the merger of three of Japan's leading optical manufacturers, Nippon Kogaku became a major producer of telescopes, microscopes, and other optical equipment, issuing camera lenses under the trademark Nikkor by 1932. At the end of World War II, only two of its manufacturing plants remained open, but within a year these were producing substantial numbers of cameras, microscopes, binoculars, eyeglasses, and surveying instruments. In 1946 the company adopted the Nikon brand name for small-size cameras, and by 1948 its first 35-millimeter camera, the Nikon I, was on the market. Within two years, Nikon cameras and Nikkor lenses were well known internationally. The company expanded overseas, opening its first American subsidiary in New York in 1953 and its first European subsidiary in 1961. In

1957 the Nikon SP camera went on the market (no. 34), followed in 1959 by the Nikon F camera, which was popular with both professional and amateur photographers. Nikon cameras were used in three U.S. Apollo missions (1971–73) to photograph the moon's surface and views of the earth from space, in the Skylab space laboratory (1973), and in the space shuttle program (1979). In 1975 the firm began producing eyeglass frames, and in 1979 commissioned HANAE MORI to design Nikon sunglasses. In the late 1970s the company introduced light, compact, automatic 35-millimeter cameras with manual override, and marketed its first compact auto-focus camera in 1983. During the 1980s cameras accounted for 65 percent of the company's business, eyeglasses for 13 percent, and microscopes and surveying instruments for 8 percent each. In 1981 the United States subsidiary was renamed Nikon Incorporated and in 1988 the parent company changed its name to Nikon Corporation to reflect its emphasis on cameras.

References: "Nippon Kogaku," *Kodansha Encyclopedia of Japan* (Tokyo, 1983), vol. 6, p. 10; "Nikon: A History of Innovation," in Nikon Corporation, *Nikon: Focusing on the Future 1990* (Tokyo, 1990), pp. 32–33; "The Nikon Years: A Calendar of Achievement," in Nikon Corporation, *Nikon '94* (Tokyo, 1994), pp. 28–29.

Nintendo Company
Established Kyoto, 1889

In 1933 the company founded by Fusajiro Yamauchi in 1889 to manufacture the Japanese playing cards called *hanafuda* became Yamauchi Nintendo Company. By 1953 it was selling mass-produced plastic playing cards, and by the end of the decade had added a popular line of children's playing cards decorated with Walt Disney characters. The company changed its name to Nintendo Company in 1963, when it began manufacturing games. In 1970 Nintendo introduced a toy gun incorporating electronic technology, the first toy with such technology in Japan. In the mid-1970s the company began developing and marketing video games for amusement arcades in North America, Europe, and Japan, its first widely popular video game being the coin-operated Donkey Kong, which it developed and distributed in 1981. Working with Mitsubishi Electric, by 1977 Nintendo had developed video games for the home. In 1980 a subsidiary, Nintendo of America, opened in New York, moving to Seattle within two years. By 1983 the company was marketing a "family computer," or video-game console for the home, elaborating on this system throughout the 1980s. The Game Boy (no. 223), a portable, hand-held game system with interchangeable software, was introduced in 1990 and immediately became a hit worldwide. The following year, Nintendo began marketing a Super NES (or Nintendo Entertainment System) for the home along with a new range of games for it.

Reference: David Sheff, *Game Over: How Nintendo Zapped an American Industry, Captured Your Dollars, and Enslaved Your Children* (New York, 1993).

Nippon Kogaku, see Nikon Corporation

Kozo Okada
Born 1934

A graphic and package designer known for his unique gift boxes in wood (no. 172) as well as for commercial packaging (no. 217), Kozo Okada was born in Osaka and attended the Kyoto University of Fine Arts. After graduating in 1958, Okada moved to Tokyo, where he worked in the advertising department of Renown, a manufacturer of clothing and table linens. In 1965 he established the Okada Design Office, which was reorganized in 1969 under the name OD, Inc. Okada is a member of the Japan Package Design Association (winning awards from the group in 1975, 1985, and 1989), as well as the Japan Sign Design Association and the Japan Graphic Designers Association.
Reference: *Package Design in Japan* (Cologne, 1989), pp. 145–63, 236.

Sinya Okayama
Born 1941

Sinya Okayama introduced an advanced postmodern idiom to Japanese design in the 1980s, creating furniture, lighting, and objects that were intended to communicate—sometimes literally—with the consumer. He became known internationally through his participation in the "Phoenix" exhibition at the Queen's Quay Terminal, Toronto, in 1984, and afterward collaborated with Alessandro Mendini, a founder of Italian postmodern design, on two projects, "Sei Mobiletti" (1986) and "7 + 7 Gioielli" (1987). Following an unusual process of design, each remained in his own country, and communicating only by letter, one designed and drew a part of an object and the other completed the design without their meeting to discuss the work. To some extent Okayama's objects all suggest the real object or object type for which they are named, such as his Top of the World shelves (1984), which resemble a snow-covered mountain with its peak in the clouds, and his Condor cupboard (1989), which extends horizontally like the wings of a bird. Certain objects relate directly to their names, such as the shelves and stools he designed in the shape of the Japanese ideograms for the words "celebration" and "wind" (nos. 228, 182). From 1961 to 1966 Okayama worked in the design department of the Mitsukoshi department store corporation, but since 1967 he has been a free-lance interior and product designer. Okayama opened his design office in Osaka in 1970, producing his first furniture and lighting fixtures under his own name in 1981.

References: New York, Gallery 91, *Interior Objects and Art Furniture by Sinya Okayama* (June 9–July 14, 1984); Tokyo, Yurakucho Asahi Gallery, *Interior Objects by Sinya Okayama* (Apr. 28–May 17, 1989).

Olympus Optical Company
Established Tokyo, 1919

Established as a microscope manufacturer in 1919, the Olympus Optical Company later added medical products to its production line, manufacturing the first gastro-camera for stomach examinations in 1951. In 1959 it introduced a pocket camera (no. 125), following it with a micro-cassette camera in 1969. Today the company is best known for manufacturing cameras (no. 234) and microscopes, but also produces medical equipment such as endoscopes, and other high-technology specialty products. Its factories in Asia, Europe, and North America employ 10,000 people worldwide. Research into new technologies is conducted at the Olympus Technology Research Institute in Japan and in three smaller centers in the United States, allowing Olympus to maintain its position as a world leader in electronic and optical technology (no. 80).
Reference: "Olympus Optical Co., Ltd.," *Kodansha Encyclopedia of Japan* (Tokyo, 1983), vol. 6, p. 100.

Plus Corporation
Established Tokyo, 1948

Founded in 1948 as a wholesale stationer, Plus Corporation was restructured and incorporated in 1959. The company expanded into the design and manufacture of office furniture, office machinery, and educational equipment, and began distributing its products on the international market, opening a branch in the United States in 1980 and one in Taiwan in 1985. In the 1980s Plus built its design reputation on stationery items and office equipment for the younger, upscale Japanese consumer, particularly through its compact, fashionable designs (nos. 177, 196).
Reference: *The International Design Yearbook* 2 (New York, 1986), pp. 439–40.

Popy, see Bandai

Makoto Saito
Born 1952

Noted for his risk-taking and the uncompromisingly individual posters that have won him international recognition and awards, Makoto Saito manipulates graphic and pictorial elements in dramatic compositions using cut and altered photographs and collages. A concept maker, Saito creates expressive images for clients who give him room to experiment: with human bones in bright blue to advertise Buddhist altars (no. 181), with shadows

wearing colored neckties for a clothing boutique (no. 165), and with a cruciform photocollage with hands, feet, ears, and face for a printing company (Taiyo). Saito was born in Fukuoka and had little formal art education. From 1976 to 1981 he worked at Tokyo's Nippon Design Center and in 1982 established his own design office. His clients have included department stores and clothing boutiques such as Alpha Cubic, Jun, Garo, and Parco. During the 1980s Saito won numerous awards, repeatedly gaining prizes from the Tokyo Art Directors Club and at the Warsaw International Poster Biennale, the Lahti Poster Biennale, and the Toyama International Poster Triennial. Saito received the Mainichi design prize in 1987.
Reference: *Makoto Saito Posters* (Tokyo, 1990).

Sanyo Electric Company
Established Moriguchi, 1947

One of the major household appliance companies in Japan, Sanyo Electric Company was founded in 1947 in Moriguchi, Osaka Prefecture, as a manufacturer of bicycle lamps. It quickly became Japan's leading producer of such lamps, and by 1952 the company had entered the consumer appliance business, first with plastic radios and then washing machines. Its line of home electrical appliances expanded to meet the high Japanese demand for such products, adding fans and heaters by the mid-1950s, and televisions, refrigerators, and air conditioners by the early 1960s. In 1958 Sanyo began mass-producing transistors, and within a year was Japan's top exporter of transistor radios. Two years later Sanyo began operating overseas, establishing an office in Hong Kong, and today it has factories in twenty-one countries. In 1963 the company developed nickel-cadmium rechargeable batteries, which allowed it to begin production of cordless electric products, lithium batteries in the late 1970s, and nickel-metal hydride batteries, which can be recharged over five hundred times, in 1990. Its interest in developing products that use alternative energy sources has also led it to explore uses of solar cells and fuel cells. In the early 1970s Sanyo began designing products that were targeted to distinct markets, and today the firm has its own Life-style Creative Center to monitor the tastes of the Japanese population through database compilations. Sanyo now designs whole series of products for specific markets, from the Robo series of electronics for children (no. 207) to the It's series of compact products for young singles (no. 221) and the Duost series of easy-to-operate machines for older consumers.
Reference: "Sanyo Electric Co.," *Kodansha Encyclopedia of Japan* (Tokyo, 1983), vol. 7, p. 19.

Toshimitsu Sasaki
Born 1949

Toshimitsu Sasaki works as a free-lance furniture and interior designer in the town of his birth, Hita, on the island of Kyushu in southern Japan. He attended the Shibaura Institute of Technology in Tokyo, but is a self-taught designer. In 1985 he was in charge of the furniture section of Seibu department store's Japan Creative campaign. Among his other clients are Yamagiwa, for which he designs lighting, and TENDO MOKKO, for which he has designed furniture (no. 147). Sasaki has won numerous awards for his designs, among them a Mainichi industrial design prize in 1978. He has served as a member of the G-Mark prize selection committee since 1987.
References: Toshimitsu Sasaki, "Designing at My Own Pace," *Designers' Workshop*, vol. 2, no. 11 (Dec. 1985), pp. 80–86; Toshimitsu Sasaki, "Jukusei shita Bessiku Dezain o Motomete" (Seeking Mature Basic Design), *Nikkei Design*, no. 46 (Apr. 1991), pp. 148–49.

Koichi Sato
Born 1944

Koichi Sato's work, noted for its highly poetic effects achieved with subtle shadings and sometimes unusual, metaphysical forms (no. 110), exploits the advanced printing techniques for which Japanese graphics are known. Sato graduated from the Tokyo National University of Fine Arts and Music in 1969 (where he later taught, from 1981 to 1986) and worked in the advertising department of Shiseido until 1971, when he opened his own office. His clients have included MATSUSHITA ELECTRIC INDUSTRIAL COMPANY (1986; no. 203) and Mitsukoshi (1987); Sato intends his work to have national and international identity, to be very traditional and at the same time, futuristic. Among the many exhibitions in which his posters have been shown are UNESCO's "20 Best Japanese Posters" (Paris, 1986), the Japan Design Committee's "Blood 6" (Tokyo, 1987), and the National Museum of Modern Art's "Graphic Design Today" (Tokyo, 1990). Sato has received the Tokyo Art Directors Club grand prize (1985) and the Mainichi design prize (1991).
References: *Koichi Sato* (Tokyo, 1990); *Best 100 Japanese Posters, 1945–89* (Tokyo, 1990), pp. 141, 147, 231, 254; Tokyo, The National Museum of Modern Art, Crafts Gallery, *Graphic Design Today* (Sept. 26–Nov. 11, 1990), pp. 86–89, 188.

Seiko Corporation
Established Tokyo, 1881

The Seiko Corporation was founded in 1881 as a watch- and clockmaking business by Kintaro Hattori, and became a public company in 1917. Renamed Hattori Seiko Company in 1983 and Seiko Corporation in 1990 (after the brand name first given its watches in 1937), the company is today the largest supplier of timepieces in the world, from the some five thousand different watch designs marketed annually to timing equipment for international sports events (Seiko was official timer of the Tokyo Olympic Games in 1964, Sapporo Winter Olympics in 1972, and Barcelona Olympics in 1992). During the 1960s Seiko was a pioneer in the use of quartz for its watches, and during the 1970s, in the use of liquid crystal displays, applying those technologies to the broad range of electronic and consumer goods that the company now markets in addition to watches and clocks (nos. 13, 176). Through its three divisions, Seiko also manufactures computers, printers, software systems, semiconductors, and other electronic components.

Reference: Seiko Corporation, *Seiko* (Tokyo, 1992).

Keisuke Serizawa
1895–1984

Keisuke Serizawa, designated a Living National Treasure in 1956 for his skill in *kata-zome* (stencil-dyeing) techniques, did not begin his training until he was in his thirties. Born the son of a dealer in kimono fabrics, in 1913 Serizawa left his hometown of Shizuoka in central Japan to study at the Tokyo College of Industrial Arts. There he concentrated on graphic design, and after graduation worked in the graphics field. A 1925 visit to Korea, where he saw the work of local artisans, and his encounter with Soetsu Yanagi and his involvement in Yanagi's folk-craft movement, proved turning points in his career. In 1928 Serizawa returned to Shizuoka to train with a local dyer and the following year exhibited his first stencil-dyed fabric. He applied his stencil-dyeing techniques not only to lengths of cloth but also to handmade paper, book covers and illustrations, screens, hangings, and kimonos (no. 60). Serizawa worked with Yanagi in planning (and acquiring objects for) the Japan Folk Crafts Museum, which opened in Tokyo in 1936. The two traveled together extensively throughout Japan and Okinawa collecting examples of regional crafts for the new museum. Serizawa also designed covers for Yanagi's Japan Folk Craft Society journal *Kogei* (Craft) from 1931. After the war Serizawa began teaching at Tokyo's Women's College of Fine Arts (1949–52), and then returned to his hometown to teach at the Shizuoka Prefectural Women's College. In 1955 he established a research center in Shizuoka for the study of stencil-dyeing techniques. Serizawa was decorated with the Order of the Sacred Treasure in 1970, and named to the Japan Academy of Arts in 1976. That same year a retrospective exhibition of his work was held at the Grand Palais in Paris, and in 1982 the Museum of Fine Arts in Boston showed his work in the exhibition "Living National Treasures of Japan."

References: Paris, Grand Palais, *Serizawa* (Nov. 23, 1976–Feb. 14, 1977); Boston, Museum of Fine Arts, *Living National Treasures of Japan* (Nov. 3, 1982–Jan. 2, 1983), nos. 93–105.

Sharp Corporation
Established Tokyo, 1912

One of Japan's largest manufacturers of home appliances, the Sharp Corporation was founded as a machine shop by Tokuji Hayakawa in 1912, and then expanded its operations to produce the mechanical Ever-Sharp pencil he had invented. After the 1923 earthquake, the firm moved from Tokyo to Osaka, where Hayakawa began to research radio technology and established the Hayakawa Metal Works. In 1925, concurrent with the first radio broadcasts in Japan, the firm produced the first Japanese-made crystal radio set, and within a year was selling radio parts to South America and throughout much of Asia. Until the firm developed a prototype television in 1951 and began to manufacture it when Japanese television broadcasting began in 1953, the company manufactured only radios and record players, but during the 1950s it began to diversify, adding washing machines, refrigerators, and other home appliances to its lines. Research laboratories for computer, semiconductor, and ultrasonic-wave technologies were established in the early 1960s, and in 1964 it developed the first transistor-diode electronic calculator (no. 62), establishing its position as a force in the field of office electronics. In 1971 Sharp received the Apollo Achievement Award from the U.S. National Aeronautics and Space Administration for developing ELSI (Extra Large-Scale Integration) chip technology for Apollo II (later used in electronic calculators and other consumer goods). As the company entered the field of office products more heavily, its name was changed in 1970 to the Sharp Corporation, producing during the 1970s photocopiers, electronic cash registers, and calculators, including one with an LCD (liquid-crystal display) in 1973 (no. 126). In 1979 the firm sold its first personal computer and word processor, and also a pocket electronic translator (no. 209); in 1980 it developed its first facsimile machine. By 1987 Sharp had developed the world's first LCD televisions, and the following year had produced one that was only about 1 1/16-inch thick. By the 1990s the company was producing LCD camcorders (no. 244) and televisions that could be wall-mounted anywhere, offered in a variety of designs to suit any style of interior decoration (no. 245).

Reference: Sharp Corporation, *Eighty Years of Sincerity and Creativity (1912–1992)* (Tokyo, 1992).

Tadao Shimizu
Born 1942

A product and interior designer, Tadao Shimizu graduated from Tama Art University in Tokyo in 1966, when he joined KENMOCHI DESIGN ASSOCIATES in Tokyo, working there until 1976. Shimizu then came to the United States, receiving a master's degree in fine arts from Cranbrook Academy of Art in Bloomfield Hills, Michigan, in 1978; working for the Burdick group in California as chief designer of furniture; and teaching design at the University of Washington (1984–87). On his return to Tokyo he assumed a post at Chiba University, where he still teaches, and established his firm, Design Studio Tad, where in his product designs he experiments with new materials and forms (no. 205). One of his other interests is design for the disabled. He has created designs for wheelchairs, which in an attempt to make them beautiful as well as comfortable, he thinks of as akin to furniture. In 1989 Shimizu won the Ministry of International Trade and Industry prize for his office furniture.

References: "Best of Category: Concepts," in *1987 Annual Design Review, Industrial Design*, vol. 34 (1987), p. 159; *East Meets West in Design: Archaeology of the Present* (New York, 1989), pp. 52–55; Masaya Yamamoto, ed., *Product Design in Japan* (Tokyo, 1990), pp. 18–25, 234.

Fukumi Shimura
Born 1924

Although Fukumi Shimura had been exposed as a youth to her mother's passion for weaving, she herself did not begin weaving until the age of twenty-nine, when circumstances forced her to seek a livelihood. Her mother had studied under the weaver Goro Aota, a member of the Japanese folk-craft movement who valued the integrity of the handspun, naturally dyed, handwoven cloth made by the rural population and by urban dwellers. Shimura had no formal training in weaving, but gathered information from her mother, from older women in nearby villages, and from traditional weavers in Kyoto. The guidance and advice of TATSUAKI KURODA, the famous Kyoto craftsman who worked in wood and lacquer, were also a major influence on her career. Having been introduced by her mother to natural dyes, Shimura continued to explore their possibilities despite the fact that chemical dyes were the preferred choice of textile professionals working in the 1950s. From 1958 through 1961, Shimura exhibited annually in the Traditional Japanese Crafts exhibitions, where she received awards for her work; she has been a juror of the exhibition since 1968, and one of the association's board members since 1978. Shimura works exclusively in silk and utilizes only natural dyes, making one-of-a-kind kimonos that reflect her exceptional sense of color and balance. She adheres to the time-honored formats of traditional textiles—plaids, checkerboard patterns, stripes, and horizontal bands—but gives expression to her modern sensibility through her use of color and texture (no. 121). Shimura was designated a Living National Treasure in 1990, and received Japan's medal of honor in 1993.

Reference: Yoshiaki Inui, *The Works of Fukumi Shimura, 1958–1981* (Kyoto, 1981).

Shounsai Shono
1904–1974

Shounsai Shono in 1967 was the first artist to be named a Living National Treasure for bamboo craft. Born in Oita Prefecture, an area known for its bamboo artisans, he apprenticed with a noted master of the craft for two years. Most of his early baskets were in the conservative Chinese (*karamono*) style, with an emphasis on precision basketry that continued to predominate throughout his career (see no. 25). Through his work in the crafts section at the Oita Prefectural industrial arts institute from 1938 and his participation in a survey of the bamboo artifacts housed in the eighth-century Shosoin imperial repository in Nara, Shono developed an expertise in judging bamboo quality, and made the most of its essential characteristics— its strength and elasticity—in his work. In 1940 he was selected to show his baskets at the government-sponsored exhibition commemorating the 2600th anniversary of the founding of Japan, and he regularly exhibited his pieces in major craft exhibitions, winning numerous prizes for his innovative creations in bamboo. In 1965 Shono entered his work in the Traditional Japanese Crafts exhibition in Tokyo and further established himself as an artist who created simple, unadorned pieces in bamboo (no. 31).

Reference: Tokyo, The National Museum of Modern Art, Crafts Gallery, *Modern Bamboo Craft* (Feb. 5–Mar. 24, 1985), pp. 35–37, 70–73, 138.

Sony Corporation
Established Tokyo, 1945

Founded in 1945 by Masaru Ibuka as the Tokyo Telecommunications Research Laboratories and, in conjunction with Akio Morita, incorporated as Tokyo Telecommunications Engineering Corporation the following year, the company known for high quality and technological innovation in audio and visual electronic equipment changed its name to Sony in 1958 after the mark on its highly successful line of transistor radios (no. 32). Ibuka, an electronics engineer, and Morita, a physicist and heir to a sake brewing concern, developed new products using imported technology after the war: in 1950 the firm marketed Japan's first magnetic tape recorder, and in 1954, under license from Western Electric, it produced Japan's first transistor, and the following year, Japan's first transistor radio, the TR-55 (no. 17). This radio was the first product to carry the Sony name and was also the firm's first export. Building its reputation on the consistent quality of its products, its emphasis on miniaturization, and its aggressive consumer-education and marketing techniques, the company created a series of innovative products and introduced them abroad. Sony developed the first all-transistor television set (1959; no. 44); the first home video recorder (1964); a color television using a single electron gun

and three-beam system (Trinitron, 1968); a portable cassette player (Walkman, 1979; no. 132) and forty-three variants and descendants of this concept (no. 222), including a flat-screen micro television (Watchman, 1982), a compact-disc player (Discman, 1984), and an "electronic book" (Data Discman, 1990); a color monitor (Profeel, 1980); and a notebook-size, portable computer with a touch-sensitive screen (Palmtop, 1990; no. 236). Sony's first products were designed largely by its engineers, but by 1960 the firm had seventeen designers. While their numbers increased over the next decades, they were scattered among, and responsible to, the company's various product divisions until 1978, when Morita created a central design department for product planning and design, which was reorganized again in 1985. Sony has production and sales facilities and subsidiaries across the globe, particularly in the United States, with, for example, a color television plant in California and a magnetic recording tape plant in Alabama. The firm entered the United States entertainment industry by purchasing CBS Records in 1988 and Columbia Pictures Entertainment in 1989, providing software programming for its audio and video products.
References: Wolfgang Schmittel, *Design Concept Realisation* (Zurich, 1975), pp. 169–96; London, The Boilerhouse, Victoria and Albert Museum, *Sony Design* (1982); Akio Morita with Edwin M. Reingold and Mitsuko Shimomura, *Made in Japan: Akio Morita and Sony* (New York, 1986); *Sony Design 1950–1992* (Tokyo, 1993).

Reiko Sudo
Born 1953

Textile designer and retailer, Reiko Sudo studied at Musashino Art University, Kodaira, and after graduating in 1975, remained in the textile department as an assistant to one of her teachers, Hideho Tanaka, while researching *tsuzure* tapestry techniques at the Kawashima Textile Institute in Kyoto (1975–77). From 1977 to 1984 Sudo worked as a free-lance textile designer for the Kanebo, Nishikawa Sangyo, and Soko companies, and in 1984, cofounded the textile firm Nuno to produce textiles of advanced design for the retail market, including her own (no. 255) and those of JUNICHI ARAI. As director of Nuno, Sudo has continued to design for other concerns, including the International Wool Secretariat, Paris (1989–92), and the Tokyo apparel company Threads (1991–93). Her work has been included in numerous textile exhibitions, including "Color, Light, Surface" at the Cooper-Hewitt Museum, New York (1990), and the Kyoto Industrial Center (1991).
Reference: *The International Design Yearbook* 5 (New York, 1990), p. 227.

Kohei Sugiura
Born 1932

To his contemporaries, Kohei Sugiura is the mystic of Japanese design. His work is imbued with a highly individualistic and spiritual world view rare among Japanese designers. Sugiura began his work as a free-lance designer after graduating from the Tokyo National University of Fine Arts and Music in 1955 with a degree in architecture. His early graphics reflected his love of music, creating record jackets and posters for events such as Igor Stravinsky's only conducting appearance in Japan in 1959, and he received the Mainichi industrial design prize for his posters in 1961. Sugiura was one of the group of designers who created the graphic program for the World Design Conference (WoDeCo) in Tokyo in 1960, and among those who organized the 1965 "Persona" exhibition in Tokyo, which won a Mainichi industrial design prize for the participants. In 1964–65 and 1966–67 Sugiura was a guest professor in architecture and visual communications at the Hochschule für Gestaltung in Ulm, West Germany, and in 1967 began to teach at the newly established Tokyo University of Art and Design. In the early 1970s he traveled extensively in India, Tibet, and other parts of Asia, and these journeys gave a new focus to Sugiura's work (no. 171). He sought out Japan's cultural and religious roots in the heritage of the Asian continent, and translated his discoveries into covers for the UNESCO publication *Asian Culture* (1972), posters for the performances in Japan of Chinese opera (1979, 1985) and Indian music (1987, 1988), postage stamps for the country of Bhutan (1983–86), and a series of books and exhibitions. In 1980 he organized an exhibition about the mandala, the complex painted representations of the Buddhist cosmos, at the Seibu Museum of Art in Tokyo, for which Sugiura also did the installation, a catalogue design, and a subsequent limited-edition book (no. 159), which are considered among the finest examples of contemporary Japanese book design. In 1985 Sugiura collaborated with Toyo Ito on the installation of an ethereal "Reflecting Space," in the exhibition "Tokyo: Form and Spirit," which toured the United States, a space that the visitor entered by passing through a seated Buddha image to experience the spiritual state of Buddhist meditation. Sugiura feels strongly that Japan's present generation has much to learn from its Asian roots. "One side of our new-found consumer culture," he says, "discards what has been created by the wisdom of generations for the sake of something 'new,' which is not nearly as good. In Asia, the old and the new still co-exist . . . and the collective wisdom of the past is re-expressed by the present generation. This is the great strength of Asia."[13]
References: Akiko Takehara, "Hito Mono Tojo: Sugiura Kohei" (Personality / Works: Kohei Sugiura), *Industrial Design*, nos. 139–40 (July 1987), pp. 64–67; Colin Naylor, ed., *Contemporary Designers*, 2nd ed. (Chicago, 1990), pp. 543–44; Mamoru Yonekura, "Kohei Sugiura," *Creation*, no. 3 (Jan. 1990), pp. 8–63; *Best 100 Japanese Posters, 1945–89* (Tokyo, 1990), pp. 63, 65, 121, 191; Richard S. Thornton, *The Graphic Spirit of Japan* (New York, 1991), pp. 93–95, 131, 212.

Tomoyuki Sugiyama
Born 1954

Tomoyuki Sugiyama, an architectural acoustician, received both undergraduate and graduate degrees from Japan University, where he now teaches at its junior college. In 1984 he worked with the sanitary fittings manufacturer Inax on developing sound-absorbing tiles, and later developed the Bubble Boy ceramic speaker (no. 192) for the company. In 1987 Sugiyama worked in the United States as a visiting researcher at the Media Laboratory of the Massachusetts Institute of Technology and in 1990 was employed by the International Media Research Foundation. Sugiyama is the author of the *Professional Acoustic Data Book*, published in 1986.
Reference: "Product Development: Inax Bubble Boy," *Axis*, no. 23 (Spring 1987), pp. 14–15, 78.

Yoichi Sumita
Born 1924

Yoichi Sumita, a graduate of the Tokyo National University of Fine Arts and Music in 1947, worked for five years for the Industrial Arts Institute in Tokyo, which led him to pursue a career in industrial design. In 1952 he moved to the Tokyo Electric Company, where he designed lighting, and from 1956 to 1963 worked in the home appliance section of Yao Electric Company. In 1963 Sumita joined Toyoguchi Design Associates (see KATSUHEI TOYOGUCHI), working principally on designs for OLYMPUS OPTICAL COMPANY and the Nippon Telecommunications and Telephone company (NTT). For Olympus he designed products such as miniaturized tape recorders (no. 80), gastro-intestinal cameras, and microscopes (for which he received the Japan Electrical Industry prize in 1975). His designs for NTT include miniature and car telephones, a number of which have received the G-Mark prize. In 1977 Sumita began to teach at the Tokyo National University of Fine Arts and Music (where he has been emeritus since 1992), and he now teaches industrial desigh at Takushoku University in Tokyo. He has also served on the board of directors of the Japan Industrial Designers Association since 1967 and as chairman from 1981 to 1985, and as a juror for the Mainichi industrial design prize (1972–80). Sumita was the general editor of the eight-volume *Complete Collection of Industrial Design* (Tokyo, 1982–84).
References: Kyo Toyoguchi, *ID no Sekai* (The World of ID) (Tokyo, 1974), pp. 34–35, 96; Kogei Zaidan, ed., *Nihon no Indasutoriaru Dezain: Showa ga unda Meihin 100* (Japanese Industrial Design: 100 Masterpieces Produced during the Showa Period) (Tokyo, 1989), p. 95; Colin Naylor, ed., *Contemporary Designers*, 2nd ed. (Chicago, 1990), pp. 544–45.

Hiroshi Tajima
Born 1922

Hiroshi Tajima became apprenticed to a master craftsman in the resist-dye (*yuzen*) technique at the age of fourteen. His training was interrupted by the war, and, after a bout with tuberculosis, Tajima resumed his creative activities in the textile field in 1958. He submitted his first resist-dyed kimono in 1959 to the annual Traditional Japanese Crafts exhibition and won one of the top awards in the 1966 exhibition. Tajima was invited to be a juror for these exhibitions from 1969, and in 1972 he became the secretary general of the Japan Craft Association. He has twice been decorated by the Japanese government for his achievements in the *yuzen* technique (no. 64), with a medal of honor in 1987 and the Order of the Rising Sun in 1993.
References: Tokyo, The National Museum of Modern Art, *30 Years of Modern Japanese Traditional Crafts* (Aug. 26–Sept. 25, 1983), p. 61; Tokyo, The National Museum of Modern Art, Crafts Gallery, *Crafts* (Tokyo, 1988), p. 152.

Akiraka Takagi
Born 1933

A lacquer artist and designer of crafts, Akiraka Takagi is a graduate of the Tokyo National University of Fine Arts and Music (1956). In 1963–64 he studied at Parsons School of Design in New York, and after his return to Japan, began to exhibit annually in the Japan New Craft exhibitions (1966–72), becoming a juror for them in 1973. The same year he opened his own office in Tokyo, where he began to design for regional craft centers, such as the Odate Craft company in Aomori Prefecture in northern Japan (no. 200). Takagi received the Kunii Industrial Art Award in 1982 for his efforts to help modernize the lacquer craft industry. He was also on the board of directors of the Japan Designer Craftsman Association (1970–76), and serves on the board of the Japan Craft Design Association, of which he was president from 1979 to 1981. Since 1977 Takagi has taught at Tama Art University in Tokyo.
References: Kogei Zaidan, ed., *Kunii Kitaro Sangyo Kogeisho no Hitobito: 35-nin no Purofuiru* (The Winners of the Kitaro Kunii Industrial Art Award: Profiles of 35 Winners) (Tokyo, 1984), pp. 131–33; Sakaiya Tetsuro, "Hand-Made by Machine," *Designers' Workshop*, vol. 5, no. 31 (Dec.1988), pp. 80–87.

Reiko (Murai) Tanabe
Born 1934

Interior designer Reiko (Murai) Tanabe graduated from the Women's College of Fine Arts in Tokyo in 1957 and joined KENJI FUJIMORI's firm, working on the design of office furniture. From 1959 to 1962 Tanabe was with the Matsuda Hirata company, working on architectural-, interior-, and furniture-design projects. It was during this time that she entered TENDO MOKKO's first furniture design competition and won a prize for her design of a plywood stool, which Tendo is still manufacturing (no. 46). In 1962 Tanabe opened her own design firm, specializing in interior design for offices and hospitals. That same year she began teaching at her alma mater, the Women's College of Fine Arts, as professor of interior design. Tanabe emphasizes individual creative expression in her work. "Universality is important. A product does not have to be 'ahead of its time' (although there have to be some that fill that role, too), but a product should be appropriate to its time and remain fresh after many generations. Individuality of design may take a long time to be recognized. In the meantime you must believe in your own individual talent."[14]
Reference: Kogei Zaidan, ed., *Nihon no Indasutoriaru Dezain: Showa ga unda Meihin 100* (Japanese Industrial Design: 100 Masterpieces Produced during the Showa Period) (Tokyo, 1989), p. 53.

Ikko Tanaka
Born 1930

Ikko Tanaka has worked in a wide field of graphics, creating not only posters (nos. 50, 142), logos (no. 134), and corporate identity programs but also packaging, book designs, exhibitions, and architectural graphics. Since the 1950s his work has won virtually every possible graphic award, from the prizes of the Tokyo Art Directors Club and Japan Advertising Artists Club to those of the New York Art Directors Club and the International Poster Biennale, Warsaw. Widely known and employed outside Japan, Tanaka is also a Japanese designer's designer, realizing books for ISSEY MIYAKE (1978), HIROMU HARA (1985), Isamu Noguchi (1985), HIROSHI AWATSUJI (1990), and others about their own work. Born in Nara and trained at the Kyoto University of Fine Arts, Tanaka worked at the Kanegafuchi spinning company (now Kanebo) in Kyoto as a textile designer (1950–52) and then joined the Sankei Shimbun in Osaka as a graphic designer for the newspaper (1952–58). In 1957 he also began working as a graphic designer for Light Publicity in Tokyo, remaining there until joining the newly formed Nippon Design Center in 1960, a job that was important to him because of the exceptional talents connected with this pace-setting firm. In 1963 Tanaka left to establish his own studio, where he continued to design the posters for the Noh performances sponsored by Sankei

Shimbun. The posters had won acclaim for their use of abstracted traditional Japanese images, including masks, calligraphy, and *ukiyo-e* prints (no. 61). "The young people of Japan would probably regard me as a fairly conservative designer," Tanaka reflected, "because I'm quite interested in the Japanese classics, and much of my subject matter and many of my themes come from that source. But of course it's still possible to be creative while working within traditional ideas."[15] From 1970 to 1980 he was responsible for several exhibition designs: for the Japanese government pavilion at Expo '70 in Osaka, the Oceanic Cultural Museum at the Ocean Expo '75 in Okinawa, and the "Japan Style" exhibition at the Victoria and Albert Museum, London, in 1980. In 1973 Tanaka began a long-term association with the Seibu Saizon Group, which includes Seibu department stores, Seiyu food stores, Seibu theater, and Seibu (now Saizon) Museum of Art. Tanaka's other principal clients include Japan IBM, Toto, Dai Nippon printing, and Morisawa.
References: *The Work of Ikko Tanaka* (Kyoto, 1975); Katsuhiro Kinoshita, ed., *The Design World of Ikko Tanaka* (Tokyo, 1987); Los Angeles, Japanese American Cultural and Community Center, *Ikko Tanaka L.A.—Graphic Designs of Japan* (June 11–July 26, 1987).

Yoshitaka Tanaka
Born 1940

Founder of the Handicapped Skiers Association of Japan (1972) and its first chairman, Yoshitaka Tanaka is a graduate of the Seibu Institute design department and worked for GK (1960–75); during that time he received a Mainichi industrial design prize for his Self-Help Environment for the Handicapped (1974) and designed his well-known Outrigger ski (no. 98). In 1979 Tanaka established his own firm, the JL Create Company. Long involved in issues concerning the physically challenged, Tanaka in 1981 established and became chairman of the National Culture Center for the Disabled, and is interested in increasing opportunities for participation in sports and recreation for people with special needs.
Reference: Takefumi Daido, "Sukijijogu no Kaihatsu" (Development of Ski Equipment for the Handicapped), *Industrial Design*, no. 95 (Sept. 1978), pp. 23–29.

Teikoku Sen-i Company
Established Tokyo, 1884

The first linen-spinning factory in Japan when it was established in the late nineteenth century, Teikoku Sen-i Company is the largest producer of linens in the country today. The company expanded into synthetics in the 1940s, experimenting with rayon and rayon-linen blend fabrics. The firm patented a new synthetic that combined linen with polyester in 1959, and has been a leader in the field ever since.

Most recently Teikoku Sen-i has worked with Dupont to develop applications for its aramide (aromatic polymide) nylon fibers for fire fighters' uniforms (no. 163).
Reference: Katsuhiko Sasada, "Kasai Yobo to Spotsuwea Yohin ni Tsukaware hajimeta Aramido" (Aramide: The Material Beginning to Be Used for Fire Protection and Sportswear), *Nikkei Design*, no. 49 (July 1991), pp. 142–46.

Tendo Mokko
Established Tendo, 1940

In 1940 in the northwestern prefecture of Yamagata, a group of carpenters and woodworkers formed a cooperative to produce wood-based products, at first making boxes for munitions and supplies for the military. Incorporated in 1942, the cooperative worked with the Industrial Arts Institute in nearby Sendai to produce decoy airplanes, using the new technology of molded plywood. After the war, with an office in Tokyo in 1947, Tendo became one of the producers of furniture for occupation housing, and by the early 1950s it had turned its attention to furniture for the domestic market. One of its first commissions came from the architect Kenzo Tange for plywood seating for the Ehime Prefectural Hall built in 1953. Plywood was then a new medium for designers, and Tendo was a pioneer in producing furniture in this material, such as the 1956 Butterfly stool by SORI YANAGI (no. 27). Tendo's policy of paying royalties for designs attracted others to the firm, and in the 1960s Tendo could boast of Japan's most important furniture designers in its ranks (see no. 59), including ISAMU KENMOCHI (no. 52), DAISAKU CHOH (no. 49), and RIKI WATANABE (no. 47), all of whose pieces are still in production. Tendo further encouraged new designers through the competition it sponsored annually from 1960 to 1967, and the company itself produced some of the winning designs, such as the plywood stool by REIKO TANABE (no. 46). In 1964 Tendo received a Mainichi industrial design prize for its leading role in furniture production. Today Tendo's mainstays are the furniture it produces for hotels, offices, theaters, museums, and libraries. The company's low chair and table (nos. 58, 48) for use in *tatami* rooms are ubiquitous elements in Japanese-style restaurants and hotels, but it continues as a leader in the production of designer furniture as well (see no. 147), with recent additions to its line by notable architects such as ARATA ISOZAKI (no. 99).
References: "Dezainaa to Meekaa wa Kyozon Kyoei de Nakereba Ikenai" (Designers and Manufacturers Must Exist Together for Mutual Prosperity), *Industrial Design*, no. 118 (July 1982), pp. 32–35; Nakamura Keisuke, "Kagu Dezain no Doko" (Trends in Furniture Design), in Kogei Zaidan, ed., *Nihon no Kindai Dezain Undoshi 1940 Nendai–1980 Nendai* (History of the Japanese Modern Design Movement, 1940s–1980s) (Tokyo, 1990), pp. 173–75.

Tokyo Telecommunications Engineering Corporation, see Sony Corporation

Tomix, see Tomy

Tomy
Established Tokyo, 1924

Tomy, a toy company, was founded by Eiichiro Tomiyama in Tokyo in 1924. In 1927 the company began exporting its metal toys, although exports did not become a major part of the firm's production until after the war, when the company established the Sanyo toy production facility to manufacture large metal toys, creating in 1953 Sanyo Industries to manage their exports. In 1963 the production company became Tomy Kyogo Company, and the sales company Tomy Company, expanding during the 1970s with offices throughout the world. The company opened a research laboratory in 1980 to design toys for disabled children, and today it regularly develops toys for visually impaired children. Tomy is active in the research and development of new toys, often using computer-related design, and incorporates new technologies in its toys whenever possible. Its Tomix division manufactures miniature models of Japanese trains (no. 253).
Reference: Tomy Company prospectus (1992).

Katsuhei Toyoguchi
Born 1905

Katsuhei Toyoguchi was an early pioneer in the field of interior and industrial design. After graduation from the Tokyo College of Industrial Arts in 1928, he became a founding member of the influential Keiji Kobo workshop, which was modeled on the Bauhaus. There, Toyoguchi developed his lifetime interest in furniture, particularly for Japanese *tatami*-floored interiors. In 1933 Toyoguchi began work at the newly established Industrial Arts Institute (IAI). From 1935 to 1937 he was the editor for the IAI monthly *Kogei Nyusu* (Industrial Art News), and he continued to be a frequent contributor to the journal throughout its existence. In 1939 his work with the IAI took him to Osaka to head the operation there, and he remained until 1943. After the war he returned to Tokyo, where he worked with the IAI project on the production of furniture for the American occupation forces housing. He was one of the interior designers that TENDO MOKKO turned to for its postwar production of furniture (no. 59). Toyoguchi retired from the IAI in 1959 to found his own firm, Design Associates, with the help of SHUTARO MUKAI. He received one of his first commissions in 1960 from the Japan External Trade Organization, to plan the display for the government's trade fair in Moscow, and Toyoguchi also designed the Japanese pavilions at the Seattle World's Fair in 1962 and at Expo '67 in Montreal. He also taught design at Musashino

Art University, Kodaira, from 1959 to 1975. Besides writing for design journals, Toyoguchi is the author and editor of several books on design, among them *Hyojun Kagu* (Standard Furniture) (1936), and *Dezain Senjutsu* (Design Tactics) (1966). He was an active member of the Japan Industrial Designers Association for its first twenty years, serving as president for seven years, and is now president emeritus. Toyoguchi has also served as a juror for the Mainichi industrial design prize (1955–70), and himself has been awarded the Kunii Industrial Art Award (1975) and the Order of the Sacred Treasure (1976) by the government of Japan.

References: Japan Industrial Designers Association, ed., *Seiichi no Kozo: Nihon no Indasutoriaru Dezain* (Structure of Dexterity: Industrial Design Works in Japan) (Tokyo, 1983), pp. 258–66; Akiko Moriyama, "Toyoguchi Katsuhei: Senzen kara sengo, hanseiki o nuki to shita Seikatsu Dezain no shisei" (Katsuhei Toyoguchi: From Prewar to Postwar, a Half-Century of Life in Design), *Nikkei Design*, no. 2 (Aug. 1987), pp. 70–77; Kunio Sano, "Hito Mono Tojo: Toyoguchi Katsuhei" (Personality / Works: Katsuhei Toyoguchi), *Industrial Design*, no. 141 (Oct. 1987), pp. 30–37.

Shigeru Uchida
Born 1943

One of Japan's most influential interior designers as well as a designer of furniture and lighting, Shigeru Uchida studied at the Kuwasawa Design School in Tokyo (1966) and established his own studio in 1970. During the 1970s he gained his reputation in exhibitions sponsored by the Japan Design Committee at the Matsuya department store (1971, 1973, 1977) and for the design of commercial interiors such as boutiques for Yoshinto (1976) and ISSEY MIYAKE (1976–82), and later for YOHJI YAMAMOTO (1983–86). In 1981 Uchida, with his associate Toru Nishioka, reorganized his office under the name Studio 80, and his interior projects since then have ranged from the Japanese government pavilion for Expo '85 in Tsukuba to the Hotel Il Palazzo in Fukuoka (1989). Uchida has designed furniture for Chairs and Pastoe, lighting for Yamagiwa (no. 185), and clocks for Alessi and Acerbis International. He has taught at the Kuwasawa Design School (from 1973) and Tokyo University of Art and Design (1974–78), and has lectured variously at Columbia University and Parsons School of Design in New York (1986) and the Domus Academy in Milan (1992). In 1987 Uchida received the Mainichi design prize.

References: *Interiors of Uchida, Mitsuhashi and Studio 80* (Tokyo, 1987); *Hotel Il Palazzo* (Tokyo, 1990).

Masanori Umeda
Born 1941

One of a number of Japanese designers who studied and practiced in Europe during the 1960s and 1970s, Masanori Umeda spent the first part of his career working in Italy. He graduated from the Kuwasawa Design School in Tokyo in 1962, and in 1967 joined the architecture and design firm of Achille Castiglioni in Milan, where he remained until 1969. Between 1970 and 1979 Umeda was a consultant designer for Olivetti, with responsibility for computer, interior, and furniture design. In 1981 he became one of the first collaborators with Memphis, a radical design group that defied convention by producing furniture and objects that were humorous and referential, brightly colored, patterned, and irregularly shaped. Borrowing motifs from Japanese popular culture, Umeda designed furniture (no. 144) and ceramics for the first three Memphis collections (1981–83). Umeda has continued to work in the postmodern style associated with Memphis, notably in a series of flower-shape chairs for the Italian firm Edra (1988–90). In 1986 he opened his first office in Japan under the name U-Metadesign and broadened his practice to include industrial- and environmental-design projects in a different stylistic language. Among his recent projects are street lighting for Iwasaki Electric (1989) and sanitary fittings for Inax (1989).

Reference: *Product Design by Masanori Umeda '80–'89* (Tokyo, 1990).

Kunibo Wada
Born 1900

Painter, illustrator, cartoonist, and graphic designer, Kunibo Wada successfully brought mainstream recognition to the images and techniques of Japanese folk woodcuts, exploiting their qualities of humor, expression, and strength in his work. Wada was born in Kotohira-cho in Kagawa Prefecture. He left to study in Tokyo and afterward joined the Tokyo Nichinichi (now Mainichi) newspaper as a cartoonist and writer of humorous essays. In 1938 Wada returned to Kagawa Prefecture, where he spent the rest of his professional career, joining the prefectural government in 1954, becoming curator of the Sanuki Folk Crafts Museum, and continuing to practice as an artist. From 1951 Wada was associated with the local confectionery firm of Arakiya, for which he designed a complete packaging program that is still in use (no. 41).

References: Japan Package Design Association, ed., *Package Design in Japan: Its History, Its Faces* (Tokyo, 1976), p. 200, nos. 706, 710, 712–13, 720, 722, 724–25; Takeo Yao, ed., *Package Design in Tokyo* (Tokyo, 1987), pp. 90–93.

Katsuji Wakisaka
Born 1944

Trained in textile design at the Kyoto School of Arts and Crafts (1960–63), Katsuji Wakisaka worked as a designer of apparel fabrics first for Itoh in Osaka (1963–65) and subsequently for the Samejima Textile Design Studio in Kyoto (1965–68). In 1968 he went to Finland, where he spent the next seven years designing furnishing and dress fabrics for Marimekko. His first designs, among which was his popular pattern of cars and trucks, Bo Boo (1975), were figurative and Western in style, and aimed at the children's market. Gradually, however, Wakisaka returned to small-scale, abstract Japanese patterns, even creating a kimono and kimono sashes for the Finnish firm. From 1976 to 1986 he worked concurrently for Jack Lenor Larsen, New York, and for the Wacoal Interior Fabric Company, Tokyo (for whom he continues to design), relying increasingly on Japanese traditional textile patterns for inspiration (no. 118). In 1980 Wakisaka's designs for Wacoal were selected by the Japan Design Committee for exhibition in the Matsuya Ginza gallery, Tokyo. Since 1988 his printed fabrics, scarfs, rugs, and T-shirts have been shown by the Kyoto Liv Art gallery.

Reference: Pekka Suhonen and Juhani Pallasmaa, eds., *Phenomenon Marimekko* (Helsinki, 1986), pp. 132–33.

Riki Watanabe
Born 1911

Riki Watanabe has been an activist among design professionals throughout his career, both through his teaching and extensive writings and as a participant in Japan's principal design organizations. A graduate of the Tokyo College of Industrial Arts, in 1936 he joined the staff of the Industrial Arts Institute (IAI). In 1949 Watanabe founded his own firm, one of the earliest independent design offices in Japan; he renamed it Q Designers in 1955, the Q to suggest quality, as well as to sound a somewhat exotic note to Japanese ears. His first success was the Rope chair (1952; no. 2), a felicitous marriage of Western (high) and Japanese (low) seating. After a trip to the United States to study American design practices in 1956, where he also realized the value of Japanese design then being praised in the West, Watanabe designed a furniture series in rattan, including the Torii stool (1956; no. 26), exhibited at the 1957 Milan Triennale, where he was awarded a gold medal. In addition to furniture (no. 47), Watanabe's designs range from cardboard toys, for which he received a Mainichi industrial design prize (1967), to a series of clocks for SEIKO CORPORATION (1980) and the interior of the new wing of the Karuizawa Prince Hotel (1982). In 1952 he was one of the founding members of the Japan Industrial Designers Association. Watanabe has also written extensively for the IAI monthly *Kogei Nyusu* (Industrial Art News), and other publications

such as the magazine *Shin-Kenchiku* (New Architecture), introducing in them the works of many European and American designers to Japan. In 1940 Watanabe began lecturing at his alma mater, and subsequently taught interior design at Tokyo University of Art and Design until his retirement in 1970. For his service to the field of industrial design Watanabe was decorated with the medal of honor in 1976 and the Order of the Rising Sun in 1984, and was the recipient of the Kunii Industrial Art Award in 1992.

References: Riki Watanabe, "Jukunen to Dezain" (Maturity and Design), *Industrial Design*, no. 109 (Jan. 1981), pp. 22–23; Japan Design Committee, *Design 19* (Tokyo, Matsuya Ginza, Sept. 3–8, 1982), n.p.; Riki Watanabe, "Watakushi ni totte no Q Dezainazu" (My Q Designers), in Kogei Zaidan, ed., *Nihon no Kindai Dezain Undoshi 1940 Nendai–1980 Nendai* (History of the Japanese Modern Design Movement, 1940s–1980s) (Tokyo, 1990), pp. 57–60; Satoshi Miyauchi, "Watanabe Riki: Okawa no Teiryu no Gotoku ni" (Riki Watanabe: Like an Undercurrent), *JSSD*, vol. 1 (1993), pp. 48–51.

Tokuji Watanabe
1934–1984

Throughout his career as an industrial designer, Tokuji Watanabe was active in promoting the development of regional manufacturing centers to produce quality goods for the domestic market. A graduate of Aichi Prefectural University of Fine Arts in 1956, Watanabe and several friends founded a design firm in Nagoya, which in 1962 he took over, renaming the firm Uni Design Office. Watanabe's designs for kitchen cutlery, sewing and garden shears, and home and office scissors won him thirty-three G-Mark citations. His products, which emphasized craftsmanship, soon became known worldwide for the beauty of their design and their high quality at an affordable price (no. 104). In recognition of his work in promoting regional manufacturing, Watanabe received the Kunii Industrial Art Award in 1981. Watanabe lectured in design at his alma mater (1975–76) and at Mie University in Tsu (1973–77), and also became a strong spokesman for environmental preservation in the face of industrial pollution, in the 1980s serving on the Nagoya environmental planning committee and the Aichi Prefecture "city beautiful" planning committee.

References: Kogei Zaidan, ed., *Kunii Kitaro Sangyo Kogeisho no Hitobito: 35-nin no Purofuiro* (The Winners of the Kitaro Kunii Industrial Art Award: Profiles of 35 Winners) (Tokyo, 1984), pp. 109–12; Kogei Zaidan, ed., *Nihon no Indasutoriaru Dezain: Showa ga unda Meihin 100* (Japanese Industrial Design: 100 Masterpieces Produced during the Showa Period) (Tokyo, 1989), p. 102; Colin Naylor, ed., *Contemporary Designers*, 2nd ed. (Chicago, 1990), pp. 599–600.

Hirosuke Watanuki
Born 1926

For Hirosuke Watanuki, his package designs reflect his own way of life. Upon his return to Japan in 1972 after fifteen years in Europe, where he was a painter, sculptor, and ceramics and glass designer, Watanuki began a career as a free-lance designer in the city of Kobe. He immersed himself in a traditional Japanese life-style, studying calligraphy and the tea ceremony, and building his own teahouse, Muho-an (The Hut without Rules). His distinctive package designs and calligraphy for the tea candies made by the Kobe confectionery firm Hon-Takasagoya since the 1980s take their brand name from Watanuki's teahouse (no. 204). Watanuki has also created textile and ceramic designs for the Muho-an line. Besides his package design, Watanuki also continues to work as a painter, printmaker, architect, and sculptor.
References: Takeo Yao, ed., *Package Design in Tokyo* (Tokyo, 1987), pp. 76–83; *El Espacio del Watanuki Hirosuke* (Kobe, 1989).

Water Studio
Established Tokyo, 1973

The design firm Water Studio, founded in 1973 by Naoki Sakai, initially specialized in textile and fashion design but has since become known for industrial and automotive design, ranging from Nissan's BE-1 and PAO automobiles and Suzuki's SW-1 motorcycle to the Olympus O-Product and Ecru cameras (no. 234) and Sharp's wall-mounted Liquid Crystal Museum television series (no. 245). A bold inventor of new forms for industrial products, Naoki Sakai studied design at the Kyoto University of Fine Arts, graduating in 1966. Known for his marketing acumen, Sakai has said that his best-selling products reflect consumer preferences that he learns on the street and then "edits."[16]
Reference: "Form Follows Fashion," *Design*, no. 511 (July 1991), pp. 12–18.

Tamotsu Yagi
Born 1949

Tamotsu Yagi has worked as a graphic designer since the age of eighteen, creating corporate identity programs for clients such as ISSEY MIYAKE and the architect Tadao Ando and advertising posters and covers for magazines. He was also part of the creative team for the Axis design building in Tokyo, and since 1986 has been a regular contributor of features to *Axis* quarterly. After working as a design consultant for Esprit Hong Kong, Yagi moved to San Francisco in 1984 to become art director for the entire Esprit organization. He worked there for the next seven years, redesigning the corporate-identity and graphic programs. In 1991 Yagi established his own firm in order to pursue areas of design that he feels are crucial for the future, including

environmental protection. He has long been an advocate for the creative applications of recycled paper, and uses it as a medium for his graphics. One of his leading Japanese clients, Akebono, maker of sweets and rice crackers, uses recycled materials for its packaging (no. 238). Yagi feels that the most effective means for promoting the use of recycled paper is to make the products beautiful and desirable for the client and the consumer. Among his other graphic activities, Yagi does book design, such as the *Amish: The Art of the Quilt* (1989), for which he was also the exhibition designer. In 1992 he collaborated with EIKO ISHIOKA on the design for an exhibition about the German filmmaker Leni Riefenstahl in Tokyo.
Reference: "Designers and Studios / Tamotsu Yagi," *Axis*, no. 42 (Jan. 1992), pp. 138–39.

Yamaha Corporation
Established Hamamatsu, 1887

Yamaha Corporation is an extremely diverse company that designs and manufactures a wide array of products ranging from semiconductors to concert pianos. Its founder, Torakusa Yamaha, built Japan's first reed organ in 1887, and ten years later established Nippon Gakki, the company's official name until it was changed in 1987 to Yamaha Corporation. Moving swiftly to the manufacture of upright pianos (1900) and grand pianos (1902), Nippon Gakki soon gained attention for the fine quality of its musical instruments by winning an honorary grand prize at the Saint Louis world's fair in 1904. Nippon Gakki continued to concentrate its efforts in the production of musical instruments throughout the next several decades, introducing the first pipe organ to Japan in 1932, and also became involved in musical education, establishing the Yamaha Music School system in 1954 and the Yamaha Music Foundation in 1966. The company's diversification began with the production of motorcycles in 1954 (which led to the founding of YAMAHA MOTOR COMPANY in 1955), and continued with the production of fiberglass-reinforced plastic bows (1959), skis (no. 98), wind instruments (no. 195), stereos (nos. 229, 237), tennis rackets, furniture (no. 199), electronic pianos (1976), and industrial robots (1984).
Reference: Roxanne Guilhamet, "In-House Rules," *Design*, no. 489 (Sept. 1989), pp. 42–43.

Yamaha Motor Company
Established Hamakita, 1955

Established by YAMAHA CORPORATION one year after it produced its first motorcycle in 1954 (no. 35), Yamaha Motor Company is now a worldwide corporation with plants in forty-five countries. In the early years, as it strove to produce new and better motorcycle technology, Yamaha developed a broad range of small-engine products for leisure and sports activities, starting with power boats and outboard motors in 1960. In addition to its

motorcycle division, Yamaha now includes a marine division, power products division, automobile engine division, and industrial machinery division. Though motorcycles still account for nearly half of all sales (no. 232), its products also include sailboats, fishing vessels, ATVs (all terrain vehicles), scooters, diesel engines, snowmobiles, golf carts, multipurpose engines, and generators.
Reference: *Invitation to Yamaha* [brochure] (1993).

Yohji Yamamoto
Born 1943

One of fashion's revolutionaries, Yohji Yamamoto was raised in postwar Tokyo by his widowed mother, a dressmaker, and remembers as a child being always surrounded by clothing. Known for his canny tailoring, which he applied to making some of the earliest Japanese women's clothes based on men's garments, Yamamoto cut uncluttered, idiosyncratic, oversize shapes and used washed, unpressed fabrics and dark colors, through which he gained his reputation (nos. 161, 239, 251). "I'm not a fashion designer, I'm a dressmaker," he said in an interview with German filmmaker Wim Wenders, whose film about Yamamoto, *Notebook on Cities and Clothes* (Aufzeichnungen zu Kleidern und Städten), was released in 1991. Eloquent and philosophic, Yamamoto "seeks to find the essence of the [garment] in the process of fabricating it," and shares with other Japanese designers a craftsman's sense of material, starting every collection with the fabric from which he then "imagines a shape."[17] Yamamoto graduated from Keio University in Tokyo with a degree in law (1966) and immediately entered the Fashion Culture Institute (1966–69), winning in 1969 two prestigious awards for design, the Soen and Endo prizes, the latter including a scholarship for study in Paris. After working for two years in Parisian haute couture, Yamamoto returned to Tokyo, where in 1972 he established his own company, Y's, with moderately priced, often playful, easy-to-wear separates for women, to which he added Y's for men in 1977. In 1977 Yamamoto presented his first collection in Tokyo, followed in 1981 and 1982 by shows in Paris and New York, and established Yohji Yamamoto Incorporated in 1984. Like ISSEY MIYAKE and REI KAWAKUBO, with whom he is often compared, Yamamoto is an artist who owns companies, freely introducing and successfully marketing radical fashion concepts. Through his own retail outlets in Japan and abroad, and shops in the Isetan, Seibu, and Takashimaya department stores in Japan, Yamamoto's sales in 1992 amounted to over one hundred million dollars. In 1982 and 1991 Yamamoto won the annual Tokyo Fashion Editors Club award and in 1986, won the Mainichi fashion prize.
References: Brenda Polan, ed., *The Fashion Year* (London, 1983), pp. 50–55; Leonard Koren, *New Fashion Japan* (Tokyo, 1984), pp. 88–99; Estor

Carla de Miro d'Ajeta, ed., *Giappone avanguardia del future* (Genoa, 1985), pp. 206–7.

Ryuichi Yamashiro
Born 1920

Ryuichi Yamashiro, admired for the lyrical quality of his graphic work, was educated at the Osaka Municipal School of Industrial Arts. His first jobs were in the advertising departments of the Mitsukoshi and then the Hankyu department stores in Osaka, where he did copywriting and layouts. Living in Tokyo in 1954, he began to work as a free-lance art director for such clients as Takashimaya department store and Toshiba; in 1960 he joined HIROMU HARA and YUSAKU KAMEKURA in founding the Nippon Design Center, which he left in 1971 to establish his own firm, Design Office R. He has received many prizes and decorations, from the first in 1954, the Asahi advertising prize, to the medal of honor in 1985. He serves on many prize juries, including that for the Tokyo Art Directors Club, and is a member of the Mainichi industrial design prize committee. Yamashiro's graphics, such as his famed tree-planting poster (1955; no. 18), often include sophisticated visual-linguistic puns or allusions to classical Japanese art and literature (no. 55). In recent years he has gained recognition for his book illustrations, including a well-known series on cats.
References: Tokyo, Ginza Graphic Gallery, *Ginza Graphic Gallery '90* [annual report] (1990), n.p.; *Best 100 Japanese Posters, 1945–89* (Tokyo, 1990), pp. 50–51, 68–69, 258.

Yoshiro Yamashita
Born c. 1932

The graphic designer Yoshiro Yamashita graduated from Nihon University, Tokyo, in fine arts in 1951, and joined the advertising group Tokyo Bijutsu. He left in 1954 to become an illustrator for the advertising department of Takashimaya department store in Tokyo, where he won the Asahi advertising prize in 1955, and the Tokyo Art Directors Club bronze prize and Mainichi prize for advertising design in 1958. In 1960 Yamashita was invited by YUSAKU KAMEKURA and HIROMU HARA to become chief illustrator for their newly established Nippon Design Center. There, Yamashita worked with Masaru Katsumie and the design team assembled to create the graphic program for the Tokyo Olympic Games of 1964 (no. 69); he was responsible for the pictograms for services and events, which became the model for subsequent ideographic programs, such as those for Expo '70 in Osaka. He also created the pictograms for the Sapporo Winter Olympics of 1972, by which time he had resigned from the Nippon Design Center to teach at Nihon University (1968).
References: Martin Krampen, "Other Pictographic Systems," *Design Quarterly*, no. 62 (1965), pp. 23–24; The International Center for the

Typographic Arts, *Typomundus 20* (New York, 1966), no. 517; Richard S. Thornton, *The Graphic Spirit of Japan* (New York, 1991), pp. 98–100.

Sori Yanagi
Born 1915

A pioneer of industrial design in Japan, Sori Yanagi initially studied oil painting at the Tokyo School of Fine Arts (1936–40), and went on to work in the studio of the architect Junzo Sakakura. In 1942 he was associated with the French designer Charlotte Perriand, who was in Tokyo counseling Japanese companies on improving product design, learning from her, as he said, "the process of design."[18] Leaving architecture in 1947 to study industrial design, Yanagi opened his own design studio in Tokyo, the Yanagi Industrial Design Institute, in 1952. The same year he was awarded first prize in the first annual Mainichi industrial design competition for his combination record player and radio (no. 1). He has won a number of prizes, both national and international, including a gold medal at the Triennale of Milan in 1957. He works in a wide range of areas and materials, including wooden furniture (no. 27), appliances, and transportation, as well as ceramics (nos. 21, 28, 43), plastics (no. 11), aluminum (no. 9), and stainless steel (no. 39). Yanagi was responsible for the torch designs for both the 1964 Olympics in Tokyo (no. 66) and the 1972 Olympics in Sapporo (no. 94). He was a founding member of the Japan Industrial Designers Association (1952) and joined the Japan Committee on International Design (now Japan Design Committee) in 1953. He taught at the Women's Art College in Tokyo in 1953 and 1954, and then in the industrial design department of Kanazawa University of Arts and Crafts, from 1954 He has written extensively about design, and his writings have been published in both Japanese and English. In 1977 he became the director of the Japan Folk Crafts Museum in Tokyo, which his father, Soetsu Yanagi, leader of an early twentieth-century folk-art movement, had founded in 1936. In 1981 Yanagi received the medal of honor for his many contributions to Japanese culture.

References: Shiro Kuramata, "Yanagi Sori ni Kiku: Waakushoppu no Jissen" (Interview with Sori Yanagi: On the Establishment of a Workshop), *Design*, vol. 3 (Mar. 1978), pp. 35–56; Tokyo, The National Museum of Modern Art, Crafts Gallery, *Contemporary Vessels: How to Pour* (Feb. 10–Mar. 22, 1982), nos. 192–99; Sori Yanagi, *Design: Sori Yanagi's Works and Philosophy* (Tokyo, 1983); Kunio Sano, "Yanagi Sori-shi ni Kiku: Akai Te-pu o Hatta no o Oboete iru yo" (Interview with Sori Yanagi: I Remember Her Sticking the Red Tape), *Industrial Design*, no. 123 (Aug. 1983), pp. 40–43; Colin Naylor, ed., *Contemporary Designers*, 2nd ed. (Chicago, 1990), pp. 618–20; Takao Takanashi, "Yanagi Sori: Dezain ni Mingei no 'Genshi' o" (Sori Yanagi: The "Roots" of Folk Arts in Design), *JSSD*, vol. 1 (1993), pp. 34–37.

Tadanori Yokoo
Born 1936

Painter, printmaker, and graphic designer (no. 85), Tadanori Yokoo led the movement toward a Pop aesthetic in Japan during the 1960s, winning the admiration and following of young graphic designers and the criticism of Japanese traditionalists who decried the vulgarity of the images he created. Yokoo became a Pop celebrity, acting in the film *Diary of a Shinjuku Thief* (1968) and the play *Chinsetsu Yumiharizuki* (Crescent Moon) by Yukio Mishima (1969) and starring in Takashi Ichiyanagi's performance, *Tadanori Yokoo Sings—An Opera* (1969). Yokoo produced posters, commercial illustrations, and book designs for clients ranging from theater and dance companies (Tenjo Sajiki, Jokyo Gekijo [nos. 86, 91], Garumella) to department stores (Seibu, Matsuya, Tokyu, Sogo) and industrial corporations (Toshiba Machine, Yamagiwa Electric, Takeda Chemical, Suntory). His design sources, which during the 1960s included matchbox covers, beer and sake labels, playing cards, and *ukiyo-e* woodblock prints, became increasingly religious and metaphysical in the 1970s with references to Hindu, Buddhist, and Zen images. Since 1980 Yokoo has largely devoted himself to painting in an expressive, painterly style focused on the human figure. A sometime collaborator of the fashion designer ISSEY MIYAKE, Yokoo provided designs for Miyake's printed fabrics (no. 136), and in 1970 he was commissioned to design the textile pavilion for Expo '70 in Osaka. Yokoo graduated from Hyogo Prefectural High School in Nishiwaki (1954) and worked as an artist at Seibundo printing company, Kakogawa (1954–56), Kobe Shimbun, Kobe (1956–59), and the National Advertisement Laboratory, Osaka (1959–60), before moving to Tokyo and the Nippon Design Center (1960–64). Yokoo established two graphic design studios during the later 1960s, Studio Illfill, with Akira Uno and Chunao Harada (1964–65), and The End Studio, with Masamichi Oikawa (1968–71). Since 1971 he has practiced independently.

References: Yokoo Tono, "Tadanori Yokoo," in S. Takashina, Y. Tono, and Y. Nakahara, eds., *Art in Japan Today* (Tokyo, 1974), pp. 178–80; Koichi Tanikawa, *100 Posters of Tadanori Yokoo* (Tokyo, 1978); Yoshiaki Tono, "Tadanori Yokoo: Between Painting and Graphic Art," *Artforum*, vol. 22 (Sept. 1983), pp. 38–43; Tadanori Yokoo, *All about Tadanori Yokoo and His Graphic Works* (Tokyo, 1989); *Best 100 Japanese Posters, 1945–89* (Tokyo, 1990), pp. 80–83, 96–97, 104–5, 126–29, 258; Mark Holborn, *Beyond Japan: A Photo Theater* (London, 1991), pp. 76–77, 118, 168, 172.

Mosuke Yoshitake
1909–1993

A graduate in metalwork from the Tokyo School of Fine Arts in 1935, Mosuke Yoshitake went to work for the Industrial Arts Institute of the Ministry of Commerce and Industry, where he rose to become head of a design department. After his retirement in 1961, he founded the design office Tokyo Craft. Yoshitake is known particularly for his design of household goods (no. 72), lamps (no. 36), kitchen wares, and other small items. Among his many prizes are the Kunii Industrial Art Award (1973) for promoting provincial crafts by using traditional metalwork techniques, the Order of the Rising Sun (1974), and the Minister of Education art prize (1976) for lifetime achievement. Yoshitake taught art education at Musashino Art University in Kodaira from 1961 to 1981. A major force behind the establishment of the Japan Craft Design Association in 1956, he became its director general in 1980. Yoshitake is the author of several books on design, including *Function of Form, Function of Beauty* (1981).

References: "Rampu Sutando" (Two Variations of Standing Lamp), *Kogei Nyusu* (Industrial Art News), vol. 25 (Oct. 1957), pp. 17–18; Tokyo, The National Museum of Modern Art, Crafts Gallery, *Contemporary Vessels: How to Pour* (Feb. 10–Mar. 22, 1982), n.p.; Japan Industrial Designers Association, ed., *Seiichi no Kozo: Nihon no Indasutoriaru Dezain* (Structure of Dexterity: Industrial Design Works in Japan) (Tokyo, 1983), p. 127; Kogei Zaidan, ed., *Kunii Kitaro Sangyo Kogeisho no Hitobito: 35-nin no Purofuiru* (The Winners of the Kitaro Kunii Industrial Art Award: Profiles of 35 Winners) (Tokyo, 1984), pp. 19–22; Kogei Zaidan, ed., *Nihon no Indasutoriaru Dezain: Showa ga unda Meihin 100* (Japanese Industrial Design: 100 Masterpieces Produced during the Showa Period) (Tokyo, 1989), p. 61.

Samiro Yunoki
Born 1922

Samiro Yunoki, born into a family of painters, entered Tokyo University in 1942 to study art history. His studies were interrupted by the war, and in 1946 Yunoki began working at the Ohara Museum of Art, located in Kurashiki, a famous center of folk crafts near Hiroshima. There, Yunoki came across a calendar with bold stencil-dyed folk patterns on handmade paper, which changed the course of his career. He became fascinated with folk arts and began reading the works of Soetsu Yanagi, leader of the folk-crafts movement. In 1947 Yunoki abandoned his university studies and his job to apprentice himself to the artist who had designed the calendar, master textile craftsman KEISUKE SERIZAWA. As part of his training, Yunoki went to live and work in a small village workshop in Shizuoka, where he learned textile-weaving and stencil-dyeing *(kata-zome)* techniques first hand. In

1949 Yunoki submitted his first piece to the exhibition sponsored by the Society for National Painting in which other members of the folk-art movement showed regularly, and has participated annually since then. His first solo exhibition was held in 1950 at the Takumi crafts gallery in Tokyo, and he has exhibited in solo and group shows over the past four decades. Yunoki won a bronze medal at Expo '58 in Brussels for his stencil-dyed wallpaper. Yunoki's work retains a strong sense of its folk-craft inspiration, with its bold colors and pictorial subjects rendered with warmth and humor (no. 150). Paralleling his artistic career, Yunoki began lecturing at the Women's College of Fine Arts in Tokyo, where his mentor Serizawa had taught, becoming full professor in 1972 and serving as president of the college from 1987 until his retirement in 1991.

Reference: *Yunoki Samiro Sakuhin-shu* (Collected Works of Samiro Yunoki) (Tokyo, 1984).

Japanese Design: A Survey Since 1950

1 In Masataka Ogawa, Ikko Tanaka, and Kazumasa Nagai, eds., *The Works of Yusaku Kamekura* (Tokyo, 1983), p. 224.

2 In W. H. Allner, *Posters* (New York, 1952), p. 50.

3 "Dai 2-kai Shin-Nihon Kogyo Dezain Nyusho Sakuhin" (Winning Designs of the 2nd Annual New Japan Industrial Design Competition), *Kogei Nyusu* (Industrial Art News), vol. 22 (Feb. 1954), p. 28.

4 Takao Takanashi, "Yanagi Sori: Dezain ni Mingei no 'Genshi' o" (Sori Yanagi: The "Roots" of Folk Arts in Design), *JSSD*, vol. 1 (1993), p. 36.

5 In *The Graphic Design of Yusaku Kamekura* (New York, 1973), p. 174.

6 In Shiro Kuramata, "Yanagi Sori ni Kiku: Waakushoppu no Jissen" (Interview with Sori Yanagi: On the Establishment of a Workshop), *Design*, vol. 3 (Mar. 1978), p. 37.

7 In Utsunomiya, Tochigi Prefectural Museum of Art, *Iizuka Rokansai: Master of Modern Bamboo Crafts* (Feb. 5–Mar. 26, 1989), p. 117.

8 Sori Yanagi, *Design: Sori Yanagi's Works and Philosophy* (Tokyo, 1983), p. 165.

9 "Sony TR Terebi 8-301 Kei" (Sony TR Television 8-301), *Kogei Nyusu* (Industrial Art News), vol. 28 (Feb. 1960), pp. 78–79.

10 Correspondence, July 5, 1992.

11 "Dezainaa to Meekaa wa Kyozon Kyoei de Nakereba Ikenai" (Designers and Manufacturers Must Exist Together for Mutual Prosperity) [Interview with Hisayoshi Sakurai and Seiki Fujikawa], *Industrial Design*, no. 118 (July 1982), p. 33.

12 Correspondence, Nov. 19, 1993.

13 In Ogawa, Tanaka, and Nagai, eds., *Kamekura*, cited above, p. 210

14 Correspondence, Dec. 1993.

15 Shigeo Fukuda, *Fukuda Shigeo Hyohonbako* (Shigeo Fukuda's Specimen Box) (Tokyo, 1978), p. 110.

16 Correspondence, Feb. 9, 1994.

17 Correspondence, Oct. 15, 1993.

18 "Domus Interviews: Shiro Kuramata," *Domus*, no. 649 (Apr. 1984), p. 60.

19 In Ginette Sainderichin, *Kenzo* (Paris, 1989), p. 34.

20 In Milan, Palazzo dell'Arte, *From the Spoon to the Town through the Work of 100 Designers* (Oct.–Nov. 1983), p. 93.

21 Masayuki Kurokawa, "Shomei Kiga mo Chiisa na Kenchikumono" (Lighting Fixtures Are Small Architectural Objects), *Designers' Workshop*, vol. 2, no. 6 (Feb. 1985), p. 32.

22 Fukuda, *Fukuda*, cited above, p. 116.

23 Correspondence, Apr. 24, 1994.

24 *Super Market by Shin Matsunaga* (Tokyo, 1992), p. 153.

25 In *Boxes by Four: Package Design* (Tokyo, 1982), p. 107.

26 "Kosei Gangu Kapusera no Kaihatsu" (The Development of Capsela—Formation Toy), *Industrial Design*, no. 87 (May 1, 1977), back cover.

27 In *Eiko by Eiko, Eiko Ishioka: Japan's Ultimate Designer* (San Francisco, 1990), pls. 236–39.

28 In *Gachi: Motomi Kawakami's Furniture* (Tokyo, 1988), p. 110.

29 In *Boxes by Four*, cited above, p. 93.

30 In Mulhouse, France, Musée de l'Impression sur Etoffes, *Les Tissus Imprimés d'Issey Miyake* (1979), p. 5.

31 In Nicholas Callaway, ed., *Issey Miyake: Photographs by Irving Penn* (Boston, 1988), p. 41.

32 In Katsu Kimura, *Box-er: Katsu Kimura's Packaging* (Tokyo, 1988), p. 130.

33 In *Boxes by Four*, cited above, p. 162.

34 Correspondence, Jan. 25, 1993.

35 *Toshiyuki Kita: Movement as Concept* (Tokyo, 1990), p. 28.

36 Correspondence, Dec. 15, 1993.

37 Correspondence, Dec. 24, 1993.

38 Toshiyuki Kita, "Borderless Design," *Design Scene*, nos. 30–31 (Jan. 1994), p. 68.

39 In *Miyake: Photographs*, cited above, p. 32.

40 In Peter Popham, "The Emperor's Clothes," *Blueprint*, no. 15 (Mar. 1985), p. 16.

41 In *Best 100 Japanese Posters, 1945–89* (Tokyo, 1990), p. 166.

42 *Asahi Shimbun*, Nov. 22, 1982.

43 In Leonard Koren, *New Fashion Japan* (Tokyo, 1984), p. 117.

44 Correspondence, Feb. 1994.

45 In *Package Design in Japan* (Cologne, 1989), p. 145.

46 In Kazuko Koike and others, *Hi-Pop Design* (Tokyo, 1985), p. 289.

47 In "Lamp: 'Japanese Men,'" *Design*, no. 182 (Nov. 1987), p. 106.

48 Correspondence, Feb. 14, 1994.

49 In Sainderichin, *Kenzo*, cited above, p. 86.

50 "Borderless Design," cited above, p. 69.

51 Correspondence, Dec. 20, 1993.

52 *East Meets West in Design: Archaeology of the Present* (New York, 1989), p. 53.

53 "Kurimoto Natsuki Urushi Kaburimono" (Lacquer Headgear by Natsuki Kurimoto), *Tokyo Shimbun*, Aug. 10, 1991, p. 3.

54 *The Textile Design of Hiroshi Awatsuji* (Tokyo, 1990), pp. 148–51.

55 In "100 Graphic Designers of the World," *IDEA*, no. 240 (Sept. 1993), p. 191.

56 Correspondence, Mar. 8, 1994.

57 In Tokyo, Axis Gallery, *Design a Dream* [brochure] (1991).

58 Reiko Sudo, "Junichi Arai: High Tech Craftsman," in Tokyo, Yurakucho Asahi Gallery, *Hand and Technology: Textile by Junichi Arai '92* (Mar. 7–25, 1992), n.p.

59 Correspondence, Dec. 27, 1993.

60 David E. Sanger, "Japan's Scratch-Pad Computers," *New York Times*, Mar. 26, 1990, p. D5.

61 Correspondence, Dec. 15, 1993.

62 Takashi Kanome, "Box XX," in Tokyo, Ginza Graphic Gallery, *Ginza Graphic Gallery '92* [annual report] (June 4–27, 1992), n.p.

63 In Sainderichin, *Kenzo*, cited above, p. 107.

64 In "Taidan: Hakai, Kaitai, Eregansu—Kawakubo Rei, Fukai Akiko" (Interview: Destruction, Demolition, Elegance—Rei Kawakubo and Akiko Fukai), *Dress Study*, vol. 24 (Fall 1993), p. 10.

Biographies

1 Reiko Sudo, "Junichi Arai: High Tech Craftsman," in Tokyo, Yurakucho Asahi Gallery, *Hand and Technology: Textile by Junichi Arai '92* (Mar. 7–25, 1992), n.p.

2 Hiroshi Shinohara, in Roxanne Guilhamet, "In-House Rules," *Design*, no. 489 (Sept. 1989), p. 45.

3 Daisaku Choh, "Interior Design and Industrial Design," in *World Design Conference 1960 in Tokyo* (Tokyo, 1960), p. 63.

4 Kitanishi Misako, "Hito Mono Tojo: Horiuchi Yoshisada" (Personality / Works: Yoshisada Horiuchi), *Industrial Design*, no. 118 (July 1982), p. 38.

5 Kiyoshi Miyazaki, "Akioka Yoshio: Jiko no Tetsugaku to Seikatsu no Hanei to shite no Dezain" (Yoshio Akioka: Design as a Reflection of Philosophy and Life), *JSSD*, vol. 1 (1993), p. 6.

6 In Tokyo, Axis Gallery, *Design a Dream* [brochure] (1991).

7 In "Zenin ga Ichinen Kakete Haizara no Dezain o shite mo ii to Omoimasu" (I Think It's Fine if the Whole Firm Spends a Whole Year on the Design of an Ashtray), *Industrial Design*, no. 113 (Sept. 1981), p. 35.

8 In "Jointo ni Yoru Pakkeeji" (Packages Based on Joints), *Designers' Workshop*, vol. 2, no. 11 (Dec. 1985), p. 13.

9 In Kogei Zaidan, ed., *Waga Indasutoriaru Dezain: Kosugi Jiro no Hito to Sakuhin* (My Industrial Design: Jiro Kosugi, the Man and His Work) (Tokyo, 1983), p. 105.

10 Miyake Design Studio, *Issey Miyake and Miyake Design Studio, 1970–1985* (Tokyo, 1985), p. 57.

11 Issey Miyake, "A Personal Statement on Fashion Design," in Phillip Dennis Cate, ed., *Perspectives on Japonisme: The Japanese Influence on America*, papers presented at the conference held on May 13–14, 1988, Rutgers, The State University of New Jersey (New Brunswick, 1989), p. 93.

12 "A Foreword to the Future," in *Hanae Mori 1960–1989* (Tokyo, 1989), p. 173.

13 Akiko Takehara, "Hito Mono Tojo: Sugiura Kohei" (Personality / Works: Kohei Sugiura), *Industrial Design*, nos. 139–40 (July 1987), p. 67.

14 Correspondence, May 7, 1992.

15 In Beryl McAlhone, "Ikko Tanaka," *Designer*, June 1980, p. 11.

16 Naoki Sakai, "Giyoho—Gidai no Dento no Imi" (The Meaning of Tradition in the Information Age), *Nikkei Design*, no. 72 (June 1993), p. 57.

17 Interview with Wim Wenders in the film *Notebook on Cities and Clothes* (1991).

18 Kunio Sano, "Yanagi Sori-shi ni Kiku: Akai Te-pu o Hatta no o Oboete iru yo" (Interview with Sori Yanagi: I Remember Her Sticking the Red Tape), *Industrial Design*, no. 123 (Aug. 1983), p. 42.

SELECTED BIBLIOGRAPHY

Antonier, Aileen, Maureen Erbe, and Roger Handy. *Made in Japan: Transistor Radios of the 1950s and 1960s.* San Francisco, 1993.

Best 100 Japanese Posters, 1945–89. Tokyo, 1990.

Blair, Dorothy. *A History of Glass in Japan.* New York, 1973.

Boston, Museum of Fine Arts. *Living National Treasures of Japan.* Nov. 3, 1982—Jan. 2, 1983.

"Design in Japan." *Design,* no. 489 (Sept. 1989), pp. 37–75.

Dietz, Matthias, ed. *Japan Design.* Cologne, 1992.

East Japan Railway Culture Foundation, ed. *Advertising Art History 1950–1990.* Tokyo, 1993.

Evans, Siân. *Contemporary Japanese Design.* London, 1991.

The Foundation Europalia International and R. DeSmet, eds. *Japanse Affiches / Modern Posters of Japan.* Tokyo, 1989.

Friedman, Mildred, ed. *Tokyo: Form and Spirit.* New York, 1986.

Fux, Herbert. *Traditionelles Kunsthandwerk der Gegenwart aus Japan.* Vienna, 1974.

Hara, Hiromu, ed. *Contemporary Book Design in Japan, 1975–1984.* Tokyo, 1986.

Interior Design in Japan. Tokyo, 1990.

Izutsu, Akio. *The Bauhaus: A Japanese Perspective and a Profile of Hans and Florence Schust Knoll.* Tokyo, 1992.

Japan Craft Design Association. *Contemporary Craft in Japan.* Tokyo, 1986.

Japan Design Committee. *Design 19.* Tokyo, Matsuya Ginza, Sept. 3–8, 1982.

Japan Design Committee. *Dezain no Kiseki* (Tracks of Design). Tokyo, 1977.

Japan Design Foundation. *Design for Every Being.* Osaka, 1992.

Japan Industrial Designers Association, ed. *Man and Tool—Discovery of New Interrelations.* Tokyo, 1973.

Japan Industrial Designers Association, ed. *Seiichi no Kozo: Nihon no Indasutoriaru Dezain* (Structure of Dexterity: Industrial Design Works in Japan). Tokyo, 1983.

Japan Interior Designers Association, ed. *Hikari to no Kakawari—Nippon no Interia Dezain* (To Live with Lights—Interior Designs in Japan). Tokyo, 1981.

Japan Package Design Association, ed. *Package Design in Japan: Its History, Its Faces.* Tokyo, 1976.

Japan Style. Tokyo, 1980.

Johnson, Chalmers. *MITI and the Japanese Miracle: The Growth of Industrial Policy, 1925–1975.* Stanford, California, 1982.

Kodansha Encyclopedia of Japan. 9 vols. Tokyo, 1988.

Kogei Zaidan, ed. *Kunii Kitaro Sangyo Kogeisho no Hitobito: 35-nin no Purofuiru* (The Winners of the Kitaro Kunii Industrial Art Award: Profiles of 35 Winners). Tokyo, 1984.

Kogei Zaidan, ed. *Nihon no Indasutoriaru Dezain: Showa ga unda Meihin 100* (Japanese Industrial Design: 100 Masterpieces Produced during the Showa Period). Tokyo, 1989.

Kogei Zaidan, ed. *Nihon no Kindai Dezain Undoshi 1940 Nendai–1980 Nendai* (History of the Japanese Modern Design Movement, 1940s–1980s). Tokyo, 1990.

Kogyo Gijutsuin Seihin Kagaku Kenkyujo, ed. *Sangyo Kogei Shikenjo 40 Nenshi* (Industrial Arts Institute's 40-Year History). Tokyo, 1976.

Koike, Kazuko, and others. *Hi-Pop Design.* Tokyo, 1985.

Koren, Leonard. *New Fashion Japan.* Tokyo, 1984.

Leitherer, Eugen, and Hans Wichmann. *Reiz und Hülle: Gestaltete Warenverpackungen des 19. und 20. Jahrhunderts.* Basel, 1987.

London, Victoria and Albert Museum. *Visions of Japan.* Sept. 17, 1991—Jan. 5, 1992.

Mainichi Shinbun-sha, ed. *Mainichi Kokoku Dezain Sho: Dai 50-kai Kinen Sakunhinshu—'82 Mainichi Dezain Sho* (Mainichi Advertising Design Prize: Works for the 50th Anniversary—'82 Mainichi Design Prize). Tokyo, 1983.

Miyazaki, Kiyoshi, and others. "Tokushu Nihon no Kogyo Dezain, 1945–1970" (Special Feature: Industrial Design in Japan, 1945–1970). *Kogei Nyusu* (Industrial Art News), vol. 39 (May 1972), pp. 2–54.

Osaka, The National Museum of Art. *Design and Art of Modern Chairs* (Isu no Katachi: Dezain kara Aato e). Aug. 19–Oct. 15, 1978.

Paris, Musée de L'Affiche. *L'Affiche japonaise: Des origines à nos jours.* Oct. 31, 1979—Jan. 13, 1980.

Posters: Japan 1800's–1900's. Nagoya, 1989.

Reischauer, Edwin O. *Japan: The Story of a Nation.* Rev. ed. New York, 1989.

Sparke, Penny. *Japanese Design.* London, 1987.

Tanaka, Ikko, and Kazuko Koike, eds. *Japan Design: The Four Seasons in Design.* San Francisco, 1984.

Tanaka, Ikko, and Shin Matsunaga, eds. *Japan Poster 1950–1988.* Tokyo, 1988.

Thornton, Richard S. *The Graphic Spirit of Japan.* New York, 1991.

Tokyo, The National Museum of Modern Art, ed. *Japanese Lacquer Art: Modern Masterpieces.* New York, 1982.

Tokyo, The National Museum of Modern Art. *Lacquer Art of Modern Japan.* Nov. 1–Dec. 9, 1979.

Tokyo, The National Museum of Modern Art, Crafts Gallery. *Contemporary Vessels: How to Pour.* Feb. 10–Mar. 22, 1982.

Tokyo, The National Museum of Modern Art, Crafts Gallery. *Crafts.* Tokyo, 1988.

Tokyo, The National Museum of Modern Art, Crafts Gallery. *Masterpieces of Contemporary Japanese Crafts.* Nov. 15, 1977–Mar. 19, 1978.

Tokyo, The National Museum of Modern Art, Crafts Gallery. *Modern Bamboo Craft.* Feb. 5–Mar. 24, 1985.

Tokyo, The National Museum of Modern Art, Crafts Gallery. *Modern Japanese Glass: Early Meiji to Present.* Sept. 22–Nov. 28, 1982.

Tokyo, The National Museum of Modern Art, Crafts Gallery. *30 Years of Modern Japanese Traditional Crafts.* Aug. 26–Sept. 25, 1983.

Tokyo, Sezon Museum of Art. *Japanese Aesthetics and Sense of Space: Another Aspect of Modern Japanese Design* (Nihon no Me to Kukan: Mo Hitotsu no Modan Dezain). Sept. 8–24, 1990.

Tokyo Art Directors Club. *Art Direction Today.* Tokyo, 1984.

Wichmann, Hans. *Japanische Plakate: Sechziger Jahre bis Heute.* Munich, 1988.

World Design Conference 1960 in Tokyo. Tokyo, 1960.

Yamamoto, Masaya, ed. *Product Design in Japan.* Tokyo, 1990.

Yao, Takeo, ed. *Package Design in Tokyo.* Tokyo, 1987.

Yoshihara, Atsushi. "Nihon no isu 100 ten" (Japanese Chairs 100 Exhibition). *Kogei Nyusu* (Industrial Art News), vol. 39 (Mar. 1971), pp. 64–69.

Japanese Design Magazines

Axis: Quarterly on Trends in Design.

Creation: Magazine of Graphic Design.

Designers' Workshop.

Design News: Magazine for Industrial Design.

Design Quarterly.

Industrial Design.

Kogei Nyusu (Industrial Art News).

Nikkei Design.

A C K N O W L E D G M E N T S

In researching and producing this book and the exhibition it celebrates, we have been assisted over the past five years by numerous colleagues, designers, institutions, and companies whom we would like now to acknowledge with our warmest thanks:

In Japan, Shukuroo Habara, professor at Meisei University and former secretary general of the Japan Industrial Designers Association, who has been a generous advisor and supporter since the inception of this project five years ago, and his colleagues Mariko Yoshida and Yasuko Nonaka, who were most helpful in searching for addresses, objects, and research materials; Professor Shutaro Mukai, who unstintingly shared his vast knowledge and insights with us, and his colleagues at Musashino Art University, who welcomed us to their rich research facilities and design collection; Naofumi Ohtsuka and Masaharu Taneichi, who offered tireless help with finding research materials and tracking objects; and Reiko Onodera, who provided photographs and slides.

For the catalogue design, Mitsuo Katsui and his staff, Masako Ishibashi, Kozo Kakei, and Risa Hirakawa; for the installation design concept, Kisho Kurokawa and his colleagues Tadao Shibata and Akira Yokoyama; for their advice and hospitality in Tokyo, the late Professor Iwataro Koike and his wife, Taka Koike; for the Japanese word processor donated to the project, Atsutoshi Nishida of Toshiba Corporation.

Kazuo Akashi of Kogei Zaidan; Nobuo Aoshima, Atsushi Ishiyama, and Masako Sone, GK; Hiroshi Arai, senior researcher, Human Environmental System Department, The National Institute of Bioscience and Human Technology, Tsukuba; Yuko Fujimoto, Japan Graphic Designers Association; Eizi Hayashi, AXIS; Kazuo Hayashi, The Museum of Modern Art, Ibaraki; Professor Toramasa Hayashi, Nagoya University of Art and Design, Nagoya; Toyojiro Hida, The National Museum of Modern Art, Tokyo; Professor Eiichi Hino, Hyogo University of Education, Kobe; Morikage Ikeda, Mainichi Shinbun Planning Division; Takuo Ikegame, Design Station Inc.; Kahoko and Kinichi Iwata; Hiraku Kido and Teisuke Mura, Japan Package Design Association; Kyoko Kinoshita; Shinji Kohmoto, The National Museum of Modern Art, Kyoto; Kazuko Koike, Kitchen; Kazuhisa Mashimo, Japan Design Foundation; Keiko Misawa, Overseas Public Affairs Office, Ministry of International Trade and Industry; Ryuichi Niimi and Toshio Takebe, Sezon Museum of Art, Tokyo; Hidemi Sakurada, D & D Studio; Hisakichi Sakurai, Tendo Company; Toshihiko Shigeyama, Kibun Foods; Reiko Sudo, Nuno; Hirohiko Takeyama, Tochigi Prefectural

Museum of Fine Arts; Yasunobu Tanaka and Yuichi Yamada, Japan Industrial Design Promotion Organization; Mariko Tsuchida, Japan Design Committee; Mioko Watanabe; Sori Yanagi.

In the United States, Gyo Aburaki and Yoichi Takaku, Yasuda Insurance Company of North America; Margaret Bemiss and Jeremy Marlton, Philadelphia; Professor Ardath W. Burks, Rutgers University, New Brunswick; Charles Burnette, Joseph Carreiro professor of industrial design, chairman, department of industrial design, University of the Arts, Philadelphia; Mary Lee Cope, Philadelphia; William B. Eagleson, Jr., honorary consul general of Japan in Philadelphia; Yoshiko Ebihara, Gallery 91, New York; Elizabeth Galloway, director, Fogg Memorial Library, Art Center College of Design, Pasadena; Satoru Gohara, Consulate General of Japan; Jun I. Kanai, Issey Miyake USA Corporation, New York; Toshihiro Katayama, director, Carpenter Center for the Visual Arts, Harvard University, Cambridge; Professor Kaori Kitao, Swarthmore College; Jack Lenor Larsen, New York; Robert Lesser, New York; Kenneth Lowengrub, New York; Tetsuya Nakao, Consulate General of Japan; Machiko Oyama and Toshihiko Okoshi, Sasakawa Peace Foundation USA, Washington, D.C.; Barbara Phillips, Allentown Art Museum; Susan Reinhold, Reinhold-Brown Gallery, New York; Jay and Kathryn Sherrerd; Kazuko Shiraishi, Consulate General of Japan; Milton Sonday, Cooper-Hewitt Museum, New York; Masahiro Suyama, LTCB-MAS; Lily Y. Tanaka; Michele N. Tanaka; Thomas Tang, export department, American Standard; Mr. and Mrs. Joseph Volk, Zip's Toys to Go, Ardmore, Pennsylvania; Yoshiko I. Wada; Motokatsu Watanabe, Information Officer, Embassy of Japan; Suzanne Wheeling; Jerry Wind, SEI Center for Advanced Studies in Management, Wharton School, University of Pennsylvania; Gerald Yoshitoshi, Japanese American Cultural & Community Center, Los Angeles.

In Europe, Emma Furlong, Design Museum, London; Kerstin Wickman, *Form* magazine, Stockholm.

At the Philadelphia Museum of Art, Caroline Demaree, Saiko Matsumaru, Emiko Mikisch, Diane Minnite, Rosemary Morrison, Diana Ott, Erisa Sekiya, Julia Smith, and Mari Yoshikawa were research assistants on the *Japanese Design* exhibition staff. The departmental staffs also contributed their support: Donna Corbin, assistant curator, European Decorative Arts; Marjorie Matthews Corr, administrative assistant, and Yan Ge, curatorial assistant, East Asian Art; Mikiko Higashi, Keiko Ikeda, Chizuko Imai, Motoko Imai, Toshiko Kono, Maxine Lewis, Masayo Nanmoto, Dr. Linda Nelson, Shoko Sekiguchi, Yumiko Tanno, and Chie Tsukahara, volunteers, East Asian Art; and Ann Lalley, volunteer, European Decorative Arts. Other Museum staff members who provided assistance were Dilys Blum and Kristina Haugland, Costume and Textiles; Anita Gilden, Gina Kaiser, and Lilah Mittelstaedt, Library; Jean Griffith and Elizabeth Woods, Publications; William Hilley, Mailroom; Lynn Rosenthal and Graydon Wood, Photography; and Terry Flemming Murphy, Rights and Reproductions.

K.B.H. / F.F.